D0765671

Springer Tracts in Advanced Robotics 80

Editors

For further volumes:
http://www.springer.com/series/5208

STAR (Springer Tracts in Advanced Robotics) has been promoted under the auspices of EURON (European Robotics Research Network)

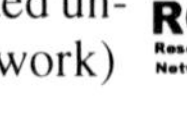

Bruno Siciliano (Ed.)

Advanced Bimanual Manipulation

Results from the DEXMART Project

Editor
Prof. Bruno Siciliano
PRISMA Lab
Dipartimento di Informatica e Sistemistica
Università di Napoli Federico II
Via Claudio 21, 80125 Napoli
Italy
E-mail: siciliano@unina.it

ISSN 1610-7438 e-ISSN 1610-742X
ISBN 978-3-642-29040-4 e-ISBN 978-3-642-29041-1
DOI 10.1007/978-3-642-29041-1
Springer Heidelberg New York Dordrecht London

Library of Congress Control Number: 2012934148

Printed on acid-free paper

Springer is part of Springer Science+Business Media (www.springer.com)

to The CC-FAB's

Foreword

Robotics is undergoing a major transformation in scope and dimension. From a largely dominant industrial focus, robotics is rapidly expanding into human environments and vigorously engaged in its new challenges. Interacting with, assisting, serving, and exploring with humans, the emerging robots will increasingly touch people and their lives.

Beyond its impact on physical robots, the body of knowledge robotics has produced is revealing a much wider range of applications reaching across diverse research areas and scientific disciplines, such as: biomechanics, haptics, neuroscience, virtual simulation, animation, surgery, and sensor networks among others. In return, the challenges of the new emerging areas are proving an abundant source of stimulation and insights for the field of robotics. It is indeed at the intersection of disciplines that the most striking advances happen.

The Springer Tracts in Advanced Robotics (STAR) is devoted to bringing to the research community the latest advances in the robotics field on the basis of their significance and quality. Through a wide and timely dissemination of critical research developments in robotics, our objective with this series is to promote more exchanges and collaborations among the researchers in the community and contribute to further advancements in this rapidly growing field.

Advances in Bimanual Manipulation edited by Bruno Siciliano provides the scientific community with the principal results of the DEXMART European project. This project brought together teams of European robotics researchers to pursue a concentrated four-year effort on the development of dexterous and autonomous dual-hand manipulation capabilities in personal and service robotics.

This volume covers a host of highly important topics in bimanual manipulation. These include issues concerned with (i) modeling and learning of human manipulation skills, (ii) algorithms for task planning, human-robot interaction, and grasping, (iii) hardware design of dexterous anthropomorphic hands. The thorough discussion, rigorous treatment, and wide span of the work unfolding in these areas reveal the significant advances in the theoretical foundation and technology basis of bimanual robotic manipulation. DEXMART culminates with this important reference to the robotics community on the current developments and new directions undertaken by this project's teams of European robotics researchers!

Stanford, California
February 2012

Oussama Khatib
STAR Editor

Preface

DEXMART (DEXterous and autonomous dual-arm hand robotic manipulation with sMART sensory-motor skills: A bridge from natural to artificial cognition) is a European research project which started in February 2008 and ended in January 2012. The ambition of the project was to fill the gap between the use of robots in industrial environments and the use of future robots in everyday human and unstructured environments, contributing to reinforce European competitiveness in all those domains of personal and service robotics where dexterous and autonomous dual-hand manipulation capabilities are required.

DEXMART contributed to the development of robotic systems endowed with dexterous and human-aware dual-arm/hand manipulation skills for objects, operating with a high degree of autonomy in unstructured real-world environments. These were the main objectives of the project:

- allow a dual-arm robot including two multi-fingered redundant hands to grasp and manipulate the same objects (different shape, dimension and weight) used by human beings;
- manipulation will take place in unsupervised, robust and dependable manner so as to allow the robot to safely cooperate with humans for the execution of given tasks;
- a robotic system able to autonomously decide between different manipulation options, and to learn new action sequences aimed at creating a consistent and comprehensive manipulation knowledge base;
- possible exploitation of high power-to-weight ratio of smart materials and structures, aimed at design of new hand components (finger, thumb, wrist) and sensors for the next generation of dexterous robotic hands.

The goal of this volume is to present the principal results of the project during the latest four years to the scientific community of people working in the field of grasping and dual arm/hand manipulation. The topics dealt with in this edited collection include:

- original approaches to interpretation, learning, and modelling, from the observation of human manipulation at different levels of abstraction;

- effective techniques for task planning, coordination and execution so as to confer to the robotic system self-adapting capabilities and reactivity to changing environment and unexpected situations, also in the case of humans cooperating with it;
- novel grasping force optimisation algorithms and posture-based control strategies for multi-fingered hands with management of redundant degrees of freedom;
- design of new actuators, as well as smart force and tactile sensors, able to overcome the limitations of current manipulation devices, and their integration in a new dexterous anthropomorphic hand;
- meaningful benchmarks for dual-hand manipulation.

The volume is based on the well-attended workshop "The DEXMART Project for Advanced Bimanual Manipulation" held on 26 October 2011 at the 11th IEEE–RAS International Conference on Humanoid Robots in Bled, Slovenia. The five presentations at the workshop are reflected into the five chapters in this collection. Noticeably, the workshop program was enriched by two plenary talks by well-known experts in the field: the cognitive aspects of manipulation were surveyed in the first presentation by Michael Beetz, while the final presentation by Oussama Khatib opened a human-centered perspective in view of the future robotics applications.

Further information about the results generated by the project can be found at *www.dexmart.eu*, including videos illustrating experiments on the available set-ups which are naturally associated to the research described in this volume: the new DEXMART hand at University of Bologna, the humanoid Rollin' Justin at DLR, the dual-robot system Adero with two KUKA lightweight arms and two Schunk anthropomorphic hands at FZI, and the mobile manipulator Jido with a KUKA lightweight arm at LAAS.

From my observing point as coordinator of this project, I trust that the results achieved by DEXMART have a great potential for European robot manufacturers, as typical assembly procedures in automotive industry require dual-arm manipulation of objects and tools similar to those generally used by production workers. The new dexterous hand will be appealing to SMEs for manipulation of work pieces of different sizes, shapes and weights currently requiring different grasping tools and frequent changes. Human–robot cooperation is to be adopted in aeronautic industry to assist humans in simple repetitive tasks, e.g. riveting and assembly. In the future, the results of the project will be useful for executing human-centered tasks in service robotics scenarios.

I hope readers will find the material contained in this volume useful for the design, modelling, planning and control of advanced bimanual robotic systems, and I would like to take this opportunity to express my sincere appreciation and warmest thanks to all those who have contributed to the success of the DEXMART project!

Naples, Italy
February 2012

Bruno Siciliano
DEXMART Coordinator

Table of Contents

List of Contributors

Giovanni Berselli
Dipartimento di Ingegneria delle Costruzioni Meccaniche Nucleari
Aeronautiche e di Metallurgia
Alma Mater Studiorum Università di Bologna
Viale Risorgimento 2, 40136 Bologna, Italy
e-mail: giovanni.berselli@unibo.it

Christoph Borst
Robotik und Mechatronik Zentrum
DLR
Münchnerstrasse 20, 82234 Wessling, Germany
e-mail: christoph.borst@dlr.de

Xavier Broquère
LAAS–CNRS
7 Av. Col. Roche, 31077 Toulouse, France
e-mail: xavier@broquere.fr

Ernesto Burattini
Dipartimento di Scienze Fisiche
Università degli Studi di Napoli Federico II
Via Cintia, 80126 Napoli, Italy
e-mail: ernb@na.infn.it

Giuseppe De Maria
Dipartimento di Ingegneria dell'Informazione
Seconda Università degli Studi di Napoli
Via Roma 29, 81031 Aversa, Italy
e-mail: giuseppe.demaria@unina2.it

Rüdiger Dillmann
Humanoids and Intelligence Systems Lab, Institut für Anthropomatik
Karlsruher Institut für Technologie
Adenauerring 4, 76131 Karlsruhe, Germany
e-mail: ruediger.dillmann@kit.edu

Pietro Falco
Dipartimento di Ingegneria dell'Informazione
Seconda Università degli Studi di Napoli
Via Roma 29, 81031 Aversa, Italy
e-mail: pietro.falco@unina2.it

Fanny Ficuciello
PRISMA Lab, Dipartimento di Informatica e Sistemistica
Università degli Studi di Napoli Federico II
Via Claudio 21, 80125 Napoli, Italy
e-mail: fanny.ficuciello@unina.it

Alberto Finzi
Dipartimento di Scienze Fisiche
Università degli Studi di Napoli Federico II
Via Cintia, 80126 Napoli, Italy
e-mail: finzi@na.infn.it

Gerhard Grunwald
Robotik und Mechatronik Zentrum
DLR
Münchnerstrasse 20, 82234 Wessling, Germany
e-mail: gerhard.grunwald@dlr.de

Katharina Hertkorn
Robotik und Mechatronik Zentrum
DLR
Münchnerstrasse 20, 82234 Wessling, Germany
e-mail: katharina.hertkorn@dlr.de

Rainer Jäkel
Humanoids and Intelligence Systems Lab, Institut für Anthropomatik
Karlsruher Institut für Technologie
Adenauerring 4, 76131 Karlsruhe, Germany
e-mail: rainer.jaekel@kit.edu

Daniel Leidner
Robotik und Mechatronik Zentrum
DLR
Münchnerstrasse 20, 82234 Wessling, Germany
e-mail: daniel.leidner@dlr.de

Vincenzo Lippiello
PRISMA Lab, Dipartimento di Informatica e Sistemistica
Università degli Studi di Napoli Federico II
Via Claudio 21, 80125 Napoli, Italy
e-mail: vincenzo.lippiello@unina.it

Martin Lösch
Humanoids and Intelligence Systems Lab, Institut für Anthropomatik
Karlsruher Institut für Technologie
Adenauerring 4, 76131 Karlsruhe, Germany
e-mail: martin.loesch@kit.edu

Emilio Maggio
Oxford Metrics Group
14 Minns Business Park, West Way, Oxford OX2 0JB, United Kingdom
e-mail: emilio.maggio@omg3d.com

Jim Mainprice
LAAS–CNRS
7 Av. Col. Roche, 31077 Toulouse, France
e-mail: jim.mainprice@laas.fr

Chris May
Lehrstuhl für Prozessautomatisierung
Universität des Saarlandes
Campus A5 1, 66123 Saarbrücken, Germany
e-mail: c.may@lpa.uni-saarland.de

Claudio Melchiorri
Dipartimento di Elettronica Informatica e Sistemistica
Alma Mater Studiorum Università di Bologna
Viale Risorgimento 2, 40136 Bologna, Italy
e-mail: claudio.melchiorri@unibo.it

Ciro Natale
Dipartimento di Ingegneria dell'Informazione
Seconda Università degli Studi di Napoli
Via Roma 29, 81031 Aversa, Italy
e-mail: ciro.natale@unina2.it

Gianluca Palli
Dipartimento di Elettronica Informatica e Sistemistica
Alma Mater Studiorum Università di Bologna
Viale Risorgimento 2, 40136 Bologna, Italy
e-mail: gianluca.palli@unibo.it

Salvatore Pirozzi
Dipartimento di Ingegneria dell'Informazione
Seconda Università degli Studi di Napoli
Via Roma 29, 81031 Aversa, Italy
e-mail: salvatore.pirozzi@unina2.it

Maximo A. Roa
Robotik und Mechatronik Zentrum
DLR
Münchnerstrasse 20, 82234 Wessling, Germany
e-mail: maximo.roa@dlr.de

Silvia Rossi
Dipartimento di Scienze Fisiche
Università degli Studi di Napoli Federico II
Via Cintia, 80126 Napoli, Italy
e-mail: srossi@na.infn.it

Fabio Ruggiero
PRISMA Lab, Dipartimento di Informatica e Sistemistica
Università degli Studi di Napoli Federico II
Via Claudio 21, 80125 Napoli, Italy
e-mail: fabio.ruggiero@unina.it

Steffen W. Rühl
FZI Forschungszentrum Informatik
Haid-und-Neu-Strasse 10-14, 76131 Karlsruhe, Germany
e-mail: ruehl@fzi.de

Florian Schmidt
Robotik und Mechatronik Zentrum
DLR
Münchnerstrasse 20, 82234 Wessling, Germany
e-mail: florian.schmidt@dlr.de

Sven R. Schmidt-Rohr
Humanoids and Intelligence Systems Lab, Institut für Anthropomatik
Karlsruher Institut für Technologie
Adenauerring 4, 76131 Karlsruhe, Germany
e-mail: srsr@ira.uka.de

Bruno Siciliano
PRISMA Lab, Dipartimento di Informatica e Sistemistica
Università degli Studi di Napoli Federico II
Via Claudio 21, 80125 Napoli, Italy
e-mail: bruno.siciliano@unina.it

Daniel Sidobre
LAAS–CNRS
7 Av. Col. Roche, 31077 Toulouse, France
e-mail: daniel.sidobre@laas.fr

Mariacarla Staffa
PRISMA Lab, Dipartimento di Informatica e Sistemistica
Università degli Studi di Napoli Federico II
Via Claudio 21, 80125 Napoli, Italy
e-mail: mariacarla.staffa@unina.it

Gabriele Vassura
Dipartimento di Ingegneria delle Costruzioni Meccaniche Nucleari
Aeronautiche e di Metallurgia
Alma Mater Studiorum Università di Bologna
Viale Risorgimento 2, 40136 Bologna, Italy
e-mail: gabriele.vassura@unibo.it

Luigi Villani
PRISMA Lab, Dipartimento di Informatica e Sistemistica
Università degli Studi di Napoli Federico II
Via Claudio 21, 80125 Napoli, Italy
e-mail: luigi.villani@unina.it

Zhixing Xue
FZI Forschungszentrum Informatik
Haid-und-Neu-Strasse 10-14, 76131 Karlsruhe, Germany
e-mail: xue@fzi.de

Franziska Zacharias
Robotik und Mechatronik Zentrum
DLR
Münchnerstrasse 20, 82234 Wessling, Germany
e-mail: zacharias@mvtec.com

Layered Programming by Demonstration and Planning for Autonomous Robot Manipulation

Rainer Jäkel, Steffen W. Rühl, Sven R. Schmidt-Rohr, Martin Lösch, Zhixing Xue, and Rüdiger Dillmann

Abstract. We propose a layered system for autonomous planning of complex service robot environment manipulation challenges. Motion planning, logic-based planning and probabilistic mission planning are integrated into a single system and planning models are generated using Programming by [human] Demonstration (PbD). The strength of planning models arises from the flexibility they give the robot in dealing with changing scenes and highly varying sequences of events. This comes at the cost of complex planning model representations and generation, however. Manually engineering very general descriptions covering a large sets of challenges is infeasible as is learning them exclusively by robot self-exploration. Thus, we present PbD for planning models together with generation of parameters from analysis of geometric scene properties to tackle that difficulty. Experimental results show the applicability of these techniques on natural learning and autonomous execution of complex robot manipulation challenges.

1 Learning and Reasoning in a Layered System for Autonomous Environment Manipulation of Service Robots

The hierarchical system for autonomous environment manipulation consists of three layers:

1. Manipulation strategy learning and execution, performing skills by constrained motion planning for manipulators in geometric scenes.

Rainer Jäkel · Sven R. Schmidt-Rohr · Martin Lösch · Rüdiger Dillmann
Humanoids and Intelligence Systems Lab, Institut für Anthropomatik,
Karlsruher Institut für Technologie, Adenauerring 4, 76131 Karlsruhe, Germany
e-mail: {rainer.jaekel,martinloesch,ruediger.dillmann}@kit.edu, srsr@ira.uka.de

Steffen W. Rühl · Zhixing Xue
FZI Forschungszentrum Informatik, Haid-und-Neu-Strasse 10-14,
76131 Karlsruhe, Germany
e-mail: {ruehl,xue}@fzi.de

B. Siciliano (Ed.): Advanced Bimanual Manipulation, STAR 80, pp. 1–57.
springerlink.com

2. Scene driven logic-based reasoning for scheduling and monitoring strategies depending on scene configurations and unfolding events.
3. Probabilistic mission-level decision making and learning thereof, selecting abstract tasks based on situations.

Within the scope of this chapter, a skill is considered being a trajectory level self-contained action, e.g. a pour-in-from-bottle or unscrew-bottle-lid. A *manipulation strategy* is a constrained motion planning representation of such a skill in the form of a graph organizing constraints in a temporal and causal manner. The strategy graph allows motion flexibility while guaranteeing important motion aspects of a skill. Qualitative properties are encoded in a quantitative manner with constraints e.g. being intervals of geometric object relations, velocities, forces etc. That way, a skill can be executed in different scenes and under different courses of events. Yet, the representation is so complex that it is infeasible to be generated manually or by self-explorative robot learning alone. Instead, natural human demonstrations of exactly these skills can be recorded, analyzed and a suitable strategy graph inferred from several demonstrations. A fine grained sensor setup, recording human manipulation skills is used in the presented system. First, demonstrations are analyzed for important geometric, temporal and force relations. Then, constraint sets are generalized and finally refined by trials in dynamics simulation. By these means, an alphabet of common manipulation skills can be created, which serve as elementary actions for higher layers of abstraction. During runtime, a collision-free motion trajectory, valid according to the strategy, given a geometric scene setup, can be planned online. Thus, in summary the manipulation strategy layer provides easy access to a discrete alphabet of complex manipulation skills during runtime situations for high level planning. It is presented in detail in Sect. 2.

Scene driven reasoning and planning manages execution of manipulation strategies in a given scene. Given a primary strategy to be executed or a more abstract task pattern, scene driven reasoning checks the applicability of a strategy on a given scene and may schedule auxiliary strategies, which transform the scene. These can be strategies which rearrange the scene in a way that a primary strategy can be applied: e.g. moving objects blocking other objects to be manipulated, out of the way. To achieve this, expressions suitable for logic-based planning are generated reflecting a scene configuration. Dynamics simulation is used to predict effects of certain strategies on the overall scene. A logic-based planner is then able to determine sequences of strategies which can transform the scene into a configuration suitable for application of a primary strategy. A monitoring process continuously checks the scene for the need of replanning. Scene driven reasoning is discussed in detail in Sect. 3.

On the most abstract level, strategic, *mission level* decision making selects coarse actions based on a situation belief. Such a coarse action is modeled being a whole strategy complex, as being executed by scene driven reasoning. Additionally, the strategic level integrates mobility and natural human-robot interaction to provide a complete autonomous behavior framework to a service robot. Strategic mission level decision making is modeled as a partially observable Markov decision process (POMDP). Description of discrete states is achieved by a filtering system which

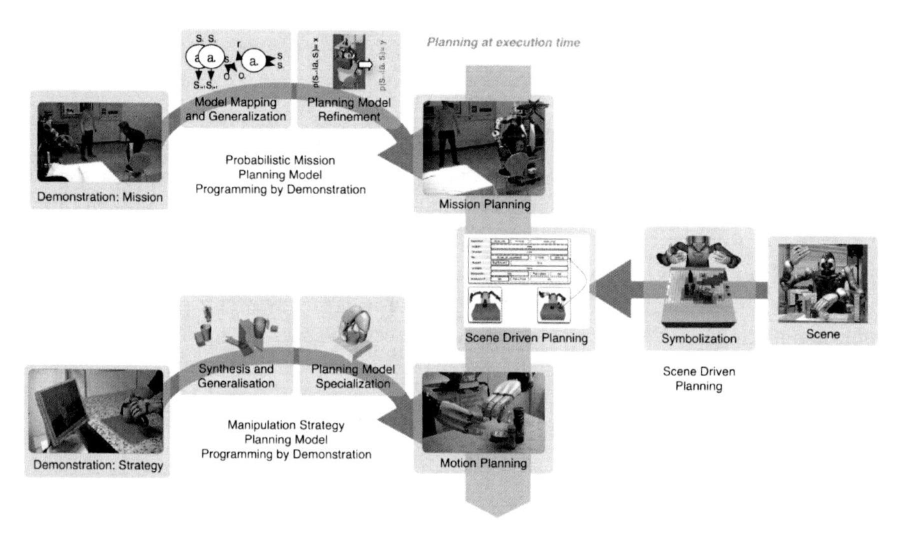

Fig. 1 A scheme of the autonomous planning system.

maps all perception component observation onto distinct state symbols. By these means, a situation belief state is updated continuously during execution time. A new, abstract action is selected by query to a pre-computed policy after the previous action terminated. However, POMDP models grow very complex for typical missions including autonomous manipulation. The model from which the policy is computed encompasses state space description and mapping, action space description, effect and observation probabilities as well as reward values. As with manipulation strategies it is infeasible to generate these models manually or to derive them from autonomous robot self-exploration. Accordingly, these models are also generated by PbD. In this case, the demonstration observation is more coarse grained, but includes more observation channels. Multiple demonstrations are analyzed, abstraction is performed, followed by model space exploration, refinement and finally policy computation. This process is described more closely in Sect. 4.

2 Programming by Demonstration of Robot Manipulation Strategies

In the human environment, the workspace of a service robot is typically restricted by self-collisions, collisions with objects and additional constraints, which are relevant to a given task. In order to execute a skill, i.e. a trajectory level self-contained action, in this restricted workspace and to be able to adapt it to different objects, local search techniques won't be sufficient since the full configuration space of the robot has to be considered. Constrained motion planning allows searching globally for a path from a start configuration to a work space goal region while considering constraints in task and joint space. The manual definition of all relevant task constraints in a subsymbolic representation, which can be used in the constrained motion planner, is time-consuming, error-prone and requires expert knowledge about manipulation and motion planning.

In the proposed DEXMART system, a skill will be represented as a *manipulation strategy*, which contains all relevant task constraints in a subsymbolic representation suitable for constrained motion planning. A symbolic action will be mapped automatically to a learned manipulation strategy during the PbD of mission planning models process, see Sect. 4. A manipulation strategy represents an action on the lowest level of abstraction and will be executed in a given scene using constrained motion planning. It will be learned based on the observation of a human teacher, who performs the manipulation task with his/her own hands in a natural way observed by different sensor systems. The basis of the flexibility of the learned planning model is the set of task constraints, which has to be obeyed to execute the task successfully. One of the key problems is to automatically deduce a minimal set of task constraints, which admits a successful execution of the manipulation task in different environments with varying objects and obstacles. Two different, complementary approaches will be discussed. In the first approach, additional, more complex human demonstrations are used to prune inconsistent constraints, following the idea of *curriculum learning*. In the second approach, the human teacher demonstrates example problems, which the robot has to solve autonomously. In this optimization process, a minimal number of task constraints is removed until a valid solution to the example problem can be found. The resulting planning model is executed in new environments using constrained motion planning. Planning times will be reduced online by learning search heuristics, which can be regarded as a specialization of the learned task constraints to the robot. Multiple experiments on the bimanual, dexterous FZI DEXMART demonstrator *Adero* with a total of 40 DOFs show the validity of the approach.

2.1 Related Work

Manipulation strategies represent a planning model for constrained motion planning. The configuration space of anthropomorphic robots with two arms and two human-like hands is high, e.g. 7 DOFs for each Arm and 13 DOFs in each hand in the FZI DEXMART demonstrator. Efficient algorithms, e.g. RRT [50] or PRM [45], to solve the path planning problem in such high dimensional spaces exist. In recent years, the algorithms were adapted to consider geometric and dynamic task constraints. In [83], different projection techniques were introduced to project a given configuration into the space defined by a set of task constraints. The projection techniques were integrated into a bidirectional RRT planner [9] to efficiently plan on constraint manifolds. In our work, an additional projection technique is introduced to consider contact constraints in the planning process. In [8], the constraint representation from [83] is used to efficiently represent goal regions for the planning process. In our work, the planning model representation was extended to represent skills requiring more than one planning step and additional constraints.

In PbD, different representations to represent motions on the trajectory level were developed, e.g. Gaussian Mixture Models (GMM) [15] and Dynamic Movement Primitives (DMP) [40]. A GMM represents a set of example trajectories as a Gaussian distribution on a temporal-spatial space. It represents explicitly the mean and variance of the set of example trajectories and the correlation between state variables. A DMP describes a set of example trajectories as the solution space of a set of differential equations. The algorithms were applied to different tasks, e.g. table tennis [62], playing pool [63], performing a chess move [16] or archery [49]. The state space is usually defined manually to capture only relevant aspects of the set of examples trajectories, e.g. in the chess example the orientation of the chess piece and the hands was ignored, which is a precondition to generalize learned knowledge to new environments. GMMs and DMPs are executed using a fast control algorithm, which allows adapting to changes in the start and goal configuration but doesn't allow for global obstacle avoidance.

Since the variety of objects is large and infinite object arrangements can be encountered in typical household settings, sufficient generalization capabilities of the robot are necessary to execute a task successfully. In contrast to a control approach with a reference trajectory, goal-directed reproduction of skills [27] offers potentially higher generalization capabilities since only the effects of the manipulation will be reproduced. The definition of the state space is the basis of the definition of the goals and task constraints, which should be reproduced by the robot. Manual definition of the state space is unfeasible for an autonomous robot and non-expert teachers. An unnecessary large state space hinders generalization since all features have to be present in the execution environment, e.g. if the manipulation was learned on a desk pad but the pad is missing in the execution environment, the skill cannot be executed. Different techniques to automatically reduce the number of features in the state space are available. In [60], task space pools are described, which consist of a set of predefined learning features, e.g. the motion of the hand relative to the object of attention, from which a set of relevant features will be chosen based on a selection criterion. Three different criteria were defined: an attention, a variance and a kinematics criterion. The criteria can be used to weight the features as well as remove features from the state space. They were applied to a single example but no data about the reduced set of features or the result in the reproduction is available.

Additionally, gazing at an object [13] and pointing gestures [10] were used to clarify ambiguities in interaction with the robot system. Since the robot has to explicitly interact with the human, a semantic representation of the learning features is required and only a small set of features can be considered. In our work, an initially large set of features, i.e. task constraints, will be incrementally pruned using additional demonstrations by the human teacher and a large set of simulation trials to deduce a minimal set of features, which admits a successful reproduction of the skill on a set of example problems.

In [82], a dexterous manipulation task, in which a lid had to be removed from a glass with the fingertips, was partially learned based on human observation. The

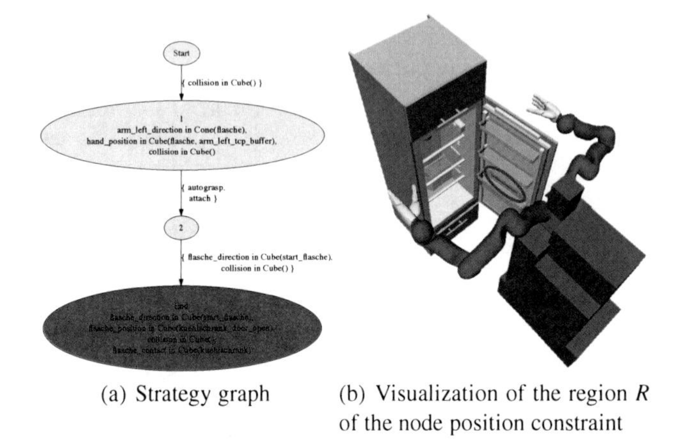

(a) Strategy graph (b) Visualization of the region R of the node position constraint

Fig. 2 Manipulation strategy to grasp a bottle and place it in a fridge door [41].

correspondence problem, i.e. the problem of mapping a motion to an actor with different geometrics, kinematics and dynamics, was considered by manually adapting the human motion until the mapping was visually valid in simulation. In contrast to this work, we consider the correspondence problem in the planning process by automatically enlarging the search space for finger and hand motions while keeping the constraints restricting the object motion.

2.2 *Manipulation Strategies*

In the human environment, an autonomous service robot has to be able to adapt a manipulation motion to a large variety of objects, different object poses and a restricted workspace. In order to plan a robot motion in a goal-directed way the goals and task constraints of the task have to be represented explicitly.

In this section, a manipulation strategy will be defined based on a set of position, orientation, direction, force, contact and force closure constraints. The manipulation strategy will be executed using a bidirectional constrained motion planner and novel projection techniques for contact and force constraints.

2.2.1 Representation

The definition of manipulation strategies is based on the definition of temporal and domain constraints. Let $\mathscr{C}_{\mathrm{R}}$ be the configuration space of the robot. The configuration space with a temporal component $\mathscr{C}$ is defined as $\mathscr{C} = \mathscr{C}_{\mathrm{R}} \times \mathbb{R}$.

A temporal constraint is defined as $(l,u) \in \mathbb{R}^2, l \leq u$. It will be obeyed in a configuration $(\theta,t) \in \mathscr{C}$ if and only if $l \leq t \leq u$.

Six different types of domain constraints exist: configuration, position, orientation, direction, force and contact constraints.

Configuration constraints restrict the joint angles of the robot with n DOFs directly. Similar to temporal constraints, they are written as $(l_j,u_j) \in \mathbb{R}^{2n}, 1 \leq j \leq n$ and will be obeyed, if and only if $l_j \leq \theta_j \leq u_j \, \forall j$.

Position, orientation, direction and force constraints will be represented as (a,b,R) with the (coordinate) frames a and b and a region $R \subseteq \mathbb{R}^3$. They restrict the value of the frame a, expressed in the frame b, to stay in the region R. Let ${}^gT_f$ be the homogenous transformation matrix, which describes the position and orientation of frame f relative to frame g. ${}^bT_a$ will be transformed into a 3D vector ${}^bt_a$ depending on the type of constraint. The constraint will be valid, if and only if ${}^bt_a \in R$. For position and force constraints ${}^bt_a$ is the translational part of ${}^bT_a$ and for orientation and direction constraints, the rotational part of ${}^bT_a$ will be converted into a scaled axis representation, see [42].

A contact constraint is defined as (a,b,R). It will be obeyed, if and only if the two unique 3D models associated with a and b are in contact and the contact normal from b to a is included in R.

The common structure of all constraints, i.e. a vector t has to be included in a predefined region R of the same dimension is exploited to define a distance measure. For each constraint c, $d(c,\theta,t)$ will be the distance to the constraint manifold described by c in the configuration (θ,t):

$$d(c,\theta,t) = \min_{y \in R} \|x - y\|$$

A manipulation strategy is defined as (N,E,C_t,C_d), where N is a set of nodes, $E \subseteq \{(u.v) \in N \times N\}$ a set of edges, C_t assigns a set of temporal constraints to each node $u \in N$ and each edge $(u,v) \in E$ and C_d assigns a set of task constraints to each node $u \in N$ and each edge $(u,v) \in E$. It can be visualized as a directed graph, which will be called *strategy graph* [43], see Fig. 2(a). Each node v represents a subgoal of the manipulation task, i.e. the goal $G(v)$ for the constrained motion planner is implicitly defined by all configurations, in which all task and temporal constraints are obeyed:

$$G(v) = \{(\theta,t) \in \mathscr{C} \,|\, d(c,\theta,t) < \varepsilon \, \forall c \in C_t(v) \cup C_d(v)\}$$

Edges represent the transition from one subgoal to the next. The motion associated with the transition is also restricted by a set of task and temporal constraints, which implicitly defines the search space for the constrained motion planner:

$$\{(\theta,t) \in \mathscr{C} \,|\, d(c,\theta,t) < \varepsilon \, \forall c \in C_t(u,v) \cup C_d(u,v)\}$$

Manipulation strategies can be regarded as a temporal constraint satisfaction problem with domain constraints, see [42].

Example 0.1. In Fig. 2(a), a manipulation strategy to place a bottle inside a fridge door is shown. The goal of the manipulation is described by a direction constraint, which restricts, that the symmetry axis of the bottle has to be aligned to the world's z axis, a collision constraint, which restricts, that no collision occurs, a contact constraint, which enforces, that the bottle bottom and the fridge door are in contact, and a position constraint, which restricts the position, where the bottle can be placed in the fridge door. The latter is visualized in Fig. 2(b) as a cubic region.

2.2.2 Instantiation

In the learning environment, for each object in the KIT ObjectModels web database [44] a set of coordinate frames is predefined, e.g. for a bottle a frame in the opening, the center of mass and the bottom is defined. In a learned manipulation strategy, which belongs to the created alphabet of common manipulation skills, all objects will be replaced by variables, e.g. *flasche_opening* will be replaced by $object_1$*_opening*. In order to execute the manipulation strategy on a robot system in a new environment, all variables have to be instantiated. The mapping of frames referring to objects in the scene will be done by the caller of the planner, e.g. the execution time system in section 4, by assigning values to the variables. The mapping will be valid, if the object contains all referenced frames.

Constraints restricting frames of the human, i.e. fingertips and wrists, will be mapped automatically to the current robot system. The *correspondence problem*, i.e. the problem how a motion can be mapped to a manipulator with different geometrics, kinematics and dynamics, is considered by relaxation of constraints restricting the motion of the human fingers or wrists. The relaxation depends on the human teacher and the robot hands. For the SAH, finger position constraints will be relaxed by 5 mm to consider the different finger widths, wrist position constraints will be relaxed by 40 mm due to the different hand sizes and wrist orientation constraints will be relaxed by 15°. Since force, contact constraints as well as constraints, which restrict the motion of the manipulated objects, the qualitative effects of the manipulation motion are not altered but the search space for finger and wrist motion is enlarged to enable the planner to find a similar robot finger and hand motion, which produces the same qualitative effects.

2.2.3 Execution

Manipulation strategies will be executed using a constrained bidirectional RRT based on CBiRRT [9]. The projection technique to project an arbitrary configuration to a constraint manifold is RGD [83], which approximates the gradient of the distance to the constraint manifold online based on a sample strategy. Although slower than Jacobian-based approaches, e.g. TS or FR [83], RGD allows considering arbitrary constraints in the planning process.

For contact constraints (a,B,R), a (special) contact projection is defined, which enforces, that the two referenced 3D models assigned to a and b will have a contact. Let $(\theta,t) \in \mathscr{C}$ be the configuration before projection. If θ is in collision, the nearest collision-free configuration is generated using a library by Min Tang based on the penetration depth [47]. The collision free configuration is incrementally refined by moving each 3D model in opposite direction of the contact normal sampled from R until a contact occurs.

Learned manipulation strategies have a unique start node $Start$ and end node End, which are connected by a single path $Start \mapsto 1 \mapsto \ldots \mapsto n-1 \mapsto End$. Each segment $u \mapsto v$ will be planned sequentially. The set of goal configurations for the constrained motion planner will be a subset of $G(v)$. Due to the interaction of constraints, $G(v)$ cannot be sampled directly. Since the constraint manifold may be thin, drawing a sample from $\mathscr{C}$ and rejecting it if not in $G(v)$ is also not sufficient. An initial configuration will be generated based on a subset of constraints using inverse kinematics. The initial configuration will be projected to the constraint manifold $G(v)$ using RGD and the contact projection. If all constraints are obeyed, the final configuration will be stored as a goal configuration. In the planning process, random samples will be drawn and the RGD and contact projection will be used to project the configuration to the constraint manifold $G(u,v)$, i.e. the search space for the manipulation planner.

If the planning process is successful, the solution is guaranteed to obey all task constraints, i.e. the arc constraints $C_d(u,v)$ will be obeyed in all points of the solution and the node constraints $C_d(v)$ will be obeyed in the final point of the solution.

2.3 *Observation*

The human teacher demonstrates the skill with his/her own hands in a sensory environment, see Fig. 3, in a natural way. The demonstrations are recorded using a stereo camera system with DragonFly II cameras, two Fastrak motion trackers, two Cyberglove II datagloves and a custom-built glove with tactile sensors based on an elastomer with changing electrical conductivity.

The vision library IVT [6] is used to detect and localize objects in the scene. In order to calculate contact points in the fingertips of the human based on the joint measurements with the datagloves, a model of the human hand is used. The hand model was generated using a high-accuracy laser scanner, see Fig. 4(a). The bone structure was added manually, see Fig. 4(b). Contacts are calculated based on a 3D visualization of the observation result using PQP [33].

Each demonstration is represented as a trajectory of two 6d poses for the left and right human wrist, two sets of 20 joint angles for each human hand, one 6d pose for each recognized object and a set of contact pairs.

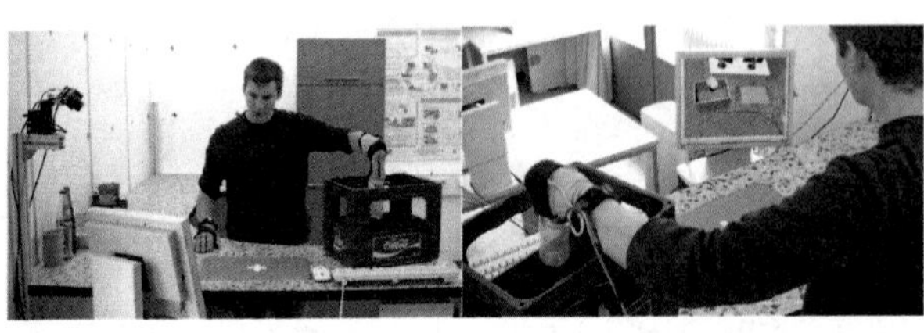

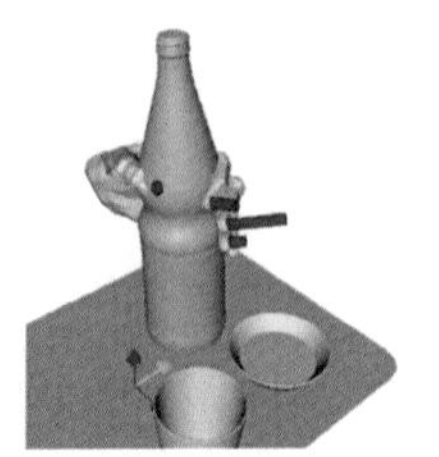

(a) Stereo camera with pan-tilt unit, two datagloves, two motion trackers and 3D visualization of observation result

(b) Tactile measurements

Fig. 3 Sensory environment: human demonstration to place a bottle in a crate.

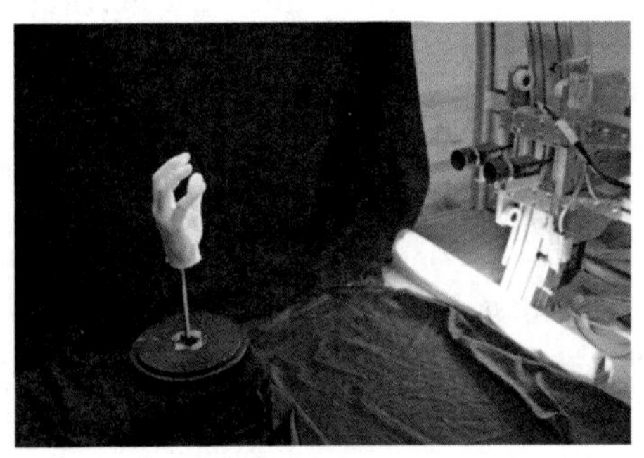

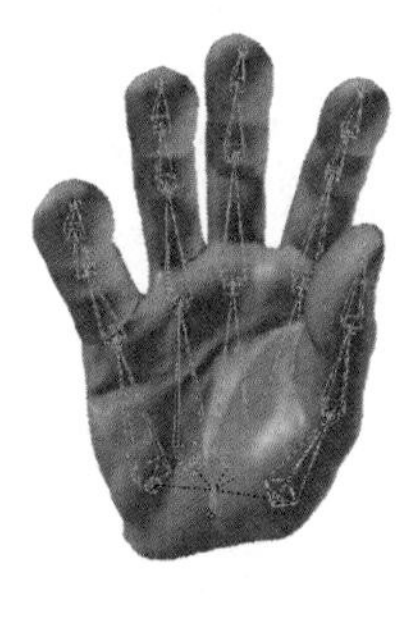

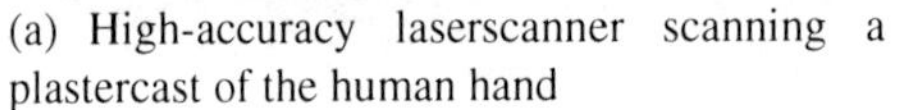

(a) High-accuracy laserscanner scanning a plastercast of the human hand

(b) Bone structure for the hand model

Fig. 4 Generation of 3D human hand model for the observation of dexterous manipulation tasks.

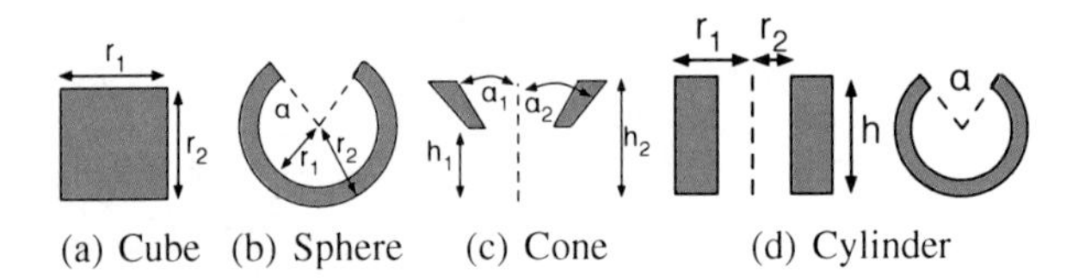

(a) Cube (b) Sphere (c) Cone (d) Cylinder

Fig. 5 Constraint volume types [43].

2.4 Preliminary Manipulation Strategy Generation

In order to learn a new skill, the human teacher demonstrates the skill multiple times in the sensory environment generating two sets of demonstrations. In the first set, the goal is to demonstrate the skill on a simple setup, e.g. with no collision and common object poses. In the second set, the teacher demonstrates the skill on more complex or different setups, e.g. with different objects of the same type. The first

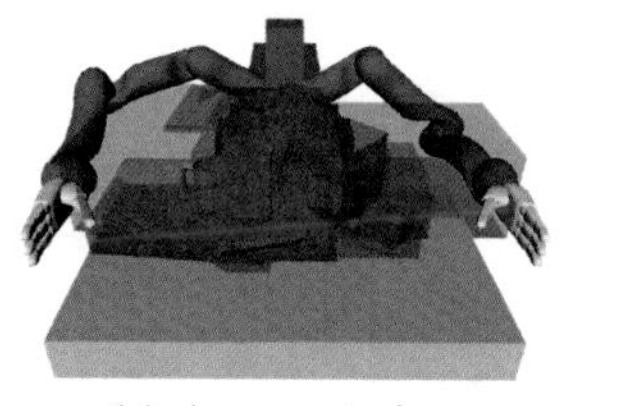

(a) Arc constraints

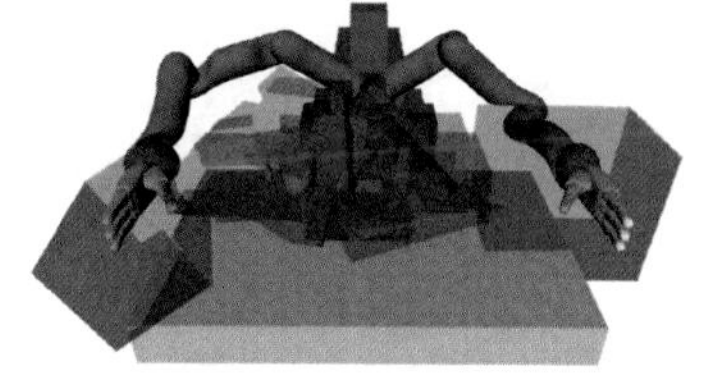

(b) Node constraints

Fig. 6 Task constraints for a (non-dexterous) pour-in task generated using the initial set of task constraints and cubic regions.

set of training data will be used to generate the initial manipulation strategy. The second set will be used to reduce the number of learning features automatically in order to improve the generalization capabilities of the skill. In the remainder of this section, the first set of training data will be used.

The demonstrations will be segmented in order to generate the structure of the initial manipulation strategy. Based on the structure and the recognized objects, the state space, i.e. the set of task constraints to learn, will be generated. For each task constraint, the parameters, i.e. the region R of minimal volume will be generated, so that all constraints are obeyed on the demonstrated trajectories.

In general, non-dexterous tasks, in which only grasping, transporting and ungrasping of objects is necessary, will be treated differently to dexterous tasks, in which force interaction with the fingertips and object motion without rigid contact, e.g. pushing, are important.

2.4.1 Segmentation

The demonstrations will be segmented based on the assumption that the velocity of fingers, hand and forces is low at important points of the demonstration. The motivation is based on the common assumption in motion planning, that the robot is at rest before and after the planning process.

Based on a simple velocity based threshold algorithm, see [29], the demonstrations are segmented. For non-dexterous tasks, grasp classification will be applied in each segmentation point and will be removed, if the grasp classification is inconsistent. In general, demonstrations will be rejected, if the segmentation is different to the segmentation of the first demonstration.

For each segmentation point, a node will be generated in the initial manipulation strategy. For two consecutive segmentation points, an edge will be generated, connecting the nodes assigned to the segmentation points.

For non-dexterous task, in each segmentation point either a grasp was detected or not. For each hand, a segment will be labelled as *grasp*, if no grasp was detected in the first segmentation point and a grasp was detected in the second segmentation point. The labels *ungrasp*, *transport* and *free* will be assigned accordingly.

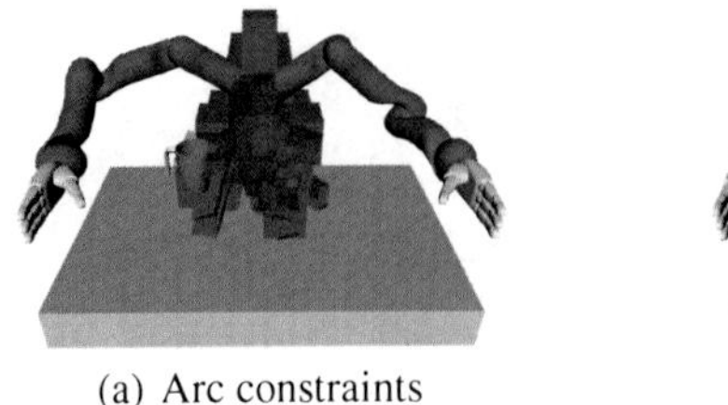

(a) Arc constraints

(b) Node constraints

Fig. 7 Reduced set of task constraints, see Fig. 6, using *Teaching*.

2.4.2 State Space

In order to generalize a skill to different environments and robots, the goals and task constraints have to be explicitly represented. In the manipulation strategy, goals and task constraints will be represented as sets of constraints. Since each constraint has to be obeyed in the execution environment and the referenced coordinate frames have to exist, one of the key problems is to generate a minimal set of task constraints, which allows executing the task successfully.

A large set of task constraints will be generated automatically, similar to task space pools [60], for each node and arc in the strategy graph. The set will be generated based on three sets of coordinate frames: A, B and C. For dexterous tasks, all sets contain the frames in the human fingertips, the human wrists and all frames (in the database) of all recognized objects. For non-dexterous tasks, A will contain the human wrist, if a *grasp* or *move* was detected on the current or ingoing arc, otherwise all frames of the grasped object will be added. B and C are equal and contain all frames of the recognized objects in the database. For all frames in B and C *start* frames will be added, which represent the value of the frame at the beginning of the current or ingoing arc.

For each constraint type and each $(a,b,c) \in A \times B \times C$ a constraint will be generated, where a is the first frame of the constraint and the second frame has the position vector of b and the rotation matrix of c. The set of constraints will be filtered to ensure that the resulting manipulation strategy can be used in the constrained motion planner, e.g. constant constraints will be removed, resulting in the initial set of constraints. For dexterous tasks, force constraints will be added constraining the force in the five fingertips of the human hand relative to the frame in the center of mass of the nearest object.

2.4.3 Parameters

For each constraint (a,b,R) in the initial set, the parameters of the region R have to be determined, so that R has minimal volume and the constraint is valid on the corresponding segments of all demonstrations. Following a bag of algorithms

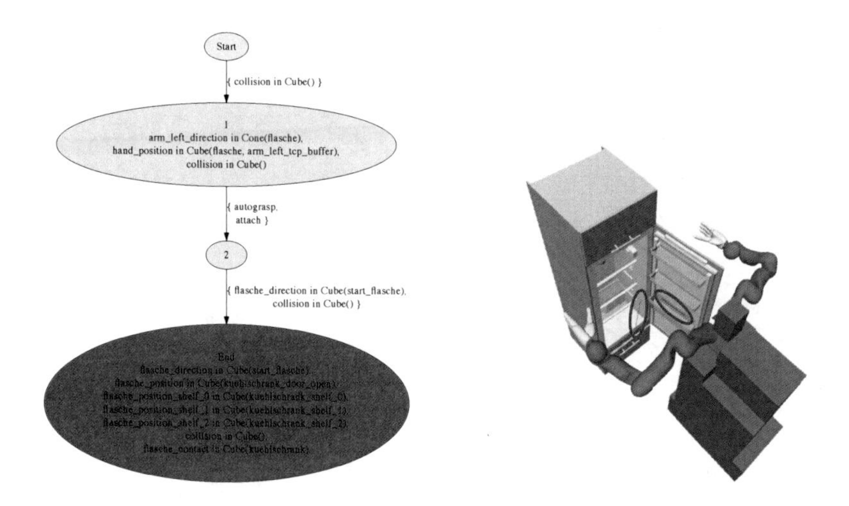

Fig. 8 Planning model to grasp a bottle and place it in a fridge door [41].

Fig. 9 Non overlapping constraints (circles) for a different door angle in the fridge experiment [41].

approach, a region with minimal volume for different region types, e.g. cube, sphere, cone and cylinder cuts, see Fig. 5, is generated and the minimal result is taken.

2.5 *Generalization*

The initial manipulation strategy is overspecialized. It contains a large number of automatically generated task constraints, which describe the subgoals and the transitions between subgoals of the skill. Since all referenced coordinate frames have to exist and all generated task constraints, even if not relevant to the task, have to be obeyed in the execution environment, generalization is limited. In Fig. 6, the initial set of task constraints in the pour-in task are shown.

Two complementary approaches to reduce the number of task constraints in the initial manipulation strategy were investigated: *Teaching* [43] and *Robot Tests* [41].

2.5.1 Teaching

Following the idea of curriculum learning [7], the second set of demonstrations, which represents human solutions to more complex problems or problems, the robot should be able to generalize to, is exploited to remove irrelevant task constraints. The second set is segmented in the same way as the first set. A constraint will be removed, if it is not obeyed on the corresponding segment on one of the second

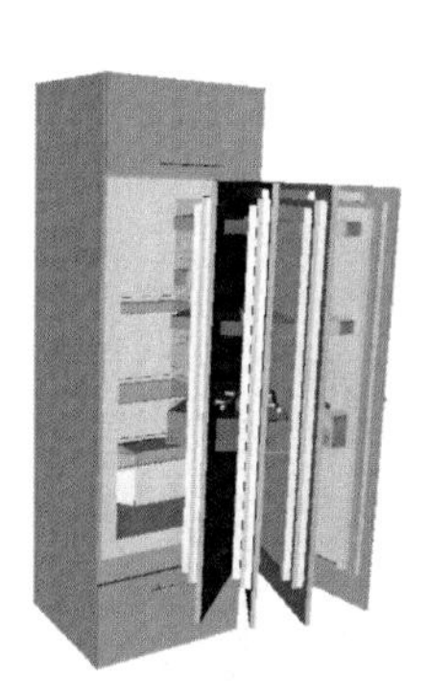

Fig. 10 Test: different door angles [41].

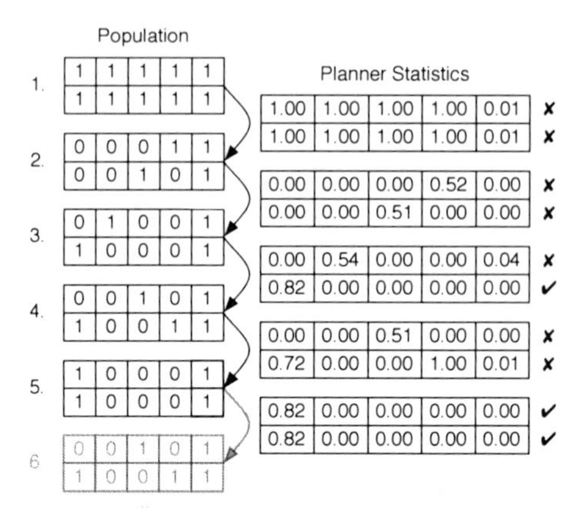

Fig. 11 Example of the evolutionary algorithm applied to the example in Fig. 8 using 2 states. Left: population in iteration 1 to 6. After iteration 1, the mutation operator is applied. Right: calculated planning statistics for each individual in iteration 1 to 5 [41].

set of demonstrations. In Fig. 7, the resulting set of task constraints after applying *Teaching* to the example in Fig. 6 is shown. More details can be found in [43].

2.5.2 Robot Tests

Since the number of human demonstrations is limited, the set of task constraints of the manipulation strategy after teaching is in general not minimal. In the example in Fig. 8, a bottle has to be grasped and moved to a position in a fridge door. Assuming that the door angles were similar during learning and teaching, the last goal of the manipulation task, i.e. that the bottle has to reside in the fridge door, is described by one position constraint, which restricts the bottle bottom relative to the fridge door, and three position constraints, which restrict the bottle position relative to the shelves. This ambiguity was not resolved automatically but greatly influences the generalization capabilities of the skill. In a new environment with different door angle, the planning process will fail since both constraints, i.e. relative to the door and relative to the shelves, cannot be obeyed at the same time, see Fig. 9. Statistics about such inconsistencies will be gathered and used to remove irrelevant task constraints automatically.

In order to further generalize the learned manipulation strategy, the human teacher defines a set of robot tests. Each robot test consists of a set of objects, a set of object poses and a mapping of objects to objects in the manipulation strategy, which is used to instantiate the manipulation strategy. In the fridge example, each robot test corresponds to a specific door angle, see Fig. 10.

The goal of the generalization algorithm is to deduce a maximal subset of constraints, which admits a successful solution to all robot tests. By using the maximal

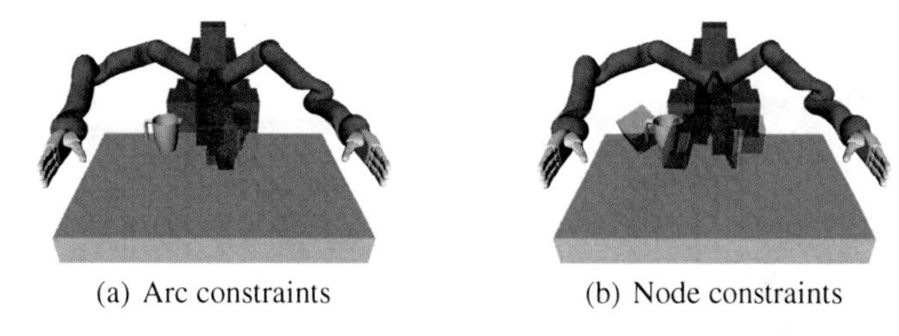

(a) Arc constraints (b) Node constraints

Fig. 12 Reduced set of task constraints, see Fig. 6, using *Robot tests*.

subset, it will be ensured, that as many as possible learned task constraints will be maintained. In order to test whether a subset of constraints fulfills the goal condition, the manipulation strategy will be used in the constrained motion planner to generate a solution to all robot tests. If no solution can be found, statistics about which constraints were inconsistent will be gathered. The resulting high-dimensional combinatorial optimization problem will be solved using Evolutionary Computation [25], which allows the computation of the goal condition to be parallelized. In Evolutionary Computation, an initial set of states, called initial population, will be incrementally modified using different operators, e.g. mutation of single states, and the value of the objective function of each modified state, called fitness, will be used to determine the follow-up population.

The manipulation strategy will be mapped to a binary state, where each bit corresponds to one task constraint, which can be deactivated potentially. In order to determine the fitness of a state, the corresponding manipulation strategy will be used in the constrained motion planner to calculate a solution for each robot tests. If the planning process fails, statistics about which constraints failed during the planning process will be gathered.

The initial population consists of n states, where each bit is set to 1, i.e. the optimization process focuses on the deactivation of constraints, which is more beneficial since only statistics about which constraints should be deactivated are available.

In each iteration, all states will be mutated using a custom mutation operator. If the planning was successful, a random deactivated constraint will be activated, i.e. the corresponding bit set to 1, to maximize the number of constraints. If planning failed, each active constraint will be deactivated according to the failure probability, which corresponds to the fraction of unsuccessful constraint evaluation on the total number of constraint evaluations in the planning process.

The fitness of each state is calculated based on the number of active constraints in each arc and node, the planning result and the progress in the planning process.

The algorithm will be stopped, if the best individual hasn't changed in 20 iterations or a time threshold was met.

The optimization algorithm was implemented using the library evolving objects [46]. The calculation of the constraint statistics was parallelized to 24 CPU cores. An example of the execution is shown in Fig. 11.

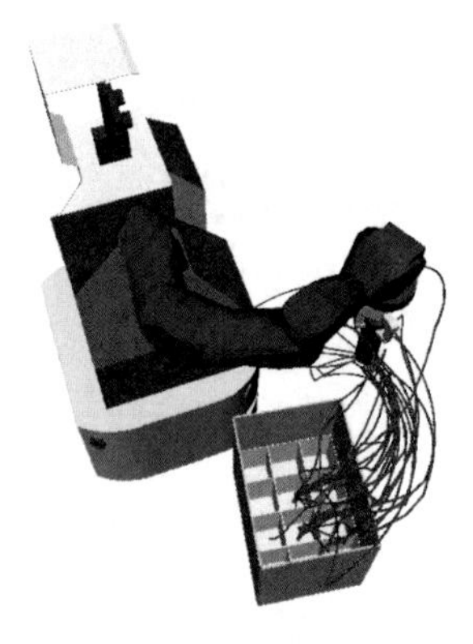

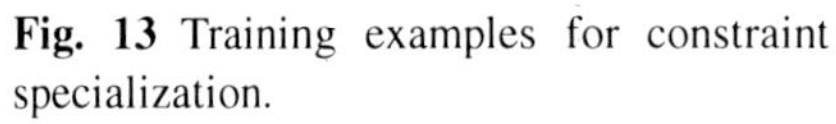

Fig. 13 Training examples for constraint specialization.

Fig. 14 Temporal alignment of paths in one cluster.

Finally, the solutions are demonstrated to the human teacher, sorted by fitness, who picks the first valid solution. In the experiments, the $1.1th$ solution was chosen on average.

In Fig. 12, the resulting set of task constraints after using *Robot tests* for the example in fig 7 is shown. More information can be found in [41].

2.6 Specialization Using Trials in Dynamics Simulation

The resulting manipulation strategy contains a maximal set of learned task constraints, which admits a successful execution of the skill in different environments and with different objects with similar geometry. It can be interpreted as a qualitative description of the goals and constraints of the given task. By using constrained motion planning, the robot infers in a goal-directed way a joint motion for which all constraints are obeyed and all subgoals will be reached. Additionally, the joint limits of the robot, obstacles in the execution environment and the geometry of the manipulated objects are automatically considered, which is an advantage compared to a low-level mapping of joint trajectories on the control level. The drawback of this flexibility are high planning times, especially for dexterous manipulation tasks, in which hand and finger motion have to be planned at the same time. In this section, the learned manipulation strategies will be specialized to the robot and task by refining each position, orientation or direction constraint based on multiple trials in dynamics simulation or on the real robot. The refined constraints will be used to execute the manipulation strategies with a fast control algorithm. Planning with the original manipulation strategy in parallel ensures, that a speed up can be guaranteed.

In the manipulation strategy, each position, orientation and direction constraint will be refined by learning a Gaussian Mixture Model (GMM) on the 3D constraint manifold. The training data is generated by executing the generalized manipulation

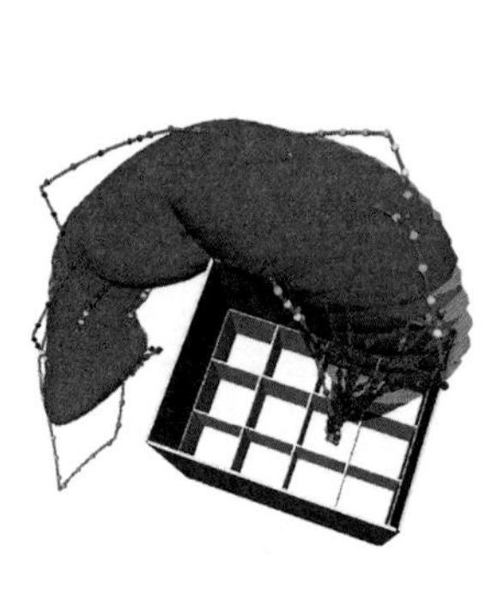

Fig. 15 Learned GMM on the temporally aligned data in Fig. 14.

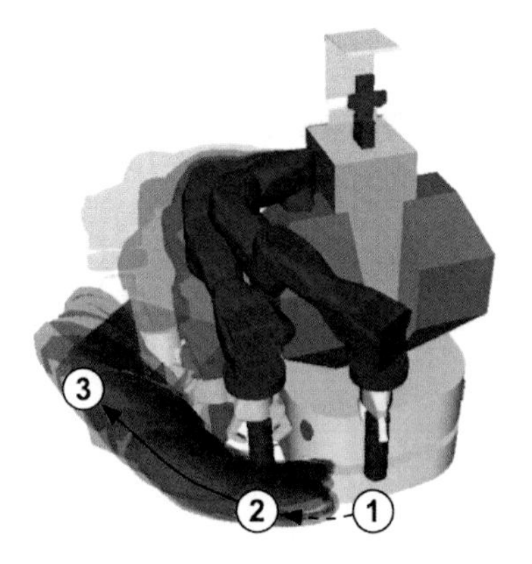

Fig. 16 The start configuration is not included in the GMM (1), a path is planned to a configuration in the GMM (2) and the controller is used to generate a motion to the goal (3).

strategy in different environments in simulation, e.g. using the object arrangements in the human demonstrations or defined robot tests, and on the real robot. Each demonstration is segmented according to the structure of the manipulation strategy. For each constraint $c \in C_d(u,v)$ of the arc (u,v), the corresponding segment is projected to the constraint manifold of c and the resulting 3-dimensional trajectory is stored as an training example. In Fig. 13, a bottle has to be placed in a crate. The resulting training examples for the position constraint, which restricts the bottle relative to the crate, are shown.

For each constraint, the training data is temporally aligned using Dynamic Time Warping (DTW) [73] and clustered using the DTW distance and hierarchical agglomerative clustering. In Fig. 14, a single, temporally aligned cluster is shown. A cluster is chosen and a GMM with fixed number of Gaussians will be generated for each position, orientation and direction constraint on the segment using the EM-algorithm. Each GMM is transformed using Gaussian Mixture Regression [14] resulting in a representation of each GMM by a mean vector and covariance matrix for each time point. Based on this representation, a set of GMMs can be executed using a simple controller, which iteratively tries to move the robot closer to the mean of all GMMs weighted with the covariance matrices in the corresponding time point, see [14]. Experimental results show, that the execution is more than 20 times faster than planning for GMMs with narrow covariance matrices but will in general fail, if the initial robot configuration is not included in the GMM, represented by the covariance ellipsoids with 3 standard deviations, see Fig. 15. Based on this observation, an additional planning step will be executed, if the initial configuration is not included in the GMM. The goal of the additional planning step contains the set of GMMs, i.e. the robot will move to a configuration, which is included in the GMM, see Fig. 16.

Table 1 Results on *robot test* approach [41].

Exp.	Tests	Constraints before/after	Chosen solution	Validation set	Success (%)
Fridge	2	5 / 2	1	10	100
Pour-in	2	174 / 60	2	15	67
Crate	3	15 / 13	1	20	100
Key	3	362 / 28	1	15	100
Cup	4	26 / 15	1	24	88
Bottle	1	16 / 8	1	20	90
Chair	1	12 / 6	1	10	100

Starting from the new configuration, the controller will be used to generate a motion for the robot system.

The controller implements a local optimization technique, which offers fast execution times for environments, which the robot system has encountered previously. Global collision avoidance, narrow passages and environments with differing object arrangements are problematic. If the controller fails, which indicates that the learned search heuristics could not be applied, the explicit knowledge about the goals and constraints of the task will be exploited to plan a motion in a goal-oriented way. By executing the controller and the constrained motion planner in parallel, the speed up for previously encountered environments as well as the generalization capabilities to novel scenes are maintained.

2.7 Evaluation

Learning and execution of manipulation strategies was evaluated on multiple tasks on the FZI DEXMART demonstrator Adero. Adero consists of two KUKA Lightweight Arm with 7 DOFs and two Schunk Anthropomorphic Hands (SAHs) with 13 DOFs. Object models were taken from the KIT ObjectModels Web Database [44]. The evaluation of the *teaching* algorithm can be found in [43]. The *robot tests* approach was evaluated on multiple examples in [41] and will be summarized here. The section will end with results about learning and execution of the dexterous manipulation task: opening a bottle with the fingertips.

2.7.1 Generalization

In [41], the *robot test* approach was applied to seven tasks: place a bottle in the fridge door, bimanual pour-in, place a bottle in a crate, pressing a key on a keyboard, grasping a cup, opening a bottle and lifting a chair. The human teacher demonstrated a set of robot tests, which was split into a training and a validation set. The training set was used to determine the maximal subset of task constraints. The resulting manipulation strategy was executed on the validation set. The results are summarized

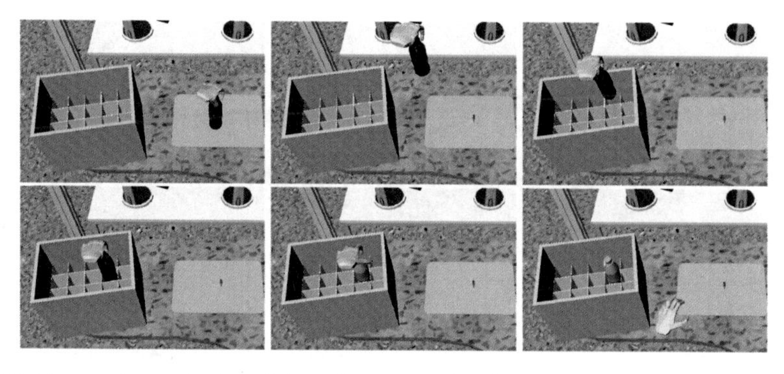

Fig. 17 Human demonstration of the task to place a bottle in a crate.

Fig. 18 Execution of planning model for placing a bottle in a crate on Adero.

in Tab. 1. The human teacher picked the 1.1th solution on average, which indicates, that human interaction is reduced to a minimum, i.e. the human is necessary to check the result for failure of implicit constraints but (s)he is not involved in finding the maximal subset of consistent task constraints. The number of constraints in all manipulation strategies could be reduced effectively and the average success rate of 92% indicates, that the heuristic to find a maximal subset of learned task constraints is valid. For manipulation strategies with a small number of constraints, the optimal solution can be found in a small number of iterations. The pour-in example showed, that a suboptimal can be found due to violation of implicit constraints.

The *bottle in crate* task will be discussed in more detail as an example. A bottle was grasped with the left hand and placed in a crate at random positions. The human demonstration is shown in Fig. 3(a) and the observation result is visualized in Fig. 17. In the learned manipulation strategy, the motion of the wrist was restricted relative to the bottle opening and bottom and the motion of the bottle opening and bottom were restricted relative to the crate. In the robot test, a bottle with different

Fig. 19 Human demonstration of bottle opening task in sensory environment and simulation environment.

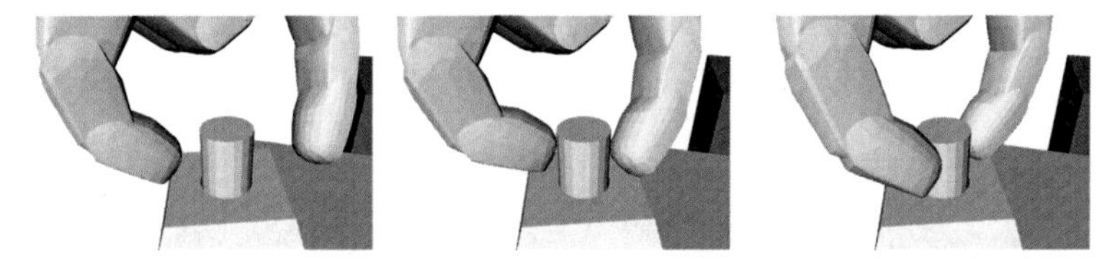

Fig. 20 Planning result for opening a bottle using physics simulation.

size was used, i.e. the bottom and opening frame were at different positions. The described constraints were inconsistent and no solution could be found to the robot test initially. In the optimization process, the constraint, which restricts the wrist relative to the bottom of the bottle will be removed since no collision-free configuration for grasping or no force-closure grasp could be found. The constraint, which restricts the motion of the bottle opening relative to the crate is also removed since no configuration, in which the bottle and the crate are in contact, could be found. After removing both constraints, the manipulation strategy could be executed successfully on the real robot, see Fig. 18.

2.7.2 Learning and Execution of Dexterous Manipulation Tasks

In this experiment, a bottle cap had to be removed by using the fingertips. In the human demonstrations, see Fig. 19, only the thumb and the index finger were used to rotate the cap. The task was considered dexterous, i.e. the motion of the human wrist and the human fingers is learned and abstracted to a set of task constraints. The constraints restricting the human wrist and fingers were enlarged to consider the correspondence problem. Since sliding contacts could not be observed with the tactile sensors force constraints, which restrict the force in the fingertips to be perpendicular to the cap surface were added in the appropriate nodes and arcs. The motion of the cap could not be tracked consistently due to occlusions and a position and orientation constraint, which restrict the cap motion relative to the bottle, were added manually. The position constraint restricts, that the cap stays at the same position during rotation. The orientation constraints restricts, that the cap is rotated between 0° and 45°. The corresponding node constraints, which represents the goal

Fig. 21 Execution of planning result for opening a bottle on Adero.

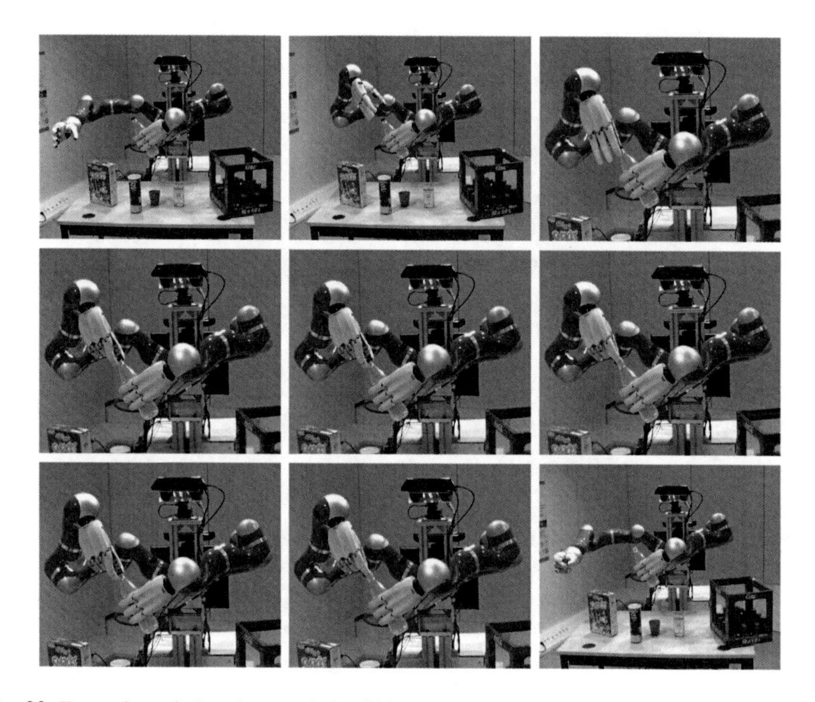

Fig. 22 Execution of planning result for lifting and opening a bottle on Adero.

orientation of the cap, restricts the cap to be 40° rotated. The resulting manipulation strategy was executed using constrained motion planning with dynamics simulation, see Fig. 20. In each planning step, i.e. each small extension of the search tree, the dynamic simulation is used to predict the movement of the cap when the motion of the fingertips is applied. The planning step will be successful if all constraints restricting the motion of the hand, the fingers, the objects and the forces are obeyed.

The manipulation strategy contained four arcs, which can be interpreted as: move the hand above the cap and open the fingers, close the fingers, rotate the cap and open the fingers. Since the rotation of the cap with the fingertips can only be executed in a small subset of the workspace of the hand, a valid hand posture has to be found in the planning process, which requires a large amount of planning time. Although dynamics simulation is used in the planning process, the finger rotation motion can be planned in only a few minutes due to the restriction of the search space based on the observation of the human. The total planning time was 5.2 minutes. The planned rotation motion is shown in Fig. 20. The whole manipulation strategy was executed on Adero, see Fig. 21.

The experimental results indicate, that dexterous manipulation skills can be learned and mapped to a robot in a goal-oriented way using a planning-based approach. For dexterous tasks, planning times are high but efficient heuristics, see Sect. 2.6, can be learned automatically based on the planning results.

3 Scene Driven Logic-Based Reasoning for Robot Manipulation Tasks

When the mission level decision making has selected a coarse action for execution, it is handled by the scene driven planning. The scene driven planning in the DEXMART system is responsible for mapping a coarse action to a sequence of symbolic actions and parameters, for which strategies exist which are able to execute the intended action in a given setup.

The need for such a mapping comes from the high level of abstraction at the mission level. The scene driven planning bridges the gap between the model of a coarse action on the one hand and the robot and scene on the other. Especially the following aspects are considered by scene level planning:

Kinematics. On the high level of abstraction, the manipulation of an object is easily described e. g. by Grasp(Object). In order to execute such a strategy, the robot has to move its gripper to a pose near the grasped object. Therefore, an arm configuration has to be calculated, which reaches that pose. This problem is known as inverse kinematics. Anyhow, such a solution may not exist. The object may be out of reach or just at a hard to reach pose in the workspace. In such a case, using the other arm of the robot is an option which must be considered, or, if the robot is mobile, moving the whole robot. Additionally, the calculated configuration must be executable without self-collisions.

Obstacles. If a robot should approach an arm configuration, it has to assure that there are no collisions between itself and obstacles in the scene. Therefore, it must model its environment and plan strategies based on that model.

Scene setup. In a case of stacked objects, if an object from the lower part of the stack is to be manipulated, the robot must not simply grasp that object. Such an action may cause the stack to collapse and damage the objects. In such a situation, the reasoning has to create a sequence of strategies, which decomposes the stack an finally allow the desired manipulation.

Since the actual scene setup is not known before execution, scene driven planning has to be done on-line. It refines an abstract description of an action to an abstraction level, where it can be evaluated, that there are configurations for the robot which could execute the associated strategy in a given scene. As pointed out, this refinement might require the scheduling of additional actions. In order to generate a sequence of actions leading to a desired goal state, a logic-based planner is employed. Queries to a classic manipulation planner ensure the feasibility of single plan steps, meaning that there is a strategy for the robot that implements the symbolic action.

The introduction of logic-based planning introduces a new challenge: the complex continuous scene in which the robot operates has to be mapped to a small but sufficiently descriptive set of symbols on which the planning operates. Since the cardinality of the symbolic space is smaller than that of the continuous space, such a mapping is an approximation of the real world. In order to verify the actual feasibility of a planned strategy, classic manipulation planner are employed to map back from the symbolic domain to the continuous. Finally, when planned strategies are executed on the robot, it has to be assured that resulting world states develop according to the generated plan. Therefore, the logic-based planner schedules monitoring assertions which are on-line monitored during the execution of strategies. In order to adapt those monitoring constraints to initially unknown actions, a supervised learning method is utilized.

Based on the analysis of the scene driven planning, we structure the component into three sub-components. The major part is the symbolization of the observed scene and possible actions. It is described in Sect. 3.2. Based on the symbolic description, planning domains have to be defined and a logic-based planer is integrated into the system as described in Sect. 3.3. Finally, the monitoring component ensures consistency between a plan and its execution. It is described in Sect. 3.4.

3.1 Related Work

The scene driven logic-based reasoning for robot manipulation tasks combines classical symbolic AI planning methods in robotics [37] with continuous planning methods [11]. Approaches from the AI world tend to ignore the problems associated with grasping and kinematics, while robotic planning algorithms lack the versatility of configurable, arbitrary goals. Different approaches, which exploit the advantages from both worlds are known in literature and discussed here briefly.

Choi *et al.* [20] propose a system, where action symbols are automatically generated by a motion planner. Point pairs are sampled from the workspace of a robot. A path planner is utilized to ensure that there is a path connecting those points. If there is one, a pick-and-place action transporting an object between those points is generated. This approach has difficulties to scale to real world complexity.

A reverse approach is taken by Dornhege *et al.* [24]. A symbolic planner plans with an action, which can approach arbitrary locations. A "semantic attachment" for such an action utilizes a motion planer to ensure the feasibility of the action. Grasping and object geometries are not considered in this approach.

The extension of a motion planner to account for re-grasp operations is proposed in [87]. The system considers kinematics, collisions and grasps, but is limited to the domain of pick-and-place with re-grasp operations. A similar approach for pick-and-place operations in complex environments is taken in [84]. In [78], a manipulation planer based on probabilistic road maps is presented. It accounts for grasps, object placement and collisions. Therefore, in the collision free space CS_{free} the subset CG containing grasps and CP containing place positions are generated and paths between their intersections searched. The approaches in this section have no outlook for higher level goals, which would be required to combine multiple operations.

For workspace discretion in the context of symbolization, we rely on the rating of workspace pose for the robot. Known research in this area includes the work of Zacharias *et al.* [92]. The inverse kinematic of a manipulator is used to sample the 3D Cartesian space around the robot. For each sampled point, different approach directions are evaluated, the amount of feasible directions defines the reachability. The reachability of all sampled points define the reachability map.

Guilamo [35] uses the forward kinematics for sampling to account for sub-optimal distribution in Cartesian space, and a prioritization is introduced. Guan [34] extends the approach for a biped robot, considering also stability.

In the area of symbolic scene description, [28] introduces "spacial relation", including a *supports* relation for two dimensional images. A qualitative description for complex mechanical systems is proposed in [48]. Concepts like impulse, collision and friction are included and it allows reasoning about the modeled scene. The model has to be created manually.

A taxonomy of monitoring systems is presented in [65]. It distinguishes between analytical and knowledge-based systems. In [59] the difference between predicted and measured joint torques under impedance control is used to detect undesired collisions. In [31], a path planer in a simulation is used to generate sensor value expectations for a mobile robot.

3.2 Symbolization

A symbolic description of the world in which a manipulation task takes place is required for logic-based planning. The generation of such a description is presented in this section. The *graspability* is introduced to reduce the continuous, 6-dimensional workspace of the robot to a small set of symbolic locations which are likely to be suitable for manipulation in context of the task. *Mechanical relations* are used to generate and transfer mechanical domain knowledge into the planning domain, for example that moving an object A, which carries another object B, will also move B. For learned actions, a free-space representation is developed, which is used as a precondition for those actions.

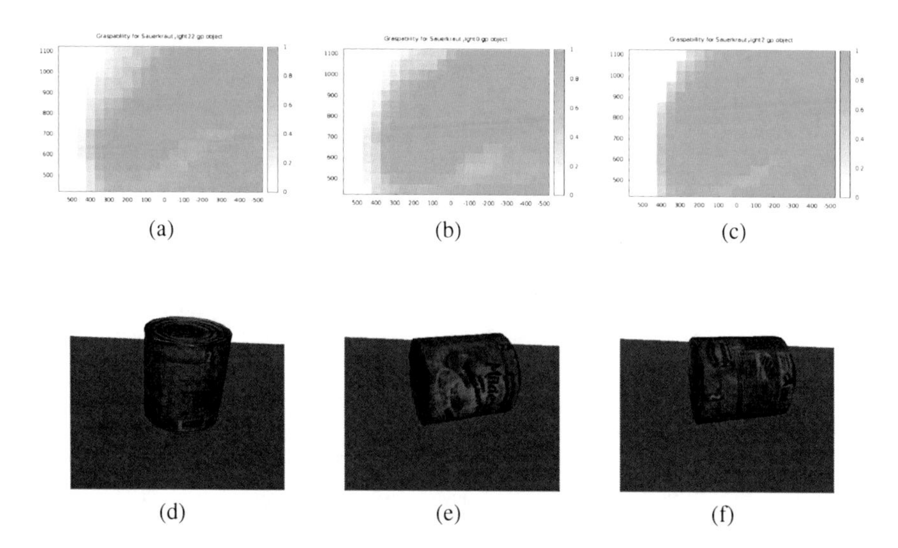

Fig. 23 Top row: Graspability for the Sauerkraut object. Three different stable planes are visualized. The bottom row shows the corresponding orientations of the Sauerkraut tin. It rendered on three different stable planes.

3.2.1 Graspability: A Symbolic Description of a Robot's Workspace

The graspability is based on the assumption that manipulation usually happens on horizontal planes. It is referred to as the "table assumption". The table assumption is justified by gravity. If an object is not manipulated, it is stored on a horizontal, flat surface. Horizontal, since otherwise the object would slip. And flat, since that way the object can be placed at different positions and yaw-orientations. Observation of humans performing manipulation tasks support the assumption that flat surfaces are preferred for placing objects. Look for example at a workbench or a working surface in a kitchen environment. Anyhow, there are some exceptions to this rule, for example a bench vise can not be modeled as a plane. Such devices usually have special semantics which must be modeled differently. A shelf on the other hand could easily be modeled as a set of planes.

From the table assumption, we can also yield constraints for the manipulated object. Since it is placed on a plane, the contact surface of the object has to be flat. This observation reduces the number of possible orientations in which an object may be placed to a set of stable planes of the object and a the rotation around the normal of such a stable plane. That rotation is referred to as yaw orientation. A set of three stable planes for a tin object is displayed in Fig. 23.

A third, weaker, assumption can be concluded from the previously described scenarios: for many tasks, the yaw orientation of an object is irrelevant. This holds particularly for a temporally stored object.

Based on the last assumption, we define the graspability $G_o(\mathbf{r})$ of an object o at the pose $\mathbf{r}$ ($\mathbf{r}$ contains the 3D Cartesian coordinates x,y,z and the orientation with roll, pitch, yaw convention) as in the following definition (from [70]):

$$
\begin{aligned}
g_o(\mathbf{r}) &= \begin{cases} 1 & \text{if o is graspable at } \mathbf{r} \\ 0 & \text{otherwise} \end{cases} \\
\tau(\gamma) &= \begin{pmatrix} \cos(\gamma) & -\sin(\gamma) & 0 \\ \sin(\gamma) & \cos(\gamma) & 0 \\ 0 & 0 & 1 \end{pmatrix} \\
G_o(\mathbf{r}) &= \int_0^{2\pi} g_o\left(\tau(\gamma) \circ T(\mathbf{r})\right) \mathrm{d}\gamma
\end{aligned} \tag{1}
$$

Herein, $\tau(\gamma)$ models the rotation of the object around the gravitational vector and $T(\mathbf{r})$ is the corresponding matrix transformation for a pose $\mathbf{r}$. Due to the integration over 2π, the yaw orientation is lost. Verbally, the graspability can be interpreted as the fraction of yaw orientations at a given pose, where an object can be grasped.

The function g_o decides, whether an object is graspable at a given position $\mathbf{r}$. This function depends on the robot's arm and hand kinematic, the robot placement and the objects geometry. Additionally, a scene setup, in which the object should be graspable must be specified. We use an empty scene with just a simple table surface under the object, the robot and the object.

For the calculation of g_o, grasp planning algorithms from [90] and inverse kinematics are employed. Since a specific pose or configuration has to be supplied they cannot be used in an equation using integration. Therefore, a discretization of the graspability is formulated. The orientation around yaw is sampled and the results are accumulated. We are interested in algorithmic exploitation of the graspability, thus a map with a discretization of x and y is generated. With those presumptions, the result is a 5-dimensional map. Its computation would be expensive, therefore we further reduce the sampled angles roll and pitch. Using the table assumption, those orientations are reduced to a finite set of stable planes for each object. The average object in our database has 12 stable planes, leading to 12 3-dimensional graspability maps. It is calculated as follows:

$$
\begin{aligned}
\bar{G}_o(\mathbf{r}) &= \frac{1}{n} \sum_{n=0}^{\lfloor \frac{2\pi}{\Delta\gamma} \rfloor} g_o\left(\tau(n\Delta\gamma) \circ T(\mathbf{r})\right) \\
\mathbf{r} &= (x, y, z, 0, 0, 0) \\
x &\in \{x_{min} + i \cdot \Delta x | i \in \mathbb{N}\} \cap [x_{min}, x_{max}] \\
y &\in \{y_{min} + i \cdot \Delta x | i \in \mathbb{N}\} \cap [y_{min}, y_{max}] \\
z &\in \{z_{min} + i \cdot \Delta x | i \in \mathbb{N}\} \cap [z_{min}, z_{max}]
\end{aligned} \tag{2}
$$

for each stable plane.

Graspability maps for a cylindrical object on three different planes are shown in Fig. 23, for a different, box shaped object, the graspability is projected into the robots workspace in Fig. 24. A typical feature is the arc around the projected shoulder pose which is caused by the length of the arm.

Fig. 24 Graspability of the ceylon tea box with the left arm.

Table 2 Time and space consumption for the graspability map.

Object	# Grasps	# Planes	# Map entries	Build time [h]
Ceylon Tea	24	6	1848	1 h
Sauerkraut	59	23	7084	6 h

In Tab. 2, the performance of the graspability calculation is evaluated. The calculation is executed for one table height. Sampling is done in a 5 cm grid, $\Delta\gamma$ is rounded to 15°. The calculation is performed on an Intel Core i7 CPU with 3GB of RAM. With 12 stable planes for a typical object, the average graspability map has 3696 entries. The calculation can be done offline and massively parallel for multiple objects.

3.2.2 Symbolic Scene Description

In the classical AI planning example domain of the "Blocks World" [72, 79] objects are connected with the *On* relation, which describes the location of an object as well as its availability for manipulation. Objects can only occur on top of other objects and only the top object of a stack can be manipulated. Real world scene configurations are by far more complex and such a simple representation is not sufficient. A more detailed description can be found in [69].

Locations. The geometrical aspects of the scene are the poses of the objects. Therefore a symbolic representation of the continuous workspace is required. We use *Location* entities to describe poses objects can potentially have. A *Location* consists of a name for the location and a 6-dimensional pose. Only poses that have a location in the scene model can be considered by the planner, thus locations for the initial position of the objects have to be provided as well as potential goal poses. Further locations, e.g. for a temporary placement of an object, are generated. Therefore, in a greedy way, locations are selected which have the highest graspability rating and are not yet in the location set already used.

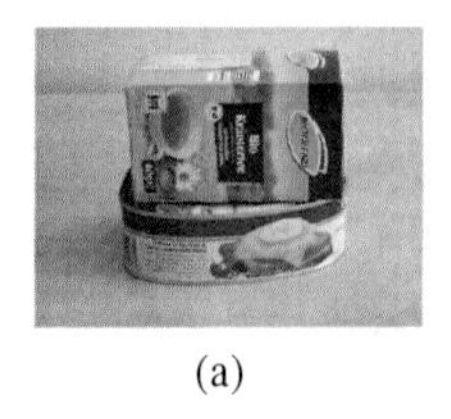

(a)

(b)

(c)

Fig. 25 Static scene relations. In (a), the tea box is on the ham tin: *Support(tin, box)*. in (b), the tin leans against the box, still *Supports(box, tin)* holds. In (c), the relation On becomes ambiguous: Is the box leaning against the box or already *on* the box? Either way: *Support(tin, box)* holds.

Objects are specified by *Object* symbols which consist of a name identifying an object in the scene and an *At* relation. *At(location)* models that the object is at the specified location.

In order to model the mechanical dependencies in a scene, the object model is extended by mechanical relations. The relations between the objects are divided into two categories. In each category, different relations are possible to determine. We selected a small set of relations based on their suitability for the given manipulation task:

Static mechanical relations. A static relation is defined by the fact that one object is exerting a force onto another object and the objects are in contact. We define the *Supports*(object1,object2) relation, meaning object2 cannot be statically at its position without object1. Typical examples are object2 resting on object1 or object2 leaning against object1. By this definition, the *Supports.* relation is anti-symmetric (*Supports(object1, object2)* $\Rightarrow \neg$*Supports(object2, object1)*). To determine the order of the objects, we assume, that the observed forces are caused solely by gravity. *Supports*(object1,object) means object2 rests on object1, in case of an *On* relation, object2 is the upper object. For the case of a horizontal force (within a threshold) between the object as in Fig. 26(b), we define that both objects support each other. This breaks the anti-symmetry of the *Supports.* relation but reflects the fact that we do not know which, if any, of the two objects can be manipulated safely.

Dynamic mechanical relations. A dynamic relation describes the consequences of the motion of one object to another. The considered dynamic relation in this chapter is *Unstabilize*(object1, object2), meaning that a small motion of object1 leads to a larger motion of object2. That motion has to be caused by gravity and thus object2 falls due to the motion of object1. Examples for the *Unstabilize* relation are shown in Fig. 26.

We use the open-source Bullet Physics Engine [12] to generate the described relations. Therefore, we have to deal with the inaccuracy from the perception. To compensate it, we run the simulation until all objects rests without motion before the tests are executed (rest phase). For this approach to be feasible, the initial scene has to be static. Anyhow, we have made this assumption on the scene for it to be able to manipulate in it, so for the scene relation generation, it introduces no new restrictions. The dynamic tests are run multiple times with different randomly

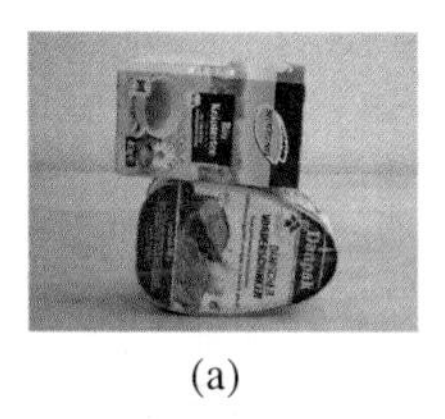
(a)

(b)

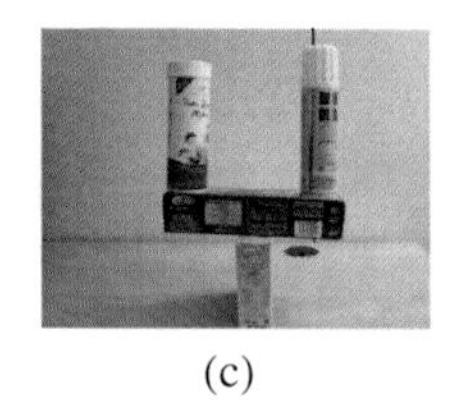
(c)

Fig. 26 Dynamic scene relations. All scenes contain unstable relations. In (a), objects are stacked but in a local unstable balance. Movement of the tin will almost certainly break the balance. The scene in (b) shows a setup, where to objects lean symmetrically against each other: *Unstablelize(box1, box2)* and *Unstablelize(box1, box2)*. Since the acting force against the objects is horizontal, *Support(box1, box2)* and *Support(box2, box1)* holds, too. A scene, where the two cylindrical objects have no physical contact but still cause a fall of the other is shown in (c).

selected parameters. The number of runs where a observed relation holds defines the believe of the system that the relation holds.

The *Unstabilize* relation is generated by applying a motion operator to the tested object. The motion operator performs a motion of the object with a distance δ at a velocity v in axis parallel directions. We use x, $-x$, y, $-y$, z as directions and random distances between 1 cm and 4 cm. There is no use to simulate a motion along the negative z axis direction, since it will move the supporting object. An object A unstabilizes another object B, if the motion of B is not zero and the difference between the applied motion operator and the motion of B is bigger than a threshold.

The presented scene description has been integrated in the scene driven planning system. Experiments are carried out on a test system equipped with a Intel(R) Core(TM)2 Duo T7800 CPU running at 2.60 GHz and 4 GB of RAM.

The shown scene in Fig. 27 contains five objects. In order to generate the scene relations, 1052 simulation steps with a simulated duration of $1/60$ s are calculated. On the test system, this process takes 33 seconds, plus 5 seconds for the initialization of the scene for the physics simulation. The 33 seconds for the scene relation generation can be split into a shot fraction (less than one second) for the static relations and the rest for dynamic relations.

3.2.3 Representation for Arbitrary Strategies

To generate a symbolic description for arbitrary strategies, they are represented by possible execution trajectories. The strategies have to be feasible in a minimal scene, containing the manipulated objects, a table and the robot.

Using the table assumption, we reduce the manipulation to a representation on the sampled surface of table. The geometric model and kinematics of the robot are used to calculate for each configuration the distance between arm and table for each cell. Those maps are fused for all configurations in the trajectory using a minimal function. A threshold, based on the obstacle height is used to create a binary

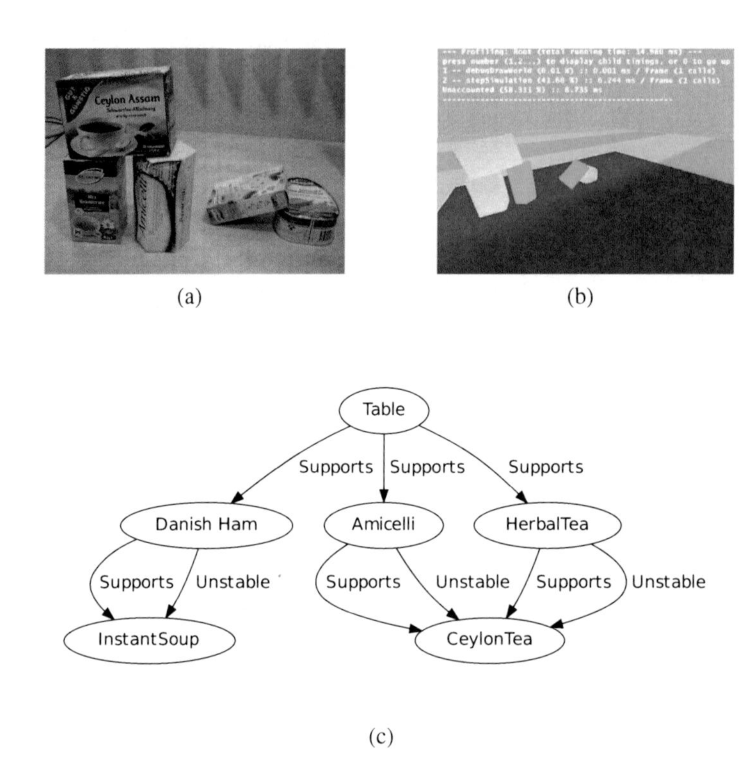

(a) (b)

(c)

Fig. 27 Evaluation of the mechanical scene relations generation. A photo of an evaluated scene is shown in (a). (b) shows its simulation in the Bullet Physics Engine and (c) a tree representation of the generated symbolic description of the scene. The table cannot be manipulated, so we do not check for dynamic relations.

representation of that map. Finally, the intersection of the location set generated by the symbolization and the set of locations in the map, which falls under the threshold is used as a precondition for that executing trajectory. Since the execution of a strategy is able to adapt the executed trajectory, a set of possible execution trajectories is used, leading to a set of preconditions, from which the planner is able to choose one with few effort to fulfill.

3.3 Changing the Course of Actions by Logic-Based Planning

A planner based on time lines is used to adapt the course of actions an keep track of the symbolic scene state. A time line is defined as a set of predicates with temporal extensions, where at every time exactly on predicate holds. Time lines are therefore a suitable representation for actions which require resources for execution as well as for state variables. To model dependencies of the predicates, Allen relations are used [2].

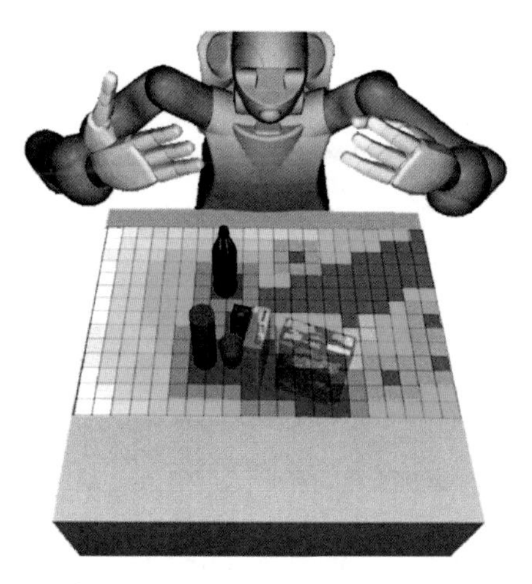

Fig. 28 Initial scene setup with the graspability projected onto the table surface. The gray area has to be free of objects in order to grasp the bottle.

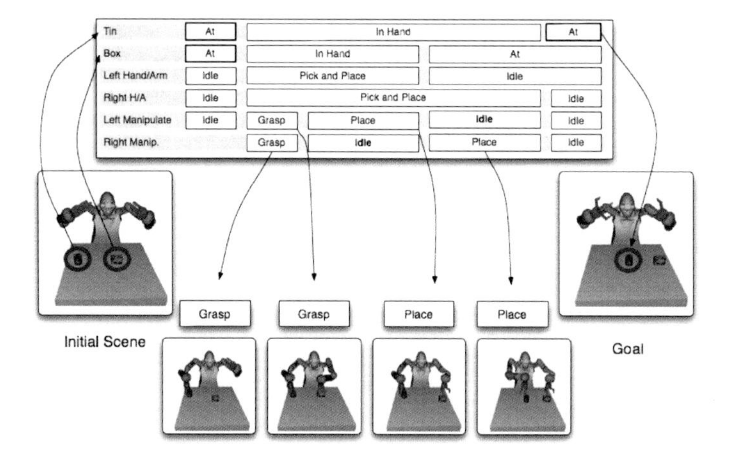

Fig. 29 An example for a plan generated by the scene driven reasoning. The goal is to place the tin in the center of the table, which is blocked by the box. The pick-and-place action to archive the goal requires the target spot to be free, therefor a second pick-and-place is planed to fulfill this condition.

The action model used for planning is divided into three abstraction layers. A fragment is visualized in the example in Fig. 29. The highest level represents the course of actions in the scene, containing objects, locations, symbolic properties such as on/off and mechanical relations, as described in Sect. 3.2.2.

On the mid level, the robot layer, robot actions on a abstraction level of pick-and-place are modeled. Here the robot is viewed as a set of abstract components, which are able to execute actions. Resource requirements and bimanual dependency is modeled on this layer.

Table 3 The proposed feature vector to monitor.

Name	Description	Dimension
$\Delta \mathbf{x_{TCP}}$	Difference TCP Pose	6
$\lvert\Delta \mathbf{x_{TCP}}\rvert$	Distance TCP Position	1
ΔF_{TCP}	Difference TCP Force/Torque measured/expected	6
$\lvert\Delta F_{TCP}\rvert$	Distance of TCP Force	1
$\Delta \tau$	Difference hand joint configuration	16
$\Delta \tau$	Difference hand joint torque	16
$\lvert\Delta \mathbf{x_{tip}}\rvert$	Distance between finger tip and thumb tip	4
Σ	Sum	50

From the robot layer, abstract actions have to be mapped to strategies. For example a pick-and-place action is mapped to a grasp, and a place strategy.

In order to apply the planer for a manipulation problem, a set of goal has to be defined, which specifies a sub-set of the desired scene state. Pick-and place operations for example lead to an "At(object,location)" predicate at the end of a plan. For the execution of learned actions, the representation presented in Sect. 3.2.3 is used to generate a set of "LocationFree(location)" predicates which have to be fulfilled, then the execution of the action can be modeled as a goal itself.

We employ the Europa PSO planning framework [21] for the implementation of the proposed planning system. Europa is a constraint based planner which implements the time line concept. The planning domain, as described in this section is implemented in the new domain description language NDDL. Static robot and action knowledge is coded manually, while the scene knowledge is compiled into NDDL online. For the execution of the planed action sequences, the framework from [71] is used.

3.4 Execution Monitoring

The symbolic scene model and planning uses an approximation of the scene and applied strategies. Unmodeled physicals effects, inaccuracy of perception and user interference may lead to divergence between the planed course of action and the eventually executed strategies. In order to act goal directed, the robot must be able to detect differences between planed actions executed strategies and react to it by generating new plans based on newer data. In this section, we present a method to classify ongoing strategies based on user classification of examples of successful and failed executions of those strategies.

In sensor data generated during execution of a grasp strategy, one observes, that some sensors determine the success state of the strategy at a point in time. Anyhow, their meaning differs in the state of the strategy. E. g. no force at a finger tip is correct in the approach phase, but would indicate an error, when the object should be graped. Further, for different grasps different forces may occur on different fingers.

Table 4 Result for SVMs on different training data bases. A grasped object is transported by the robot. The action fails, if the object is drop. The success rate describes the share of sample, where the SVM classified the situation correctly.

Objects trained	1	5
Grasps trained	2	13
Samples	244	795
Rate on training data	100 %	100%
Rate on test data	45%	93.4%
Number of support vectors	43	159
Training time	3.7 s	16.9 s
Demonstration time	10 min	50 min

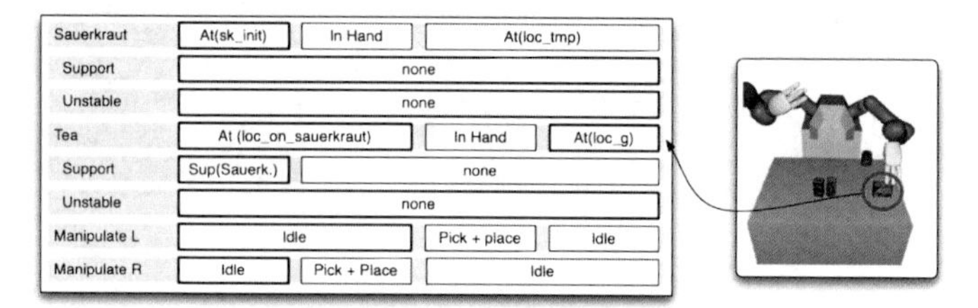

Fig. 30 Sketch of the generated plan for the problem of moving the box from a stack. At first, the tin on top is removed to a generated position by the robot.

For meaningful interpretation of sensor values, we require a segmentation of the action, and a flexible evaluation of the current sensor state.

A segmentation of the action sequence is provided by the planner. Further segmentation of strategies could be provided by learning methods [57].

The selected features use impedance control to execute the strategies on the robot. External forces cause offsets from commanded positions and configurations. The approach would also be applicable to different hardware configurations using force and torque sensors. The evaluated features are shown in Tab. 3.

For the generation of execution examples, the robot executes a strategy in different setups multiple times. A user classifies the execution into "successful" and "failure". In the case of a failure, he has to specify the time, when the failure occurs.

The classified data points of the examples are then used to train a support vector machine. We use the implementation from LIBSVM[18].

The proposed method has been evaluated on a set of 6 objects. The results of two experiments are summarized in Tab. 4. The evaluated scenario is grasping different objects. Segmentation is provided by the logic-based planner, failure cases are generated by the experimenter manually removing the object either form the table or from the gripper. In the first experiment, a single object is used to train the svm. As seen in the table, this is insufficient for correct classification (45% accuracy on test data). Training on 5 objects leads to a better performance of 93% percent correctly classified data points. This experiment still includes one untrained object. The training time takes about 10 minutes per object, where the user has to supervise the robot.

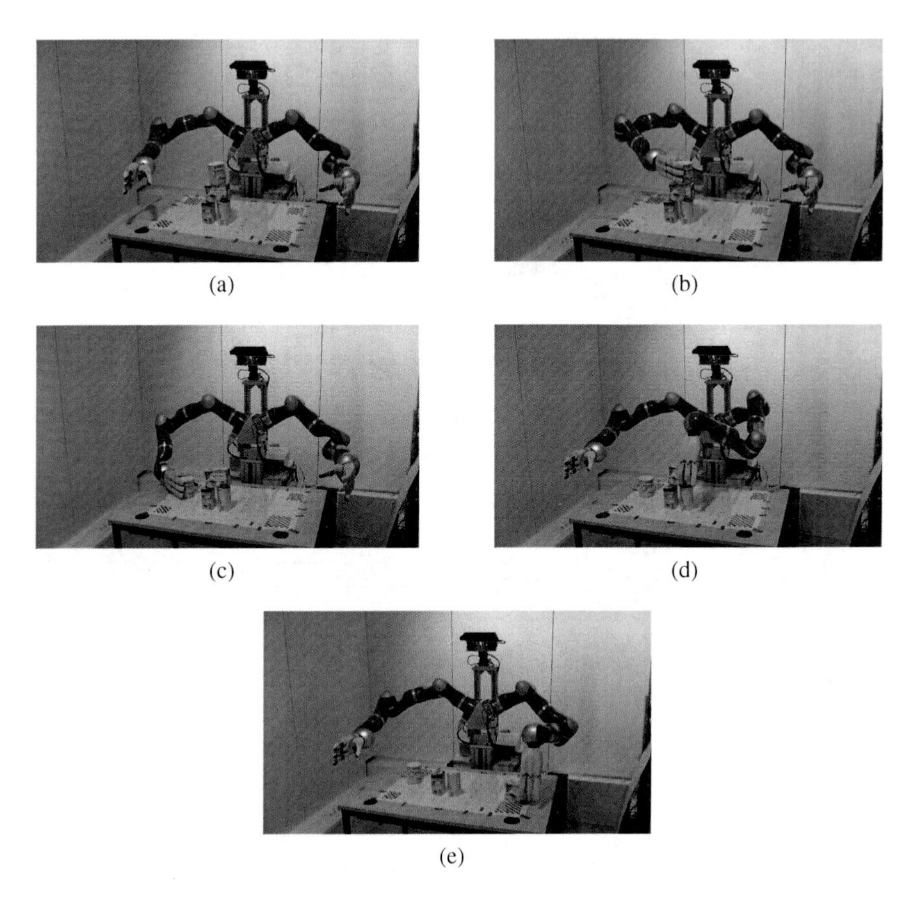

(a) (b) (c) (d) (e)

Fig. 31 Successful execution of the plan on the bimanual demonstrator. The tin is grasped (Fig. 31(b)) and placed to a temporary location (Fig. 31(c)). Then the box can be grasped (Fig. 31(d)) and transported to the goal location (Fig. 31(e)).

3.5 Evaluation

In this section, the subsystem for scene driven logic-based reasoning is evaluated. The evaluation scenario is a stack of different objects as depict in Fig. 31(a). The task is to move the box from the middle of the stack to a predefined position. In the experiment, suitable mechanical relations are generated within 30 seconds. They are consistent with the experimenters perception of the scene.

Based on that description, the logic-based planner is able to find a plan as shown in Fig. 30. It takes 15 seconds, where 8 seconds are spent in grasp and path planning, leaving 7 seconds for the logic-based planning. The execution on the demonstrator is shown in Fig. 31.

In a second experiment on the same setup, the experimenter removes the tin object from the top of the stack at the beginning of the execution. The monitoring

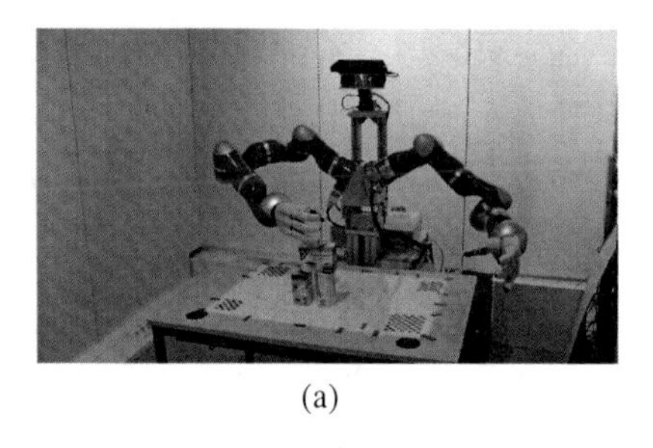

(a)

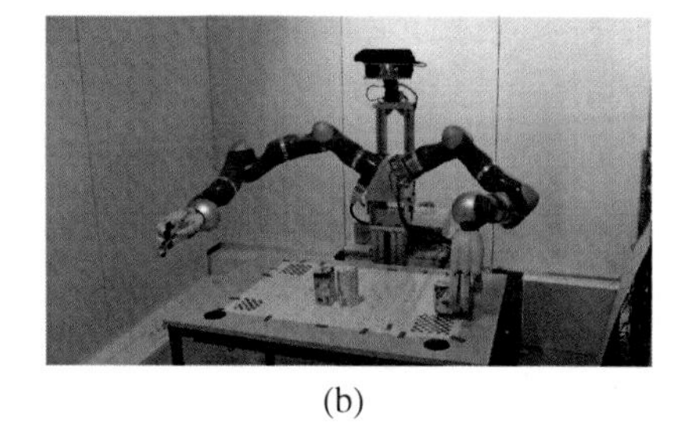

(b)

Fig. 32 Execution with errors: The same setup and plan as in Fig. 29 and Fig. 30 are used. When the robot starts executing the generated plan, the experimenter removed the top tin. The execution of the grasp strategy fails, which is detected by the monitoring. A new plan is generated, which achieves the goal, the execution of a pick-and-place action, directly.

unit is able to detect a flaw in the execution. After replanning, the desired strategy can be executed directly on the robot.

4 Programming by Demonstration of Probabilistic Decision Making in Robot Missions

To achieve true autonomy, an anthropomorphic service robot has to be able to select manipulation tasks and subsumed strategies *proactively* depending on a situation. Consequently, on the highest level of reasoning, a decision making system has to assess the current state of the world continuously and permanently, selecting new tasks from the task repertoire, when a previously executed task has terminated - which may include just idling. Such a level of abstraction, called *mission* or *strategic* level here, typically uses symbolic representations to model states, actions and general planning in literature.

Symbolic representations are able to group and classify complex characteristics of world and robot into easily processable chunks, for which powerful planning algorithms exist. Furthermore, vastly different aspects, e.g. human-robot interaction and object manipulation, can be modeled in the same representation and thus be directly related during planning computations. Therefore, symbolic representations are suitable especially for the abstract, strategic level.

POMDPs are a framework for symbolic decision making, which in contrast to some other paradigms, allows considering quantified uncertainty in situation assessment and robot action effect prediction. By these means, more robust action selection can be performed in the face of real world uncertainty.

To compute decision making policies in that framework, explicit mission models containing these properties are necessary. However, representing a mission setting with sufficiently grounded symbols as well as appropriate quantification of real world uncertainties leads to complex models, generation of which is manually or with explorative learning alone infeasible.

Instead, as with manipulation strategies, recording and analysis of human mission demonstrations is performed, which, in combination with some explorative learning refinement steps, generates a full mission model.

In summary, strategic mission level decision making has to provide symbolic action selection while considering situative uncertainty and covering multiple, distinct skills domains. Mission model generation is made feasible utilizing PbD.

4.1 Related Work

Information processing architectures for autonomous service robots have been investigated in detail. Strictly hierarchical, three-layer architectures with a low-level reactive skill control layer, an intermediate sequencing layer and a deliberative strategic planning layer are suitable for autonomous robot control [30]. In three-layer architectures, the lowest control level can mediate a vast range of different perception and actuation components which are coordinated by the sequencer [1]. Planning in the deliberative layer then just has to schedule abstract functionalitities. On anthropomorphic and humanoid service robots, tight integration between layers is achieved by sharing many properties of models and environment information [3].

POMDPs are an action planning framework for autonomous agents considering uncertainty both in measurements of the current state of the world as well as predicted results of actions [4]. Employing an action selection policy, reflecting these uncertainties together with quantified mission objectives (rewards), an agent can be able to assess risks and opportunities to optimize long-term success [80]. For discrete space and time POMDP representations, so called policies, can be computed from an explicit model representation including discrete sets of true world states S, action choices A, measurements M, stochastic action effect probabilities T, stochastic measurement-state correlations O and mission rewards/costs R [17]. Computing exactly optimal policies is highly intractable, thus approximate techniques have to be applied, with point-based approaches being most promising [66]. Good approximations can be achieved computing the policy preferably only in areas of all potential subjective situation experiences an agent might encounter (belief states), which are highly likely to happen according to a model. Such an approach is taken by SARSOP [51], the algorithmic procedure also used to compute a policy from a model in the presented system.

With approximate policy computation, real world application is feasible, however modeling real task and mission domains is still challenging. A manually crafted model setup for an autonomous service robot, though without manipulation, is presented in [67]. State and action grounding along skill domains to derive symbol-world relations in a suitable and well established manner along perception and actuation skill capabilities is of foremost importance for technical robust real world applications [38]. Manipulation is a skill domain, especially challenging, with low-level modeling explored in [23] and [39]. Furthermore, striving for information gain, realized by actions explicitly exploring and gathering more knowledge about world states to reduce uncertainty and apply subsequent actions in a more directed manner, is a crucial concept in POMDPs, which should be reflected in model design [81].

Persistent knowledge storage, organisation and inference beyond short term, specific mission models can enhance robot capabilities, enabling easier generation of more complex planning models. Description logic (DL) based ontologies are the terminology best investigated organising abstract level, symbolic, general knowledge covering all aspects of service robot reasoning. Such reasoning systems can be hierarchically organized themselves and utilized for online planning [85]. Integration with functional skill-level components, using different, more specific algorithmic methods e.g. considering uncertainty is feasible [86]. DL-ontology based reasoning can easily be modularized and made portable over different robot architectures and planning paradigms [52]. Scalability to robotic domains beyond service robots has also been investigated [64], [19].

PbD of symbolic, abstract task and mission representations (in contrast to trajectory level representations discussed in Sect. 2) has been investigated for various task sequencing and planning methodologies. Symbolic Hierarchical Task Networks (HTNs) can be learned from segmentation of human demonstrations of manipulation tasks [93]. Hierarchical PbD systems can integrate manipulation-motion-level learning with learning of symbolic sequences [61]. Visual interaction and grouping of abstract subtasks can help to organise representations suitable for logic-based planning [26]. Furthermore, logic-based planning representations can be inferred from observed sequences of human task representations and in turn used for flexible task execution [88]. Finally, DL-ontologies and abstract-level learning from human demonstrations can be integrated, using strengths of both paradigms complementary [86].

4.2 Execution Time Decision Making System

By utilizing a proven three-layer architecture design, the autonomous execution reasoning system is able to integrate skill level, sequencing and task reasoning as well as mission-level decision making in a clearly organized manner. The skill domain of autonomous object manipulation is tightly integrated with natural human-robot interaction and mobility on the strategic level. All layers share information from basic environment perception, of which localization of small objects [5], furniture objects [58], known walls (map) and forces in arm and hand are relevant for autonomous manipulation. Object models and symbols for common reference are shared among manipulation strategies, scene driven reasoning and mission-level decision making as shown in Fig. 33.

Nonetheless, each layer performs different kinds of planning on these situation models. Manipulation strategy execution utilizes geometric models directly for collision-free constraint based motion planning. Scene driven reasoning performs

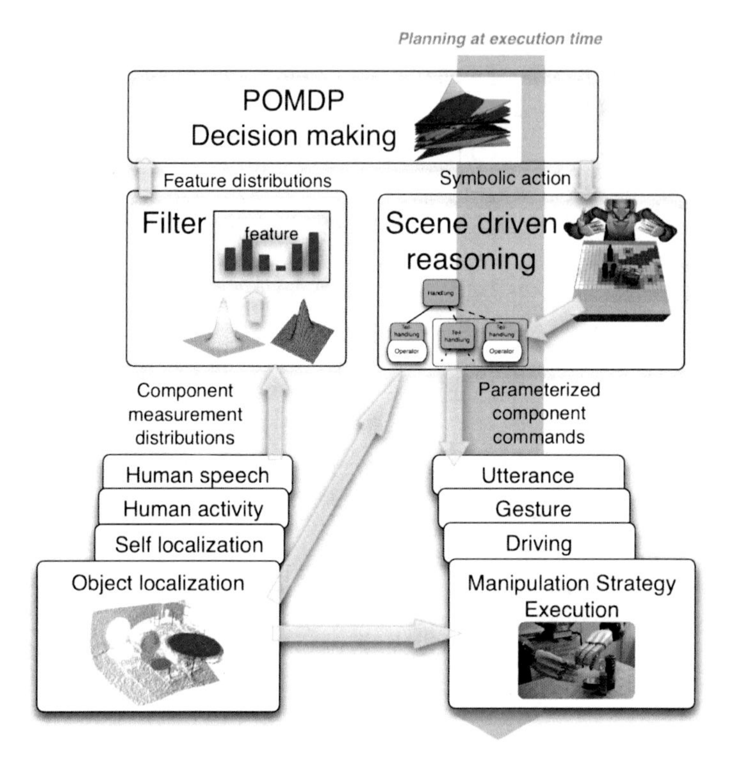

Fig. 33 The execution-time three-layer architecture.

moderate abstraction on the geometric representation to derive predicates usable for logic-based planning.

In turn, being on the highest level of abstraction and integrating distinct skill domains, the grade of abstraction is highest for mission-level action selection. Yet, states, actions and measurements have to be defined unambiguously and clearly grounded in quantifiable world properties. Furthermore, to consider real world uncertainties, reflected by probabilities given by low-level perception components, those probabilities have to be passed up to the most abstract belief state.

Therefore, a state description has to be grounded on modes of perception, provided by available skill components. Concerning just object manipulation, these are object localization, self-localization, force detection and state of actuator joints in the given system. For each of these skill domains $d_k \in D$, sets of quantified values are delivered, denominating certain properties of the world, typically by real numbers in continuous domains: $d_k := \{val_1 ... val_n\}$. Based on that, a *feature* is defined here as a discretizing function f_{feat_j}, mapping a specified, fixed input space of domain members onto a set of discrete, symbolic categories: $f_{feat_j} : (val_{x_1} \in d_k, ..., val_{x_n} \in d_k, ..., val_{y_1} \in d_l, ..., val_{y_n} \in d_l) \rightarrow \{c_1, ... c_m\}$. This set of categories forms a *feature state space*, $feat_j = \{c_1, ... c_m\}$. Symbolic state grounding can then be achieved by mapping each distinct combination of feature categories onto precisely one state, called *feature-state map*: $f_{fsmap} : (c_{z_1} \in feat_1, ..., c_{z_p} \in feat_p) \rightarrow s_i$. That mapping is surjective, but not injective and a unique combination of categories must never map onto more than one distinct state. By these means, the whole space of values, the robot is not agnostic about, can be mapped to one discrete set of states S.

However, with perception skill components, delivering estimations of measurement uncertainty derived from sensor and algorithmic characteristics, domain members are not described by crisp values, but probability distributions. Continuous values are typically described by parametric (e.g normal) or non-parametric sampling (e.g. particle) distributions. Accounting for this, the feature definition above has to be extended considering two aspects: 1) not just values, but probability distributions have to be discretized, and 2) Bayesian filters which combine measurement distributions with prediction distributions may occur in a stacked way, thus conditional independence of stacked filters has to be assured.

Discretization can be achieved by approximating distributions by Gaussian Mixture Models and then integrating individual Gaussians over discrete regions by state-of-the-art numerical methods [32]. Alternatively, sampling based methods can approximate a discrete representation by summing up samples in regions. Prediction (transition) models may be already applied within a skill component (e.g. a Kalman filter for self-localization) with the distribution then already being a filtered, though continuous belief. Or in contrast, an additional predictive element $p(c_t|u_{t-1}, c_{t-1}) \in T_{external}$, which is conditionally independent from any in-skill prediction, is included after observation discretization into $b_{skill}(c_t)$:

$$b_{feat}(c_t) = \alpha\, b_{skill}(c_t) \left(\sum_{c_{t-1}} T_{external}(c_t, u_{t-1}, c_{t-1})\, b_{feat}(c_{t-1})\right) \quad (3)$$

As a result, for each feature, a discrete probability distribution, a *feature belief* over all categories $b_{feat}(E_s) := \{p(c_1), ..., p(c_n)\}$ with $\sum_i p(c_i) = 1$ is obtained by such a *feature filter model*. The feature-state $c_{i \simeq feat_i, j} \rightarrow s_k$ mapping can then be used to compute the state belief:

$$b_{filter}(s_k) = \prod_{i=1}^{|feat|} \sum_{j, c_{i,j} \in s_k} p(c_{i,j}) \quad (4)$$

A discussion of implemented, practical feature filter models can be found in [75].

While the filter structure, especially the set of input $\{val_i\}$ is highly reusable, more specific state grounding, the precise discretization $\{f_{feat_j}\}$ has to be determined for each mission model. Automating the latter step by means of PbD is discussed in the next section.

Furthermore, by using this low-level uncertainty information preserving filter concept, the high-level POMDP observation model may not be well grounded anymore. While it is always just an approximation of real world observation uncertainty, grounding can be achieved by deriving an average observation uncertainty model from detailed analysis of probabilities delivered by a skill component.

Table 5 Evaluation of *filter*POMDP against other behavior control methods [75].

Method / Type	FSM	MDP	POMDP	fPOMDP
Correct fetch/put	8	12	12	16
Incorrect fetch/put	2	4	6	2
Reassurance	15	7	10	5

Such an analysis and model was derived for a furniture localization system, described in [58].

Action grounding is more straight forward: symbolic actions represent primary strategies, transformed into HTN sequences by scene driven reasoning. Cost (negative rewards representing effort and duration) as well as effect probabilities can be derived from typical execution behavior as discussed in the next section. The available action repertoire is formed by all learned strategies and their execution in typical scenes.

Evaluation of *filter*POMDP against other behavior control methods like finite state machine (FSM) and MDP on the same robot in the same setting has indicated superior robustness of the *filter*POMDP [75], see Tab. 4.2.

4.3 Probabilistic Mission Planning Model PbD

As with manipulation strategy PbD, mission model PbD consists of: recording of demonstrations, abstraction and analysis, generalization and model refinement. Finally, based on a generated model, policy computation is performed which can subsequently be used for execution time decision making.

In contrast to manipulation strategy PbD, mission recording is not performed in a specialized sensor setup, but solely robot-based as shown in Fig. 34. As human motion observation is more coarse grained than with manipulation strategy PbD, marker-less, robot-based tracking is sufficient. Furthermore, missions encompass more agent mobility in space, thus the robot follows demonstrating humans with its head —and potentially also with its mobile base.

The human is tracked using the MS Kinect sensor and the NITE body-tracking framework [68], providing $h.pose$. Object locations of small objects and furniture objects are tracked using execution-time perception as mentioned in the previous section, providing $obj.poses$. Robot head movement always tries to keep the human demonstrating the robots role and surrounding objects in its view. Additionally to human pose, a body movement activity classification systems interprets the human body pose constantly, providing $h.activity$, as discussed next.

This leads to the following recording data point vector relevant for object manipulation: $rec(t) = h.pose(t), obj.poses(t), h.activity(t)$.

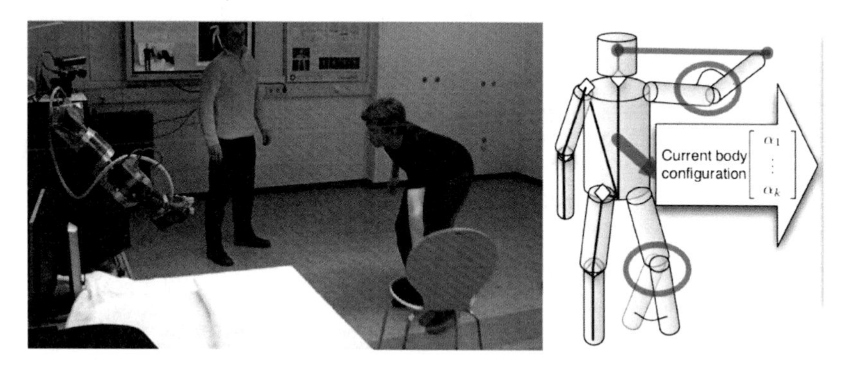

Fig. 34 Mission demonstration recording setup (left) and body model data output (right).

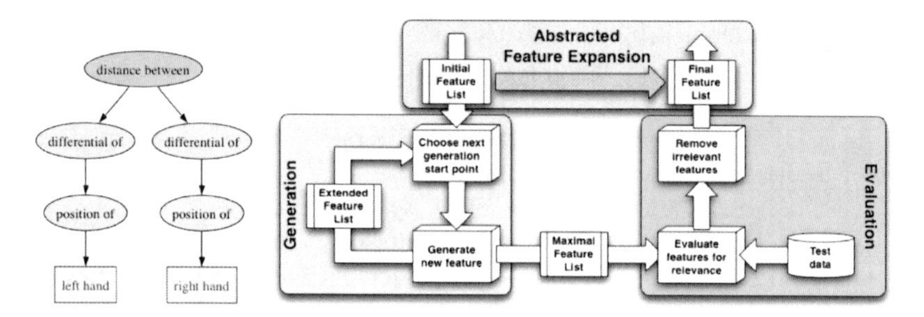

Fig. 35 Feature Extraction Modules for human body activity classification training scheme (right) and example (left).

4.3.1 Human Activity Recognition

Human body movement activity recognition labels a time series of human skeleton kinematics (see Fig. 34) with probabilities of affiliation to a certain, symbolic type of activity. There are basically two stages for this process: training and usage. During training, a human demonstrates a set of exemplary motions, some reflecting activities to be trained and some reflecting counterexamples [53].

A set of over 300 features, which are derived from recorded skeleton poses, is used to train the classifiers. A feature is represented as a tree of *Feature Extraction Modules (FEMs)*, where each FEM represents a certain derivation technique (e.g. difference, derivation, calculation of mean value etc.). An example for such a feature tree is shown in Fig. 35 (left). The initial feature set is determined by an automatic, iterative feature space exploration (depicted in Fig. 35 (right)), which is used to find potentially relevant features in a domain.

Naturally, the initial feature set contains many features which are not relevant to describe a certain activity besides the relevant features. Thus, extraction of a small subset of features, relevant for classification within a set of activities, which can be used with classification learning techniques like Support Vector Machines (SVMs) is critical. The employed approach combines the *Fast Correlation based Filter* [91], which determines a small subset of relevant features by calculating features with a high correlation to the class and a low correlation to each other, with an interactive approach involving user hints to guide the feature selection process [54].

Finally, given the refined set of features, classifiers can be trained on the recorded data with manually labeled activities. Depending on the complexity of the movements which define the activity, different classifiers can be used. For complex activities which involve a sequence of very diverse motions, HMMs are better suited, while for other activities, SVMs perform better.

To allow for later reuse of recognizers for activities in other settings, the training process treats the learning problem as a one-class problem (i.e. each classifier is trained to recognize the probability that a movement belongs to one specific activity), instead of a multi-class problem (where the recognizer would return a probability distribution over all learned activites for the current movement). This way, classifiers for different activities can be combined independently.

Activity classifier usage then takes place during recording of human mission demonstrations, thus leading to two distinct demonstration stages: first classifier training and then mission training. Manipulation strategy training can be performed in another, distinct stage. Activity classifiers can be reused for demonstrations of multiple missions, though. The most likely symbolic human activity at a given time can then be considered as a representative of the agents manipulation task (activity) performed. Subsequently, a mission demonstration recording time series $obs(t)$, called *trace*, is obtained with environment situation E_s and agent activity G_a: $obs(t) = (E_s, G_a) = ((h.pose(t), obj.poses(t)), h.activity(t))$.

4.3.2 Demonstration Based State and Action Grounding

With traces E_s, G_a of several demonstrations being available, the next step has to determine feature and action mappings, thus providing state and action grounding as well as preliminary state and action space determination.

State grounding generates feature mappings for observed human poses onto discretized robot self-localisation categories as well as mappings from observed object poses onto discretized object regions: $f_{feat_{pose}}(h.pose) \rightarrow c_x^{self-pose}$ and $f_{feat_{objs}}(obj.poses) \rightarrow c_x^{objposes}$. Category sets and category limits in the domain value space suitable for a demonstrated mission are determined. Furthermore, a limited automatic choice of input domain values relevant for a feature can be performed.

Like in any following step, a set of recording *traces*, $Demo := Obs_1, ..., Obs_n$ of n different demonstrations of possible courses of events in the corresponding mission is required.

First, data preparation takes place, beginning with dimensionality reduction of $h.pose : (x, y, z, r, p, j) \rightarrow (x, y, \theta)$. Object pose dimensions, on the other hand, will be automatically assessed in the following steps —e.g. for a chair, always standing in a mission, only (x, y, θ) are relevant for state distinction. Then, temporal and spacial interpolation of pose data balances frequency irregularities in trace data points,

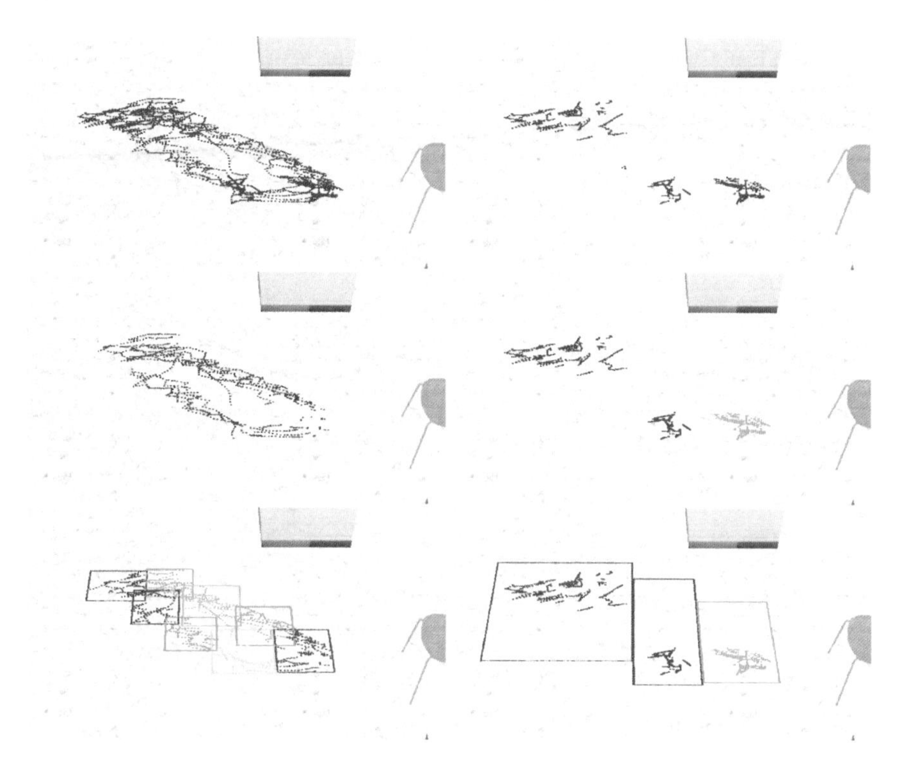

Fig. 36 Exemplary $f_{feat_{self-pose}}$ generation: demonstration data (top left), filtered standing (top right), filtered walking (middle left), DBScan stationary clusters (middle right), k-means transition clusters and limits (bottom left) and final, merged region categories after BSP (bottom right).

resulting from perception component temporal peculiarities. Subsequently human and object velocities and accelerations can be computed.

Next, derived attributes can be computed for data point components. Concerning human mobility state, two thresholds $thres_s, thres_w$ are used to distinguish between *standing*: $v(h.pose) < thres_s$, *walking*: $v(h.pose) \geq thres_w$ and *moving*: $thres_s \leq v(h.pose) < thres_w$. As limitations in object localization may indicate moving object poses temporally shifted, an additional *object moving* attribute is derived from the beginning of corresponding human activities. Accordingly, objects can be tagged being *manipulated* (both moving or stationary) at a given data point.

Based on these attributes, data points of all demonstrations can be selected for further processing, labelled Obs_{proc}, using motion or activity filters ψ_{filter}.

Spacial clustering, being the most crucial process step for state grounding follows subsequently. To determine category limits and category *sets* automatically, clustering algorithms which do not only compute cluster limits, but also identify suitable cluster numbers k have to be applied. Furthermore, relevant groups of points representing moving stages of human and objects have geometric properties differing significantly from stationary groups.

Consequently, different types of clustering methods are applied in a complementary manner. Primary clustering applies DBScan [22] on points filtered by stationary motion filters as defined previously. Checking the continuous DBScan ε-environment parameter over a large value range $0 < \varepsilon < max_{\varepsilon}$ gives a maximum range plateau corresponding to one specific number of clusters $k_i = dbs(\varepsilon)$: $\varepsilon_{min}(k_i) < \varepsilon \leq \varepsilon_{max}(k_i)$ which corresponds to the optimum k_{opt}. In practise, clustering has to be performed for discretized samples of the ε-range with the maximum plateau being evaluated on that discretized range.

In contrast to compact stationary pose aspects, pose point groups representing motions (see Fig. 36), e.g. transition regions of human or objects, are not suitable for DBScan clustering, but instead k-means [55] and EM clustering. As those are fixed-k clustering methods, secondary clustering has to be performed on the motion-filtered data for ranges or k, estimated based on primary clustering k_{opt}. For each k, clustering results are evaluated on compactness and separability, using a balanced vote including DB-index [22], XB-index [89] and SD-index [36] metrics.

After determining clusters and thus category sets, category limits are computed using Binary Space Partitioning on the clustered points (see Fig. 36). Subsequently, activities occurring during the majority of data points in certain categories can be used to further enhance the description or, if not present in the majority of points, split up categories into pragmatically more complete ones. As an example, a manipulation action might take place only in one spacially distinct part of a previously clustered single region.

Finally, with category sets and category limits on domain input values defined, f_{feat_j} definitions can be exported and subsequently used for demonstration segmentation as well as execution-time belief state computation.

Generating f_{feat_j} can also be interpreted as a switch in point of view: $h.pose$ is transferred into robot self-pose pragmatics, thus the robot puts itself into the position of the demonstrating human in further interpretations of traces using the set of f_{feat_j}. Objects are then also interpreted along characteristics as relevant from the robots point of view.

Action grounding concerning manipulation has to map observed human activities to manipulation strategy skills executable by the robot. As mission demonstrations and strategy demonstrations are not immediately correlated, a dedicated process stage has to create such a connection.

Basically, object-relative trajectories acquired during strategy training are matched with object-relative trajectories which are classified as certain symbolic activities during mission demonstration recording. A finite set of learned manipulation strategies with given demonstration trajectories relative to a set of objects and a finite, trained set of human activity classifiers are assumed as input and output domains. Subsequently, a mapping is computed for one specific set of mission demonstration recordings. Thus, mappings from classifiers to strategies are not universally valid, but in different missions, distinct strategies may be suited to take the role of a human activity classifier as the latter is only coarsely defined and not object-relative. Furthermore, with finger joint angles not available from coarse grained tracking, relevant trajectories can only regard the hand center.

First, all trajectories of a strategy demonstration are normalized temporally, concerning data points $|\mathbf{x}_i|$ and in relation to the manipulated object. Next, a Gaussian Mixture Model (GMM) is computed for the trajectory bundle according to the approach presented in [16]. Accordingly. the GMM is transformed into a Gaussian Mixture Regression (GMR) representation, suitable for distance metric computation. Any recorded mission demonstration activity has to be normalized in respect to each GMR-represented strategy it is compared with so that both contain the same number of data points.

Based on the GMR representation Φ, a suitably normalized mission activity trajectory Ψ and a weight matrix ω, the Mahalanobis [56] distance metric between the bundle and the trajectory can be determined:

$$D(\Phi,\omega,\Psi) = \frac{\sum_{i=1}^{n} \omega_i \sqrt{(\psi_i - \mu_i^{\Phi})^T * \Sigma_i^{-1} * (\psi_i - \mu_i^{\Phi})}}{\sum_{i'=1}^{n} \omega_{i'}}. \tag{5}$$

The weight matrix ω determines how much individual points x_i contribute to the overall trajectory distance metric. Four different varieties have been considered:

1. Equally distributed weights
2. Constraint density weighting
3. Object distance weighting
4. A combination of constraint and object weights

The combination of constraint variance and object distance (and normalizers α):

$$\omega_i = \alpha_c det(\Sigma_i)^{-1} + \alpha_o d(\psi_i - pose(obj))^{-1} \tag{6}$$

has empirically delivered superior results on typical simple manipulation activities as it preferably considers trajectory parts which are especially relevant to the object manipulation action.

Another approach of comparing mission demonstration trajectories directly with strategy graphs, without the intermediate GMR representation, to achieve matching of more complex and bi-manual manipulation actions, is currently investigated.

Again as in state grounding, action pragmatics change from observing a different agents action to assessing the agents own skill action being performed in a certain situation. Therefore, after state and action grounding is completed, the robot is able to assess demonstration traces for abstract model generation from a point of view as if it had executed a sequence itself. This concept of a student putting himself/herself into the teachers point of view is also crucial in humans learning from demonstrations of human teachers.

4.3.3 Segmentation and Demonstration Model Generation

With f_{feat_j} available and a generic $f_{fsmap} : (c_{z_1} \in feat_1, ..., c_{z_p} \in feat_p) \rightarrow s_i$ for any category combination, a mapping $s_t = \phi(obs(t)), \phi : f_{fsmap_{mission}}(f_{feat_1}, ..., f_{feat_n})$ can be applied for trace segmentation [76]:

$$Q'(t) = \begin{cases} \emptyset, & \phi(obs_{t-1}) = \phi(obs_t) \\ s_t, & \phi(obs_{t-1}) \neq \phi(obs_t),\ s_t = \phi(obs_t). \end{cases} \tag{7}$$

By removing $\emptyset$ from Q', e.g. $Q' = (\emptyset, \emptyset, s_3, \emptyset, s_5, \ldots)$, a discrete, abstract time sequence of states is derived $Q = (s_{x_1}, \ldots, s_{x_n}) = (s_3, s_5, \ldots)$.

Next, abstract actions correlated to state transitions are determined. Concerning object manipulation, three types of actions are possible: 1) mobility (goto) only, 2) manipulation only, 3) mobility with concurrent manipulation. Therefore, it is determined if $c_i \in s_t$: *self-pose* $\neq c_j \in s_{t+1}$: *self-pose* and if true, a mobility action a_{mobil} is generated. In case a mapped manipulation strategy was performed while s_t occurred, a manipulation action a_{manip} is generated. When both cases hold, a mobile manipulation action is compiled: $a_{mobilmanip}$, containing both corresponding *goto* and manipulation actuation parameters in a task sequencer description.

Fully observable recordings are assumed here, although an extension was developed, accounting for recording errors with HMM-based smoothing, using causal meta-models reflecting impossible and unlikely transitions and observations.

A preliminary state space S_D, including all non robot-specific states inferable from demonstrations can then be accounted for: $\forall s_i \in Q_1 \ldots Q_n : s_i \in S_D$. In the same manner, a preliminary action set A_D is compiled: $\forall a_k \in Q_1 \ldots Q_n : a_k \in A_D$. Subsequently, absolute frequencies of transitions are reflected in the *counting transition model* TC_D which is initialized to 0:

$$\forall Q_1..Q_n : \forall s_t \in Q_i : TC_D(s_{t-1}, a_t, s_t) + 1. \tag{8}$$

Stochastic effects, observable in demonstrations, can then be derived by considering frequencies of outcomes of each pair (s,a):

$$\forall (s,a,s') : T_D(s,a,s') = \frac{TC_D(s,a,s')}{\sum_{s'_i} TC_D(s,a,s'_i)}. \tag{9}$$

Finally, the preliminary reward model R_D has to account for goals and costs. As action costs are known only from background knowledge and goal values computed from costs on causal paths, goals can only be flagged at this point. Accordingly, for each final pair, a goal is flagged $(s_r, a_k)_{x_n} \in Q_d; R_D(s_r, a_k) = 1$.

To account for missing courses of events which could potentially occur in a mission, but are missing from a set of demonstrations, techniques to automatically explore the transition space of the preliminary model based on state and action similarities have been developed. Transition hypotheses beyond the observed ones are generated, which in turn are validated by further demonstrations requested verbally by the robot from human demonstrators. A detailed discussion is beyond the scope of this chapter, however.

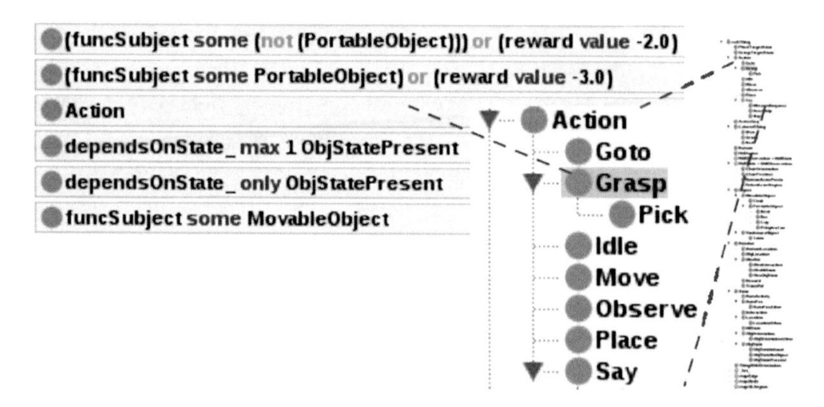

Fig. 37 An ontology excerpt, with reward model costs for manipulation of certain object types.

4.3.4 Mission Model Refinement

Robot specific model aspects cannot be inferred from human —an agent with different capabilities— demonstrations. Such aspects include:

- robot capability specific error transition effects $\in T$,
- resulting additional error states $\in S$,
- subsequent, generic recovery actions $\in A$,
- information gain actions $\in A$,
- information gain transitions $\in T$,
- action cost penalties $\in R$,
- skill component perception characteristics O.

Merging these aspects from a background knowledge source can transform a generic preliminary mission model as generated from demonstration analysis S_D, A_D, T_D, R_D into a final, robot specific POMDP mission model S, A, M, T, R, O [74]. As discussed in the related work section, ontologies are a technique well investigated to organize general-purpose, symbolic knowledge about the world. Therefore, an ontology described by the *OWL 2* language, based on the description logic expression system SROIQ is employed. Consistency checking as well as inference is done by the reasoner *Pellet* and the primary query interface provided by the RDF-System *Jena*.

A combination of three basic description types encodes the ontology: TBox, ABox and RBox. Abstract concepts, their class hierarchy, including both physical objects, e.g. types of manipulable objects, and model specific aspects, e.g. state definitions, are encoded by the TBox as shown in Fig. 37.

The ABox models specific instances, e.g. actual manipulable objects *object: model:Chair(model:WoodenChair)* and as with the following example, relations between physical and model specific properties:

model:ObjStatePresent(model:ObjStatePresentChair)
model:subject(model:ObjStatePresentChair, model:WoodenChair)

Further dependencies and relations beyond basic ones have to be modeled using Horn clauses called *DL-Safe-Rules* in an RBox, being a DL extension and expressed by the SWRL language. The following example relates a model specific aspect (action) with a tangible object class:

Grasp(?action) $\wedge$ ObjStatePresent(?os) $\wedge$ subject(?os, ?obj) $\wedge$ object(?action, ?obj)
$\Rightarrow$ dependsOnObjState(?action, ?os)

By these means, an ontology structure can be created, relating physical world properties, reoccurring in multiple mission with model specific, intangible aspects like states, actions and observations. Hierarchy allows inferring shared properties of similar types of classes —e.g. for all types of chairs handled in manipulation.

Furthermore, as mission models are both symbolic and numeric, quantities reflecting properties in T, O and R have to be stored in and inferred from the knowledge base. To be able to infer quantities for new instances from their parent classes und thus exploit the relationship concept in the acquisition of new knowledge, such numeric information can be "tagged" to subjects and objects. A relation defined as $Rel(x,y,v)$ with x being the subject, y the object and v the value. An exemplary transition frequency $t(s,a,s'), freq(t)$, expressing the probability of the robot being stuck at the origin is 0.1 when trying to reach position PullStartPos from LookPos, can be modeled thus as: TransRel(TransRelId0); subject(TransRelId0, GotoPullStartPos)
state0(TransRelId0, LookPos); state1(TransRelId0, PullStartPos);
intValue(TransRelId0, 9)

TransRel(TransRelId1); subject(TransRelId1, GotoPullStartPos)
state0(TransRelId1, LookPos); state1(TransRelId1, LookPos)
intValue(TransRelId1, 1)

There are two directions for knowledge transfer: a) knowledge added to the ontology from analysis of mission demonstrations or explorative refinement, and b) knowledge inferred from the ontology to complete a model.

To account for a), actions can be verbally commented during demonstrations, e.g. while grasping a chair with the utterance "this is furniture manipulation", and thus this new activity can be connected to a matching parent class, e.g. *furniture manipulation*. Generalized transition knowledge, stored as relations to that parent class can be extended, from transition knowledge derived from the new action. On the other hand, concerning b), further information concerning this new action can be derived from related actions in the ontology while generating model parameters. Error states like "jamming manipulator and object during manipulation due to execution errors", not occurring during human demonstrations can be inferred from related action objects. Exploiting transitive relations in the ontology, further instances can be inferred and previously listed aspects as recovery actions, information gain actions and coarse transition probability estimates derived.

Costs can be derived from length and effort associated with an action. Goal reward values have to consider the effort (sum of costs) of action sequences, accomplishing certain goals, otherwise the robot might never risk the associated costs. Costs of sequences of actions a_i occurring in demonstrations and leading to flagged

Table 6 Execution with POMDP (P) and FSM (F), performance times (Mi, Av, Mx) and failures (Fl).

	Mi.P	Av.P	Mx.P	Fl.P	Mi.F	Av.F	Mx.F	Fl.F
Mi1	4:25	4:50	5:50	1/10	4:40	5:10	5:35	2/10
Mi2	4:05	4:25	5:00	1/10	4:10	4:35	4:50	0/10
Mi3	5:35	6:00	7:25	2/10	5:50	6:20	7:00	3/10

goals are added up with the scaled absolute value replacing the previous flag-value in the preliminary model: $R(s_r, a_k) = \nu \sum_i |R(s_i, a_i)|$ with $\nu > 1$.

Observation model probabilities in certain situation can be derived from empirical perception skill analysis and stored and retrieved from the ontology in the same manner as transition models.

After robot-specific knowledge is incorporated, the demonstration model is turned into a complete final model.

Techniques were developed for further refinement of robot-specific transition probabilities using geometric analysis of action trajectories (both concerning mobility and manipulation strategies), followed by execution trials in dynamics simulation, detailed description of which is beyond the scope of this presentation, however.

4.3.5 Results

In an evaluation of model generation —without automatic grounding though— three missions sharing common background information were studied [76]. Assessment explored the execution performance of automatically generated POMDP models with manually expert-crafted FSM controllers. Each mission included mobility, simple manipulation actions and human-robot interactions. Demonstration courses of events had to follow a script while execution-time interacting human behavior was sampled randomly from script behavior frequencies. Real, robot based recordings of ten demonstrations of each mission were used for model generation, resulting in

1. Mi1: $|S| = 500$, $|A| = 12$, $|M| = 18$
2. Mi2: $|S| = 50$, $|A| = 6$, $|M| = 12$
3. Mi3: $|S| = 350$, $|A| = 14$, $|M| = 19$

as well as T, R, O.

Real, robot based autonomous execution of policies, which were computed with SARSOP from these models, was performed successively with both POMDP (P) and FSM (F) ten times and performance times (Mi, Av, Mx) and failures (Fl) were recorded, see Tab. 6.

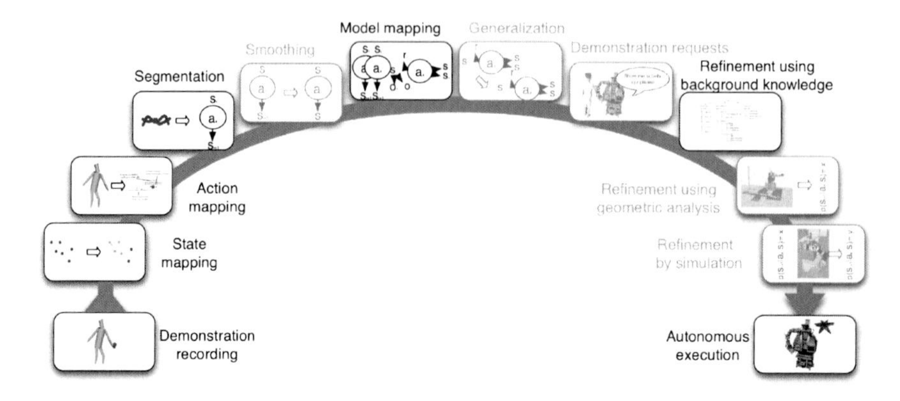

Fig. 38 A scheme of PMPM-PbD with components not discussed here, greyed out.

On average, learned POMDP performance was superior to delicately handcrafted FSM.

In summary, a process to compile abstract POMDP strategic action selection models by means of PbD as shown in Fig. 38 tackles the challenge of model generation complexity.

5 Conclusion

The proposed layered system for autonomous environment manipulation of service robots will be summarized based on the execution of a complex manipulation task with focus on Manipulation strategy learning and execution, see Sect. 2, and scene driven logic-based reasoning, see Sect. 3. In the manipulation task, the robot has to *prepare a drink for the human*, i.e. the action sequence consists of opening a bottle, pouring liquid into a cup and placing the bottle into a crate. For each action, a manipulation strategy was learned and automatically assigned, see Sect. 4.3.2. Assuming that the manipulation task was commanded by the Execution Time Decision Making System, see Sect. 4.2, the execution starts with the robot configuration and scene shown in Fig. 39(a). In the given scene, the workspace is restricted by self-collisions and collisions with multiple obstacles in different arrangements, which makes the execution of predefined trajectories unfeasible. Additionally, the execution of the pour-in manipulation strategy is impossible due to collisions with objects near the cup and the robot has to actively change the course of actions and schedule auxiliary actions to rearrange the scene in a way, that the learned pour-in manipulation strategy can be applied.

For the scene driven logic-based reasoning, the symbolic scene description is generated, see Fig. 39(b), and the preconditions of the learned manipulation strategies are calculated by projecting example trajectories on the table surface. Since the

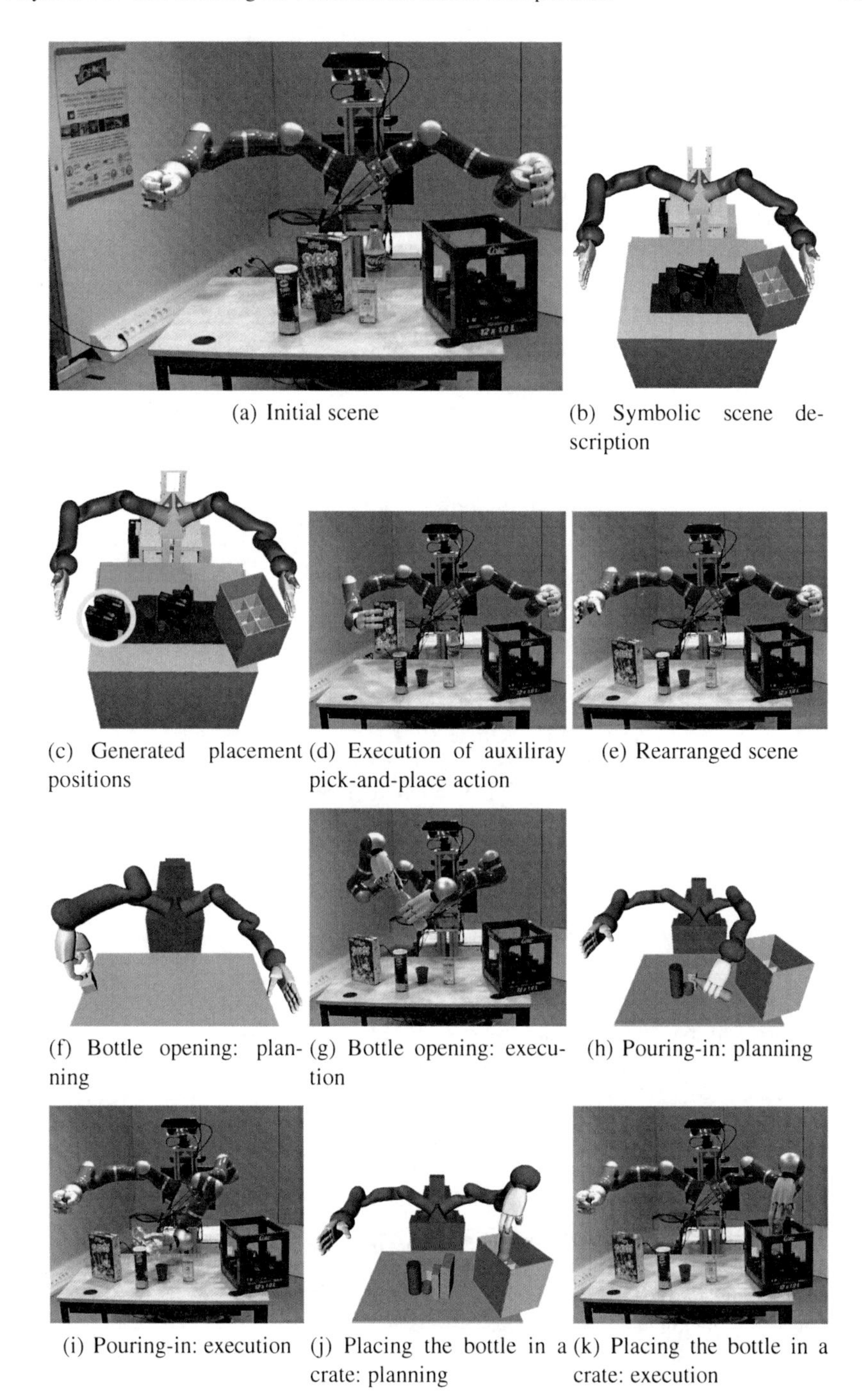

(a) Initial scene
(b) Symbolic scene description
(c) Generated placement positions
(d) Execution of auxiliray pick-and-place action
(e) Rearranged scene
(f) Bottle opening: planning
(g) Bottle opening: execution
(h) Pouring-in: planning
(i) Pouring-in: execution
(j) Placing the bottle in a crate: planning
(k) Placing the bottle in a crate: execution

Fig. 39 Execution of a complex manipulation task with the proposed layered system for autonomous environment manipulation of service robots.

cereals box blocks a part of the projected map, the symbolic planner evaluates the *mechanical* relations to determine, that the cereals box can be moved, and decides, that the cereals box has to be placed somewhere else on the table. Multiple placement positions are generated using the offline-computed graspability database, see Fig. 39(c), and finally an auxiliary pick-and-place action is introduced to grasp the cereals box, move it to a placement position and ungrasp it, see Fig. 39(d).

Since the preconditions of all actions are established, see Fig. 39(e), the learned manipulation strategies will be executed one after another using constrained motion planning. First, the manipulation strategy to lift a bottle and remove the bottle cap, which was learned in Sect. 2.7.2, is planned, see Fig. 39(f), and executed, see Fig. 39(g). Second, a manipulation strategy to use the grasped bottle to pour-in liquid into the cup on the table is executed. The manipulation strategy was learned based on multiple human demonstrations and generalized using *teaching* and *robot tests*, see the experiments in Sect. 2.7.1. Since the cereals box was removed, the constrained motion planner successfully finds a robot trajectory, see Fig. 39(h), which is executed, see Fig. 39(i). Finally, the bottle is regrasped and the manipulation strategy to place the bottle into the crate is planned, see Fig. 39(j), and executed, see Fig. 39(k), which was learned in the experiments in Sect. 2.7.1.

Acknowledgements. The research leading to these results has been supported by the DEXMART Large-scale integrating project, which has received funding from the European Communitys Seventh Framework Programme (FP7/2007-2013) under grant agreement ICT-216239. The authors are solely responsible for its content. It does not represent the opinion of the European Community and the Community is not responsible for any use that might be made of the information contained therein. In addition, the support of the German SFB 588 "Humanoid Robots" granted by DFG is gratefully acknowledged.

References

1. Alami, R., Chatila, R., Fleury, S., Ghallab, M., Ingrand, F.: An architecture for autonomy. International Journal of Robotics Research 17, 315–337 (1998)
2. Allen, J.: Maintaining knowledge about temporal intervals. Communications of the ACM 26, 832–843 (1983)
3. Asfour, T., Regenstein, K., Azad, P., Schröder, J., Vahrenkamp, N., Dillmann, R.: Armar-III: An integrated humanoid platform for sensory-motor control. In: 6th IEEE-RAS International Conference on Humanoid Robots, Genova (2006)
4. Aström, K.J.: Optimal control of Markov decision processes with incomplete state estimation. Journal of Mathematical Analysis and Applications 10, 174–205 (1965)
5. Azad, P., Asfour, T., Dillmann, R.: Combining apperance-based and model-based methods for real-time object recognition and 6D localization. In: IEEE/RSJ International Conference on Intelligent Robots and Systems, Beijing (2006)
6. Azad, P., Gockel, T., Dillmann, R.: Computer Vision: Das Praxisbuch. Elektor-Verlag, Aachen (2007)
7. Bengio, Y., Louradour, J., Collobert, R., Weston, J.: Curriculum learning. In: 26th Annual International Conference on Machine Learning, New York (2009)

8. Berenson, D., Srinivasa, S., Ferguson, D., Collet, A., Kuffner, J.: Manipulation planning with workspace goal regions. In: IEEE International Conference on Robotics and Automation, Kobe (2009)
9. Berenson, D., Srinivasa, S., Ferguson, D., Kuffner, J.: Manipulation planning on constraint manifolds. In: IEEE International Conference on Robotics and Automation, Kobe (2009)
10. Breazeal, C., Berlin, M., Brooks, A.G., Gray, J., Thomaz, A.L.: Using perspective taking to learn from ambiguous demonstrations. Robotics and Autonomous Systems 54, 385–393 (2006)
11. Brock, O., Kuffner, J., Xiao, J.: Motion for manipulation tasks. In: Siciliano, B., Khatib, O. (eds.) Springer Handbook of Robotics, pp. 615–645. Springer, Heidelberg (2008)
12. Bullet: The bullet physics library, http://bulletphysics.org
13. Calinon, S., Billard, A.: Teaching a humanoid robot to recognize and reproduce social cues. In: 15th IEEE International Symposium on Robot and Human Interactive Communication, Hatfield, UK (2006)
14. Calinon, S., Billard, A.: A probabilistic programming by demonstration framework handling skill constraints in joint space and task space. In: IEEE/RSJ International Conference on Intelligent Robots and Systems, Nice (2008)
15. Calinon, S., Guenter, F., Billard, A.: Goal-directed imitation in a humanoid robot. In: IEEE International Conference on Robotics and Automation, Barcelona (2005)
16. Calinon, S., Guenter, F., Billard, A.: On learning, representing, and generalizing a task in a humanoid robot. IEEE Transactions on Systems, Man, and Cybernetics, Part B 37, 286–298 (2007)
17. Cassandra, A.R., Kaelbling, L.P., Littman, M.L.: Acting optimally in partially observable stochastic domains. In: 12th National Conference on Artificial Intelligence, Seattle, WA (1994)
18. Chang, C., Lin, C.J.: LIBSVM: a Library for Support Vector Machines (2001)
19. Choi, D., Kang, Y., Lim, H., You, B.J.: Knowledge-based control of a humanoid robot. In: IEEE/RSJ International Conference on Intelligent Robots and Systems, Saint Louis, MO (2009)
20. Choi, J., Amir, E.: Combining planning and motion planning. In: IEEE International Conference on Robotics and Automation, Kobe (2009)
21. Daley, P., Frank, J., Iatauro, M., McGann, C., Taylor, W.: Planworks: A debugging environment for constraint based planning systems. In: International Conference on Knowledge Engeneering in Planning and Scheduling, Monterey, CA (2005)
22. Davies, D., Bouldin, D.: A cluster separation measure. IEEE Transactions on Pattern Analysis and Machine Intelligence 1, 224–227 (1979)
23. Deckers, P., Dollar, A.M., Howe, R.D.: Guiding grasping with proprioception and Markov models. In: Robotics: Science and Systems, Workshop on "Robot Manipulation: Sensing and Adapting to the Real World", Atlanta, GA (2007)
24. Dornhege, C., Gissler, M., Teschner, M., Nebel, B.: Integrating symbolic and geometric planning for mobile manipulation. In: IEEE International Workshop on Safety, Security and Rescue Robotics, Denver, CO (2009)
25. Eiben, A., Smith, J.: Introduction to Evolutionary Computing. Springer, New York (2003)
26. Ekvall, S., Aarno, D., Kragic, D.: Task learning using graphical programming and human demonstrations. In: 15th IEEE International Symposium on Robot and Human Interactive Communication, Hatfield, UK (2006)
27. Erlhagen, W., Mukovskiy, A., Bicho, E., Panin, G., Kiss, C., Knoll, A., van Schie, H.T., Bekkering, H.: Goal-directed imitation for robots: A bio-inspired approach to action understanding and skill learning. Robotics and Autonomous Systems 54, 353–360 (2006)

28. Freeman, J.: The modelling of spatial relations. Computer Graphics and Image Processing 4, 156–171 (1975)
29. Friedrich, H., Dillmann, R., Rogalla, O.: Interactive Robot Programming Based on Human Demonstration and Advice. In: Noltemeier, H., Christensen, H.I. (eds.) Dagstuhl Seminar 1998. LNCS (LNAI), vol. 1724, pp. 96–119. Springer, Heidelberg (1999)
30. Gat, E.: On three-layer architectures. In: Kortenkamp, D., Bonnasso, R.P., Murphy, R. (eds.) Artificial Intelligence and Mobile Robots, pp. 195–210. AAAI, MIT Press, Cambridge, MA (1997)
31. Gat, E., Slack, M., Miller, D., Firby, R.: Path planning and execution monitoring for a planetary rover. In: IEEE International Conference on Robotics and Automation, Taipei (2003)
32. Genz, A.: Numerical computation of rectangular bivariate and trivariate normal and t probabilities. Statistics and Computing 14, 151–160 (2004)
33. Gottschalk, S., Lin, M.C., Manocha, D.: OBBTree: A hierarchical structure for rapid interference detection. In: 23rd International Conference on Computer Graphics and Interactive Techniques, New Orleans, LA (1996)
34. Guan, Y., Yokoi, K.: Reachable space generation of a humanoid robot using the Monte Carlo method. In: IEEE/RSJ International Conference on Intelligent Robots and Systems, Beijing (2006)
35. Guilamo, L., Kuffner, J., Nishiwaki, K., Kagami, S.: Efficient prioritized inverse kinematic solutions for redundant manipulators. In: IEEE/RSJ International Conference on Intelligent Robots and Systems, Edmonton (2005)
36. Halkidi, M., Vazirgiannis, M., Batistakis, Y.: Quality scheme assessment in the clustering process. In: Zighed, D.A., Komorowski, J., Żytkow, J.M. (eds.) PKDD 2000. LNCS (LNAI), vol. 1910, pp. 265–276. Springer, Heidelberg (2000)
37. Hertzberg, J., Chatila, R.: AI reasoning methods for robotics. In: Siciliano, B., Khatib, O. (eds.) Springer Handbook of Robotics, pp. 207–223. Springer, Heidelberg (2008)
38. Hoey, J., von Bertoldi, A., Poupart, P., Mihailidis, A.: Assisting persons with dementia during handwashing using a partially observable Markov decision process. In: 5th International Conference on Computer Vision Systems, Bielefeld (2007)
39. Hsiao, K., Kaelbling, L.P., Lozano-Pérez, T.: Grasping POMDPs. In: IEEE International Conference on Robotics and Automation, Roma (2007)
40. Ijspeert, A.J., Nakanishi, J., Schaal, S.: Movement imitation with nonlinear dynamical systems in humanoid robots. In: IEEE International Conference on Robotics and Automation, Washington, DC (2002)
41. Jäkel, R., Meißner, P., Schmidt-Rohr, S.R., Dillmann, R.: Distributed generalization of learned planning models in robot Programming by Demonstration. In: IEEE/RSJ International Conference on Intelligent Robots and Systems, San Francisco, CA (2011)
42. Jäkel, R., Schmidt-Rohr, S.R., Lösch, M., Dillmann, R.: Representation and constrained planning of manipulation strategies in the context of Programming by Demonstration. In: IEEE International Conference on Robotics and Automation, Anchorage, AK (2010)
43. Jäkel, R., Schmidt-Rohr, S.R., Lösch, M., Kasper, A., Dillmann, R.: Learning of generalized manipulation strategies in the context of Programming by Demonstration. In: 10th IEEE-RAS International Conference on Humanoid Robots, Nashville, TN (2010)
44. Kasper, A.: KIT ObjectModels Web Database, http://wwwiaim.ira.uka.de/ObjectModels
45. Kavraki, L., Svestka, P., Latombe, J.C., Overmars, M.: Probabilistic roadmaps for path planning in high-dimensional configuration spaces. In: IEEE International Conference on Robotics and Automation, Minneapolis, MN (1996)

46. Keijzer, M., Merelo, J.J., Romero, G., Schoenauer, M.: Evolving objects: A general purpose evolutionary computation library. In: Collet, P., Fonlupt, C., Hao, J.-K., Lutton, E., Schoenauer, M. (eds.) EA 2001. LNCS, vol. 2310, pp. 231–242. Springer, Heidelberg (2002)
47. Kim, Y.J., Otaduy, M.A., Lin, M.C., Manocha, D.: Fast penetration depth computation using rasterization hardware and hierarchical refinement, Technical Report, Department of Computer Science, University of North Carolina (2002)
48. Kleer, J.D., Brown, J.S.: A qualitative physics confluences. Artificial Intelligence 24, 7–83 (1984)
49. Kormushev, P., Calinon, S., Saegusa, R., Metta, G.: Learning the skill of archery by a humanoid robot Icub. In: 10th IEEE-RAS International Conference on Humanoid Robots, Nashville, TN (2010)
50. Kuffner, J.J., LaValle, S.M.: RRT-Connect: An efficient approach to single-query path planning. In: IEEE International Conference on Robotics and Automation, San Francisco, CA (2000)
51. Kurniawati, H., Hsu, D., Lee, W.: SARSOP: Efficient point-based POMDP planning by approximating optimally reachable belief spaces. In: Robotics: Science and Systems, Zurich (2008)
52. Lemaignan, S., Ros, R., Mösenlechner, L., Alami, R., Beetz, M.: ORO, a knowledge management platform for cognitive architectures in robotics. In: IEEE/RSJ International Conference on Intelligent Robots and Systems, Taipei (2010)
53. Lösch, M., Schmidt-Rohr, S., Knoop, S., Vacek, S., Dillmann, R.: Feature set selection and optimal classifier for human activity recognition. In: 16th IEEE International Symposium on Robot and Human Interactive Communication, Jeju Island, Korea (2007)
54. Lösch, M., Schmidt-Rohr, S.R., Dillmann, R.: Making feature selection for human motion recognition more interactive through the use of taxonomies. In: 17th IEEE International Symposium on Robot and Human Interactive Communication, München (2008)
55. MacQueen, J.: Some methods for classification and analysis of multivariate observations. In: 5th Berkeley Symposium on Mathematical Statistics and Probability, Berkeley, CA (1967)
56. Mahalanobis, P.C.: On the generalised distance in statistics. Proceedings National Institute of Science, India 2(1), 49–55 (1936)
57. Meeussen, W., Rutgeerts, J., Gadeyne, K., Bruyninckx, H., De Schutter, J.: Contact-state segmentation using particle filters for programming by human demonstration in compliant-motion tasks. IEEE Transactions on Robotics 23, 218–231 (2006)
58. Meißner, P., Schmidt-Rohr, S.R., Lösch, M., Jäkel, R., Dillmann, R.: Robust localization of furniture parts by integrating depth and intensity data suitable for range sensors with varying image quality. In: 15th International Conference on Advanced Robotics, Tallinn (2011)
59. Morinaga, S., Kosuge, K.: Collision detection system for manipulator based on adaptive impedance control law. In: IEEE International Conference on Robotics and Automation, Taipei (2003)
60. Mühlig, M., Gienger, M., Steil, J., Goerick, C.: Automatic selection of task spaces for imitation learning. In: IEEE/RSJ International Conference on Intelligent Robots and Systems, Saint Louis, MO (2009)
61. Mühlig, M., Gienger, M., Steil, J.J.: Human-robot interaction for learning and adaptation of object movements. In: IEEE/RSJ International Conference on Intelligent Robots and Systems, Taipei (2010)

62. Mülling, K., Kober, J., Peters, J.: Simulating human table tennis with a biomimetic robot setup. In: Doncieux, S., Girard, B., Guillot, A., Hallam, J., Meyer, J.-A., Mouret, J.-B. (eds.) SAB 2010. LNCS, vol. 6226, pp. 273–282. Springer, Heidelberg (2010)
63. Pastor, P., Kalakrishnan, M., Chitta, S., Theodorou, E., Schaal, S.: Skill learning and task outcome prediction for manipulation. In: IEEE International Conference on Robotics and Automation, Shanghai (2011)
64. Patrón, P., Miguelañez, E., Petillot, Y.R., Lane, D.M.: Fault tolerant adaptive mission planning with semantic knowledge representation for autonomous underwater vehicles. In: IEEE/RSJ International Conference on Intelligent Robots and Systems, Nice (2008)
65. Pettersson, O.: Execution monitoring in robotics: A survey. Robotics and Autonomous Systems 53, 73–88 (2005)
66. Pineau, J., Gordon, G., Thrun, S.: Point-based value iteration: An anytime algorithm for POMDPs. In: International Joint Conference on Artificial Intelligence, Acapulco (2003)
67. Pineau, J., Thrun, S.: High-level robot behavior control using POMDPs. In: AAAI Workshop on Cognitive Robotics, Edmonton (2002)
68. Primesense Nite Middleware, http://www.primesense.com/en/nite
69. Rühl, S.W., Hermann, A., Xue, Z., Kerscher, T., Dillmann, R.: Generating a symbolic scene description for robot manipulation using physics simulation. In: Multibody Dynamics, Brussels (2011)
70. Rühl, S.W., Hermann, A., Xue, Z., Kerscher, T., Dillmann, R.: Graspability: A description of work surfaces for planning of robot manipulation sequences. In: IEEE International Conference on Robotics and Automation, Shanghai (2011)
71. Rühl, S.W., Xue, Z., Zöllner, J., Dillmann, R.: Integration of a loop based and an event based framework for control of a bimanual dextrous service robot. In: IEEE International Conference on Robotics and Biomimetics, Guilin, China (2009)
72. Russell, S.J., Norvig, P.: Artificial Intelligence: A Modern Approach, 2nd edn. Prentice Hall, Upper Saddle River (2003)
73. Sakoe, H., Chiba, S.: Dynamic programming algorithm optimization for spoken word recognition. IEEE Transactions on Acoustics, Speech and Signal Processing 26, 43–49 (1978)
74. Schmidt-Rohr, S.R., Dirschl, G., Meissner, P., Dillmann, R.: A knowledge base for learning probabilistic decision making from human demonstrations by a multimodal service robot. In: 15th International Conference on Advanced Robotics, Tallinn (2011)
75. Schmidt-Rohr, S.R., Knoop, S., Lösch, M., Dillmann, R.: Bridging the gap of abstraction for probabilistic decision making on a multi-modal service robot. In: Robotics: Science and Systems, Zurich (2008)
76. Schmidt-Rohr, S.R., Lösch, M., Jäkel, R., Dillmann, R.: Programming by Demonstration of probabilistic decision making on a multi-modal service robot. In: IEEE/RSJ International Conference on Intelligent Robots and Systems, Taipei (2010)
77. Siciliano, B., Khatib, O. (eds.): Springer Handbook of Robotics. Springer, Heidelberg (2008)
78. Simeon, T., Cortes, J., Sahbani, A., Laumond, J.P.: A manipulation planner for pick and place operations under continuous grasps and placements. In: IEEE International Conference on Robotics and Automation, Washington, DC (2002)
79. Slaney, J., Thiebaux, S.: Blocks world revisited. Artificial Intelligence 125, 119–153 (2001)
80. Sondik, E.J.: The optimal control of partially observable Markov decision processes, Ph.D. thesis, Stanford University (1971)
81. Sridharan, M., Wyatt, J., Dearden, R.: Hippo: Hierarchical POMDPs for planning information processing and sensing actions on a robot. In: International Conference on Automated Planning and Scheduling, Sydney (2008)

82. Steffen, J., Elbrechter, C., Haschke, R., Ritter, H.: Bio-inspired motion strategies for a bimanual manipulation task. In: 10th IEEE-RAS International Conference on Humanoid Robots, Nashville, TN (2010)
83. Stilman, M.: Task constrained motion planning in robot joint space. In: IEEE/RSJ International Conference on Intelligent Robots and Systems, San Diego, CA (2007)
84. Stilman, M., Schamburek, J.U., Kuffner, J., Asfour, T.: Manipulation planning among movable obstacles. In: IEEE International Conference on Robotics and Automation, Roma (2007)
85. Suh, I.H., Lim, G.H., Hwang, W., Suh, H., Choi, J.H., Park, Y.T.: Ontology-based multi-layered robot knowledge framework (OMRKF) for robot intelligence. In: IEEE/RSJ International Conference on Intelligent Robots and Systems, San Diego, CA (2007)
86. Tenorth, M., Beetz, M.: KnowRob — Knowledge processing for autonomous personal robots. In: IEEE/RSJ International Conference on Intelligent Robots and Systems, Saint Louis, MO (2009)
87. Vahrenkamp, N., Scheurer, C., Asfour, T., Kuffner, J., Dillmann, R.: Adaptive motion planning for humanoid robots. In: IEEE/RSJ International Conference on Intelligent Robots and Systems, Nice (2008)
88. Veeraraghavan, H., Veloso, M.M.: Learning task specific plans through sound and visually interpretable demonstrations. In: IEEE/RSJ International Conference on Intelligent Robots and Systems, Nice (2008)
89. Xie, X., Beni, G.: A validity measure for fuzzy clustering. IEEE Transactions on Pattern Analysis and Machine Intelligence 13, 841–847 (1991)
90. Xue, Z., Kasper, A., Zöllner, J., Dillmann, R.: An automatic grasp planning system for service robots. In: 14th International Conference on Advanced Robotics, München (2009)
91. Yu, L., Liu, H.: Efficient feature selection via analysis of relevance and redundancy. Journal of Machine Learning Research 5, 1205–1224 (2004)
92. Zacharias, F., Borst, C., Hirzinger, G.: Capturing robot workspace structure: Representing robot capabilities. In: IEEE/RSJ International Conference on Intelligent Robots and Systems, San Diego, CA (2007)
93. Zöllner, R., Pardowitz, M., Knoop, S., Dillmann, R.: Towards cognitive robots: Building hierarchical task representations of manipulations from human demonstration. In: IEEE International Conference on Robotics and Automation, Barcelona (2005)

Observation and Execution

Christoph Borst, Franziska Zacharias, Florian Schmidt, Daniel Leidner, Maximo A. Roa, Katharina Hertkorn, Gerhard Grunwald, Pietro Falco, Ciro Natale, and Emilio Maggio

Abstract. Assistive robotic systems in household or industrial production environments get more and more capable of performing also complex tasks which previously only humans were able to do. As robots are often equipped with two arms and hands, similar manipulations can be executed. The robust programming of such devices with a very large number of degrees of freedom (DOFs) compared with single industrial robot arms however is laborious if done joint-wise. Two major directions to overcome this problem have been previously proposed. The programming by demonstration (PbD) approach, where human arm and recently also hand motions are tracked, segmented and re-executed in an adaptive way on the robotic system and the high-level planning approach which tries to generate a task sequence on a logical level and attributes geometric information as necessary to generate artificial trajectories to solve the task. Here we propose to combine the best of both worlds. For the very complex motion generation for a robotic hand, a rather direct approach to assign manipulation actions from human demonstration to a human hand is taken. For the combination of different basic manipulation actions the task constraints are segmented from the demonstration action and used to generate a task oriented plan. This plan is validated against the robot kinematic and geometric constraints and then a geometric motion planner can generate the necessary robot motions to fulfill the task execution on the system.

Christoph Borst · Franziska Zacharias · Florian Schmidt · Daniel Leidner · Maximo A. Roa · Katharina Hertkorn · Gerhard Grunwald
Robotik und Mechatronik Zentrum, DLR, Münchnerstrasse 20, 82234 Wessling, Germany
e-mail: {christoph.borst,florian.schmidt,daniel.leidner, maximo.roa,katharina.hertkorn,gerhard.grunwald}@dlr.de, zacharias@mvtec.com

Pietro Falco · Ciro Natale
Dipartimento di Ingegneria dell'Informazione, Seconda Università degli Studi di Napoli, via Roma 29, 81031 Aversa, Italy
e-mail: {pietro.falco,ciro.natale}@unina2.it

Emilio Maggio
Oxford Metrics Group, 14 Minns Business Park, West Way, Oxford OX2 0JB, United Kingdom
e-mail: emilio.maggio@omg3d.com

B. Siciliano (Ed.): Advanced Bimanual Manipulation, STAR 80, pp. 59–122.
springerlink.com

1 Introduction

One of the major ideas followed in DEXMART was to make use of the observation of human manipulation actions to guide an optimize the execution of the same actions on a bimanual robotic platform. This principle to use humans to demonstrate actions to be later adapted to robot programs is well known as the PbD paradigm. In DEXMART this principle has been studied on different levels of abstraction. On lower levels of trajectory observation a special focus is set on the observation of finger movement during the task execution. For similar kinematic constraints of human and robot hand such generalized manipulation trajectories might be tried directly to program a robotic hand.

For dissimilar kinematics such approach needs a proper mapping to the different workspace of the robotic device. Such a mapping is difficult to be found for arbitrary kinematics. Therefore we tried to extract invariant task constraints from demonstrations in various setups to store the essentials of the task in newly developed representations. This abstract information is then combined with the kinematic constraints of the robot to be successfully executed on the robot.

This chapter is organized in two parts. The first is focusing on the observation part of human hands combining the observation of finger movements with tactile information essential for the successful task execution. Models that allow the adaption to slightly changing situations regarding the kinematics as well as geometry of the object due to uncertainties are presented. In cases where the mapping of the trajectories to heavily changing kinematics is needed a model based approach using the extracted task constraints combined with intelligent planning to optimally using the kinematics of the robot is used. This is presented in the second part of the chapter where the execution of constrained tasks with a limited robotic manipulation device is the main focus.

2 Human Observation

The robotic systems of the next decade will be, potentially, a part of everyday life as our appliances, servants and assistants, as our helpers and elder-care companions, assisting surgeons in medical operations, intervening in hazardous or life-critical environments for search and rescue operations, and operating in field areas like forestry, agriculture, cleaning, mining, freight transport, construction and demolition, and so on. In this scenario, bringing a robot to the same manipulation skills as those of human beings is recognized as the crucial issue for transferring the robots from industry to the service robotics application domain. Several researchers work towards this objective within the DEXMART project. It attempts to extend a bridge from research on natural cognition to research on artificial cognition, as it will primarily contribute to the development of robotic systems endowed with dexterous and human-aware dual-arm/hand manipulation skills for objects, operating with a high degree of autonomy in unstructured real-world environments.

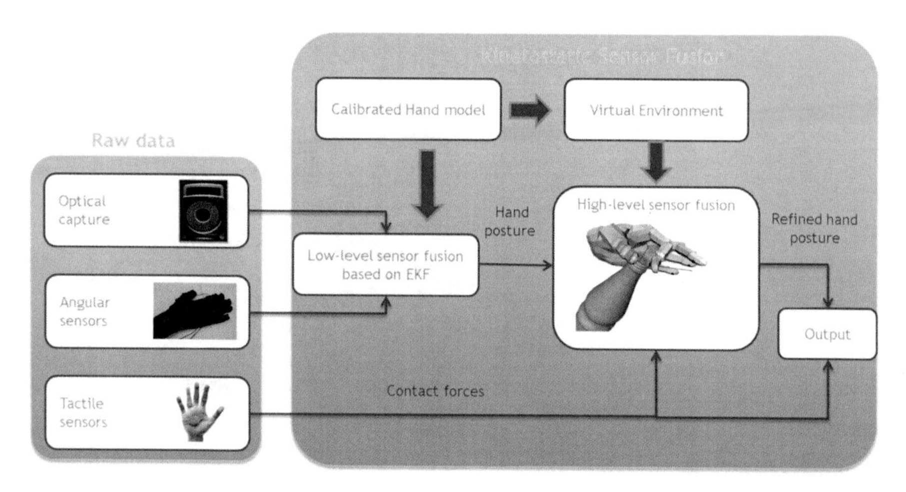

Fig. 1 Architecture of the sensor fusion system used for observation.

The approach followed in the project to pursue this challenging goal is based on PbD-like strategies, which require development of original methods for interpretation, learning, and modelling, from the observation of human manipulation at different levels of abstraction [9]. At the state of the art, the observation of the human hand motion during the execution of complex manipulation tasks, is a very difficult problem, handled through two main approaches

- optical position measurements based on motion capture systems, which require many markers and many cameras, for trying to reduce the marker occlusion phenomenon, very frequent in hand tracking problems;
- direct angular measurements, which require complex and expensive sensorized gloves.

In this chapter, a novel architecture, based on sensor fusion of kinetostatic data, is proposed that overcomes the limitations of the classical systems. As shown in Fig. 1, the system is constituted by a low-level sensor fusion module that estimates the hand posture and a high-level module that exploits the knowledge of fingertip contact forces to refine the initial estimation. The low-level module is in charge to observe human hand motion combining, through a Bayesian senor fusion technique, both the classic approach described above with a significantly decreased hardware complexity, i.e. using a small number of markers (typically only three markers per finger) and cameras, and only three low-cost angular sensors per finger, specifically designed and realized for this purpose. Usually, extracting the joint angles requires reliable marker data due to the high number of human hand DOFs.

Figure 2 shows the marker sets required by the proposed low-level algorithm. Evidently, on some occasions the positional data of some of the markers will not be available due to occlusions. This lack of data could lead to unconstrained kinematics and finally to unreliable hand pose estimates. To cope with this challenge the proposed approach not only exploits the additional information coming from the

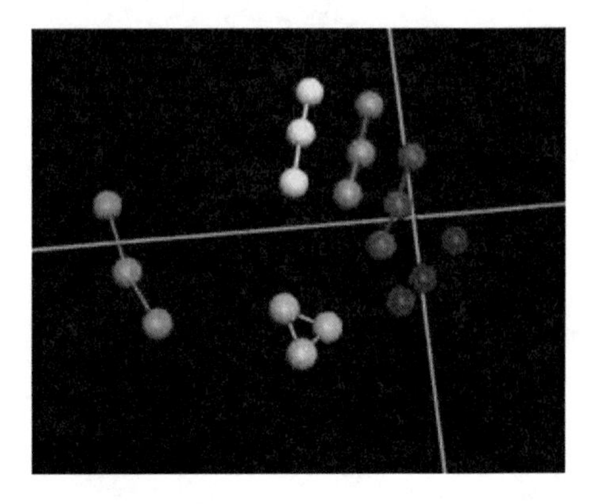

Fig. 2 Marker set used in the project for the entire hand.

data glove, thus filling the data gaps, but also usefully exploits the kinematic model of the hand. The algorithm is constituted by two steps. The first is devoted to estimate the constant kinematic parameters exploiting the recursive nature of the open kinematic chains. The second step consists of estimating the joint angles through a finger-centralized sensor fusion algorithm which takes into account also the marker slipping over the glove surface.

The approach followed within the low-level sensor fusion component presents three key innovations:

- design and realization of an optoelectronic low-cost data glove;
- estimation of the model parameters by methods that exploit the recursive nature of open kinematic chains;
- real-time joint angle tracking by Bayesian sensor fusion algorithms for nonlinear systems.

Until now, only few papers have addressed the problems of kinematic model parameter and joint angle estimation in human complex manipulation tasks; none of these papers uses approaches based on the sensor fusion. In [29] a protocol to determine the link structure of human hand using motion capture data is proposed. [52] describes a global optimization method for off-line assessing joint angles in human hand and for calibrating instrumented glove from motion capture system measurements. The papers [6] and [57] use, respectively, a deterministic and a stochastic global optimization algorithm to determine the centers and the axes of rotation for fingers in a simplified model of the human hand. In [8] an anatomic-based cost metric is proposed to identify a model of the carpometacarpal joint of the thumb, that is suitable for measuring mobility. [24] and [7] try to estimate the whole body motion from the measurement by a Kalman-like approach, but without focusing on the hand motion. [16] is one of the first works that address the real-time finger tracking problem; it uses a Kalman approach to track the marker positions for an augmented reality application.

Concerning the realization of the low-cost data glove, the optoelectronic technology has been selected not only for cheapness sake, but also for its typical interesting properties such as immunity to electromagnetic field, low power consumption and lightness. The data glove proposed is equipped with sensing elements whose developments is based on the use of angle-varying radiation pattern of common LEDs (Light Emitting Diodes) and responsiveness pattern of PDs (PhotoDetectors). The effectiveness and advantages of using a solution that exploits this property of optoelectronic components has already been shown in [34], [33] and [5], where the measurement of different physical variables is proposed.

The high-level sensor fusion module aims at improving the observation of the human hand motion, exploiting the measurements of fingertip contact forces and a virtual environment. The main idea of the proposed algorithm is to compare the fingertip contact information, obtained by commercial tactile sensors, with the contact information computed in a virtual environment, that reproduces the real one. In case the estimation of the joint angles and the relative pose between the hand and the object are accurate, the contact information in the virtual and in the real environment are fitting, i.e the contact-consistence condition is satisfied. On the other hand, when the two sources of information are not consistent, a correction of the hand posture is carried out. The correction is constituted by two steps. The first step consists in computing, on the basis of the geometry of the grasped object and of the hand posture, the fingertip position and orientation, such that the contact-consistence condition is satisfied. The second step finds the posture of the hand (i.e. position, orientation and joint angles) such that the end-effectors assume the poses computed in the first step. To tackle this Inverse Kinematics (IK) problem, a Jacobian-based technique known as CLIK (closed-loop inverse kinematics) has been used, which is suitable for the on-line implementation of the correction algorithm. Since the starting point of the IK algorithm is given by the data from the sensors, only a local correction is required and Jacobian based methods, that are fast to find local minima, are particularly suitable to this application. It is important to emphasize that in the inverse kinematics algorithm, the hand is modeled not as five independent kinematic chains, but as a "kinematic tree" with a root and five branches. The root is composed by six 1-DOF joints, describing the pose of the hand in space and the branches are the five serial chains describing the five fingers. The correction brings two advantages: improve of the accuracy of the hand observation and guarantee of the coherence between hand posture and measured force in the virtual environment, which is very important for PbD applications. In order to correctly tune the parameters of the CLIK algorithm, the results of the study found in [17] on the stability of CLIK algorithms have been exploited.

3 The Sensory Environment

The measurement system adopted for tracking human hand motions uses two different types of sensing systems. The first one is a commercial optical motion capture system. The second one is a sensorized glove equipped with three angular sensors per finger and with three reflective markers per finger. Marker positions are measured through a Vicon optical motion capture system (see Fig. 3), composed by four

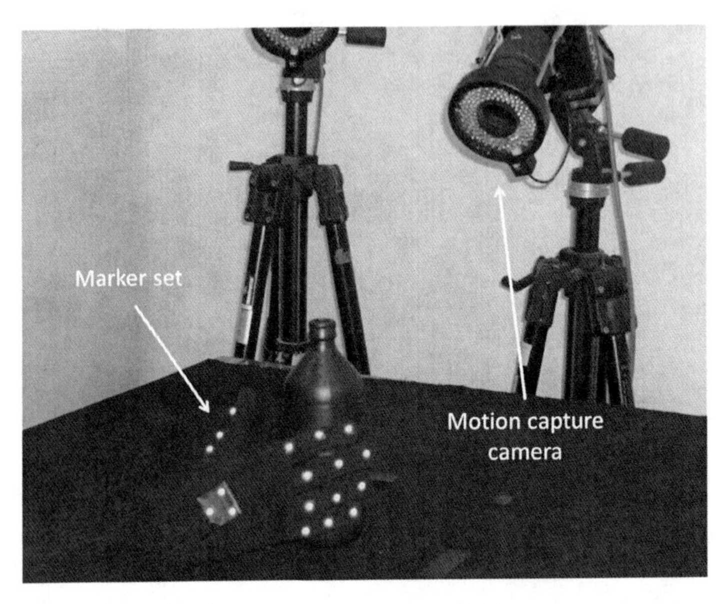

Fig. 3 The motion capture system: cameras and data glove.

Fig. 4 Picture of a single sensing element of the data glove.

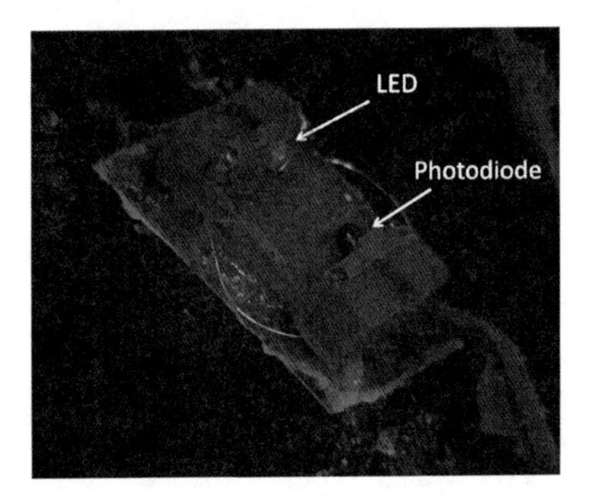

CMOS cameras, a workstation and a host PC, on which a real-time engine and the software for calibration and management are installed. The workstation is connected to a PC through an Ethernet cable, with TCP/IP protocol. The data glove has been principally designed to be used in a sensor fusion system and the concept is presented in Sect. 3.1. The calibration curve of the sensor can be estimated according to the specific procedure described in [11]. Both the motion capture system and the data glove work with a sampling rate of 60 Hz.

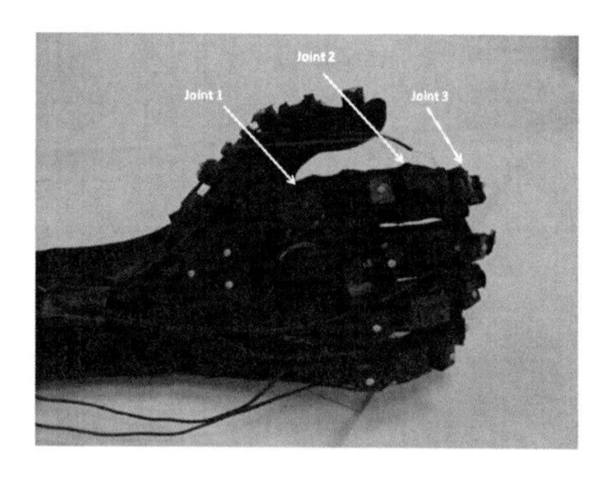

Fig. 5 Picture of the data glove prototype with angular sensors and markers used for motion capture system.

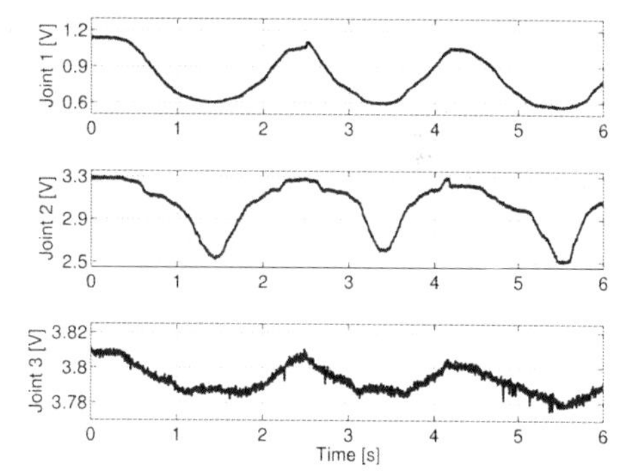

Fig. 6 Typical measured voltages on joint angular sensors pointed out in Fig. 5 for a repeated pick and place task.

3.1 Data Glove Design

A prototype of a data glove has been realized installing sensing elements based on optoelectronic technology (see Fig. 4) on the joints of a commercial neoprene glove. More details on this technology can be found in [11]. Figure 5 shows a picture of the realized prototype. Figure 6 reports typical measured voltages for all index finger joints in a simple pick and place task. In order to calibrate the data glove, a recursive inverse kinematic algorithm has been used. It elaborates measurements from a motion capture system and exploits the recursive nature of the open kinematic chains to calculate the joint angles. ¿From the user point of view, a calibration session consists of observing, by an optical motion capture system, repeated flexion-extension motion of all the fingers. It is important to underline that the result of the calibration procedure depends on the shape and dimension of the user's hand. Then, for a different performer, a new calibration procedure is required.

Figures 7, 8, 9 report examples of calibration curves for the index finger joints of the data glove prototype, which have been obtained through the calibration procedure which can be found in [11]. Obviously, the range of angular motion is different for each joint and also the sensitivity.

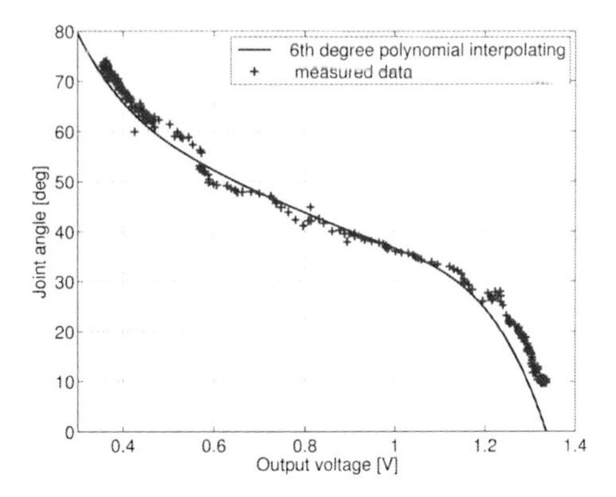

Fig. 7 Calibration curve of the sensor on joint 1 of the index finger.

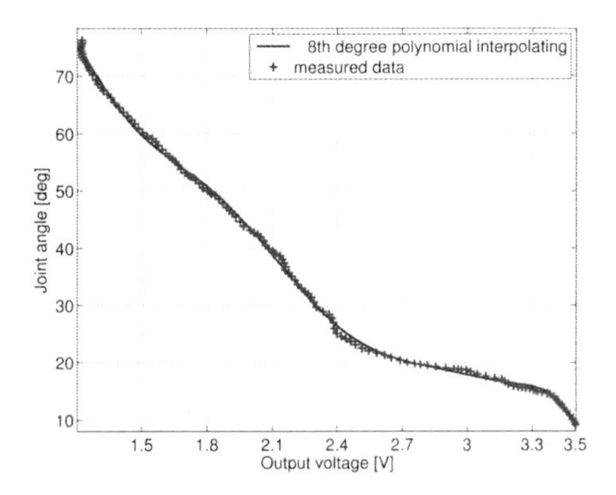

Fig. 8 Calibration curve of the sensor on joint 2 of the index finger.

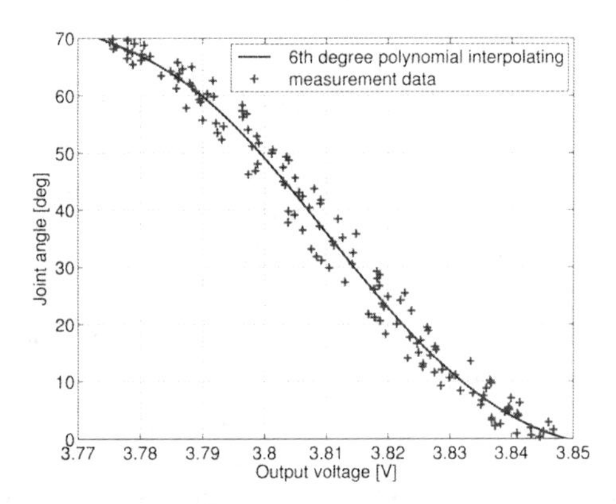

Fig. 9 Calibration curve of the sensor on joint 3 of the index finger.

3.2 Tactile Sensors

The glove is equipped with tactile sensors is based on the Pressure Profile FingerTPS, see Fig 10. In each fingertip, a tactile sensor pad is used to measure the

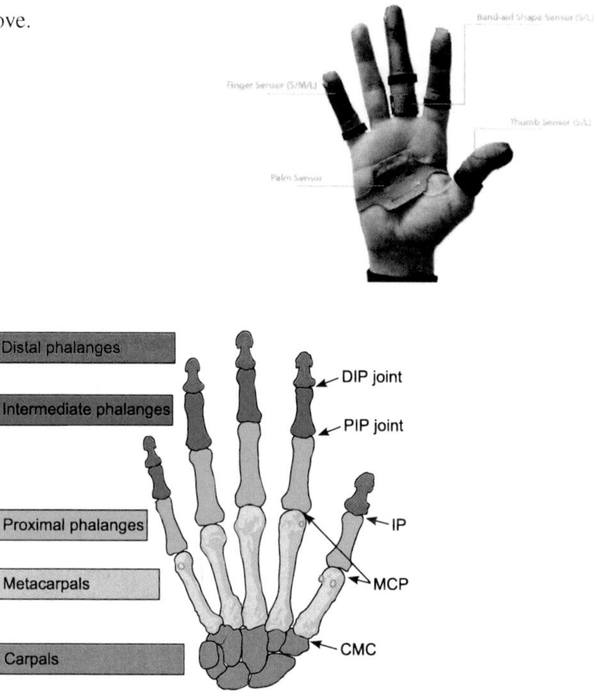

Fig. 10 Tactile sensors installed into the data glove.

Fig. 11 Human hand skeleton and articulations.

force intensity applied to an object. Due to the size and location of the pads, the human operator has to consider the sensor pads and adapt the manipulation motion in order to make consistent force measurements. An additional wrist sensor is available but has not been used in this work. The system is calibrated using a provided dynamometer. The repeatability is $< 4\%$ of the full scale range, which is 4.55 to 22.73 kg [40].

4 Kinematic Model of the Hand

In order to reconstruct the motion of the human hand, a kinematic model has to be selected. Then, its parameters have to be estimated, including the joint angles necessary to animate the model. The articulations of the human hand are more complex than the comparable articulations of other animals. In fact, the skeleton only consists of 27 bones, 14 for the fingers, five metacarpal forming the palm and eight carpal bones in the wrist (Fig. 11). With respect to the main goal of the DEXMART project, i.e., improve robotic manipulation capabilities, a necessity arises for a model that can reproduce the vast majority of the manipulation tasks and that

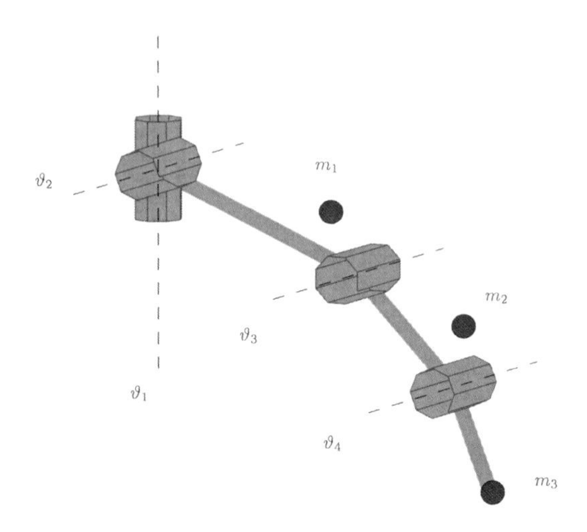

Fig. 12 Kinematic model and marker set for one finger.

Table 1 Denavit-Hartenberg table for all the fingers.

link	a	d	ϑ	α
1	0	0	ϑ_1	$\frac{\pi}{2}$
2	a_2	0	ϑ_2	0
3	a_3	0	ϑ_3	0
4	a_4	0	ϑ_4	0

can be measured with the motion capture technology presented in the previous sections. Unfortunately, a model that emulates in toto human hand is not desirable as the subtle movements of the carpal bones are difficult to measure using non-invasive techniques. However measuring at such a details is not required as state of the art robotic hands are usually much simpler (in terms of kinematics) than the human ones; therefore an approximated kinematic model might suffice. The next section defines such a model.

4.1 Model Definition

A universally recognized as an accurate model of the human hand is the one proposed in [39], which allows describing also palm arching movements. A simplified version of the model is adopted in this chapter assuming a rigid palm. In detail, for each finger, the selected 4-DOF kinematic model is depicted in Fig. 12, where also the markers attached to each bone are reported. The above kinematic model assumes that the first two joints of each finger are two consecutive pin joints with orthogonal axes. Another key assumption of the adopted kinematic model is that the flexion axes of each finger are all aligned. The method adopted to define the kinematic model, i.e. the relationship between the joint angles and the fingertip poses, is the

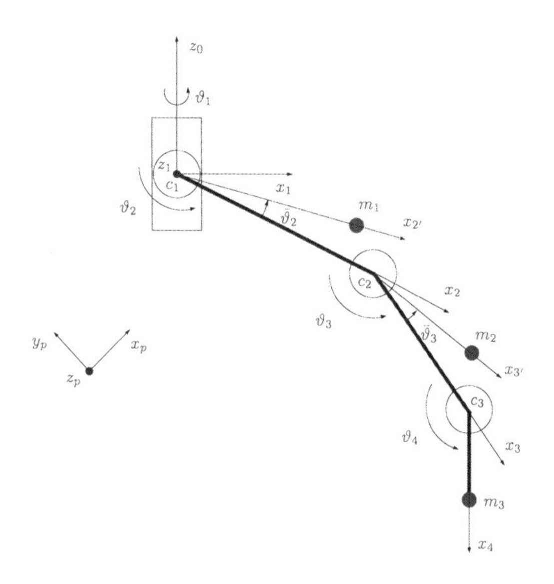

Fig. 13 Denavit-Hartenberg frames and markers attached to the finger.

so-called Denavit-Hartenberg technique [12], which is widely used both in robotics and biomechanics. The reader unfamiliar with this method can refer to, e.g., [46]. The resulting Denavit-Hartenberg (D-H) parameter table is the one in Tab. 1. The D-H reference frames fixed to the links of the finger are depicted in Fig. 13, together with two intermediate frames used in the calibration algorithm.

4.2 Modeling Marker Local Motion

When the hand flexes, due to skin stretch, glove sliding over the skin, and muscles deformations, the distance between markers and the center of rotation of the articulations changes, i.e., the markers "slide" over the bones while the hand moves, and this might cause large residual errors during calibration and fitting with the optical system.

The results in Fig. 14(a)–(c) form a range of motion trial show the motion behavior of three different markers. Each graph displays time instances of a component of the fitting residual vector plotted against its maximally correlated joint angle. The plots clearly show that a dependency between marker motion and joint angles in θ indeed exists. In one case (Fig. 14 (a)) a linear function could possibly approximate the relationship between the variables. This is valid as long as the joint angle range is small. In other cases a non-linear mapping function might be more appropriate (Fig. 14 (b)). Finally, as shown by residual-parameter correlation matrix in Fig. 14 (d) marker motion can be highly correlated with more than one joint parameter (i.e., matrix rows with more than one red cell). Consequently a model with a single input cannot provide good predictions (Fig. 14 (c)).

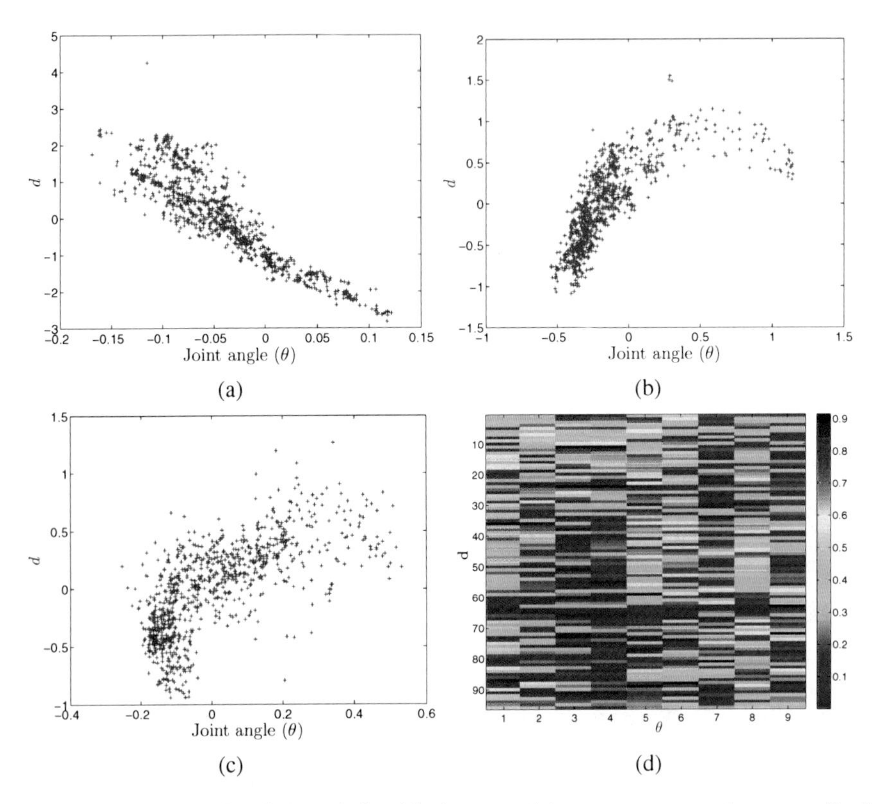

Fig. 14 Visual analysis of the relationship between joint parameters and unnormalised marker to reckon residuals. (a)-(c): sample residual components (in mm) plotted against the maximally correlated joint parameter (in radians). (d): correlation coefficient absolute values.

The analysis of the residual error shows that it may be possible to improve the predictive power of the kinematic model by explicitly accounting for marker movement. To this extent we propose to model $\mathbf{l}_i$ the position of the i-th marker in the local coordinate system of its parent segment as a parametric function of some joint angles θ and a set of regression parameters $\mathbf{w}_i \in W$. In particular we chose a linear form, $\mathbf{l}_i = F(\theta)\mathbf{w}_i$, where

$$F(\theta) = \begin{bmatrix} \phi_x(\theta)^T & \mathbf{0} & \mathbf{0} & 0 \\ \mathbf{0} & \phi_y(\theta)^T & \mathbf{0} & 0 \\ \mathbf{0} & \mathbf{0} & \phi_z(\theta)^T & 0 \\ \mathbf{0} & \mathbf{0} & \mathbf{0} & 1 \end{bmatrix}$$

contains the regressor vectors $\phi(\theta)$ for each of the three positional coordinates. In our implementation the regressors are simple polynomial components. The marker residual used by the optimization procedure becomes

$$\mathbf{d}_i = \mathbf{m}_i - S_i(\theta)\mathbf{l}_i = \mathbf{m}_i - S_i(\theta)F(\theta)\mathbf{w}_i. \tag{1}$$

Algorithm 1. Active inputs pre-selection

1: INPUT: $\{\theta, \mathbf{d}^{(u)}\}$
2: OUTPUT: set of active inputs A_u
3: Compute the correlation $c_j^{(u)}$ between $d_k^{(u)}$ and $\theta_{j,k}$ (a row in Fig. 14 (d)).
4: **for all** j **do**
5:    **if** $c_j^{(u)} > T$ **then**
6:       Add j to the set of active inputs A_u.
7:    **end if**
8: **end for**

where $\mathbf{m}_i$ is the measured marker position and $S_i(\theta)$ is the function dependent on the kinematic chain that maps a 3D point in the local coordinate system onto the coordinate system of the world. Note that, if the polynomials are zero-order (i.e., $F(\theta) = I$) Equation (1) reduces to a standard kinematic model.

In general for a specific residual not all the joint angles in θ convey useful information to improve the predictive power of the model. As shown in Fig. 14, each residual is usually correlated to a limited number of joint angles and in some cases non-linear components may not be necessary. At the same time favouring simpler models with a low number of extra parameters might help preventing model overfitting problems. To this end we implemented a model selection procedure which selects only those parameters which contribute to a significant residual reduction.

We treat the marker motion modeling as a post-progessing step to the standard calibration procedure. Given a calibrated standard model and a range of motion trial we treat the residual components independently. To the purpose of this explanation it is convenient to define $\mathbf{d}_k = [\mathbf{d}_{1,k} \dots \mathbf{d}_{n_m,k}]$ as the concatenation of the residuals at frame k and an index $u = 1, \dots, U$ for the single residual components $d_k^{(u)}$. Also, we define the 1D polynomial function $g^{(u)}(\theta, \mathbf{w}^{(u)}) = \phi^{(u)}(\theta)^T \mathbf{w}^{(u)}$ modelling the marker motion for the component u. For each residual the model selection procedure is composed of two steps: (i) first we analyse the correlation between θ_k and $d_k^{(u)}$; those parameters with correlation larger than a threshold T are selected as active inputs for the skin model function $g^{(u)}$ (Algorithm 1); (ii) then we initialise the set $\phi^{(u)} = \{1\}$ with the zero order regressor only and for each input we add higher order regressors (i.e., θ, θ^2, θ^3, etc.) in a greedy fashion (Algorithm 2); the greedy procedure comes to a halt when none of the more complex models under test improves the performance with respect to the current best model.

The performance of two models g' and g'' on the data $\mathbf{d}$ is compared by computing the Bayes factor

$$\frac{p(\mathbf{d}|g')}{p(\mathbf{d}|g'')} = \frac{\int p(\mathbf{d}|\mathbf{w}', g')p(\mathbf{w}'|g')d\mathbf{w}'}{\int p(\mathbf{d}|\mathbf{w}'', g'')p(\mathbf{w}''|g'')d\mathbf{w}''}. \tag{2}$$

As in (1) we use a Gaussian noise model with independent marker residuals. Thus we can write the likelihood as

Algorithm 2. Greedy selection of the polynomial components

1: INPUT: $\{\theta, \mathbf{d}^{(u)}, A_u\}$
2: OUTPUT: set of regressors $\phi^{(u)}$
3: Initialize zero order model: $\phi^{(u)} = \{1\}$
4: Regress model $g^{(u)}(\theta, \mathbf{w}^{(u)})$ to the data $\mathbf{d}^{(u)}$.
5: **repeat**
6: $\phi^{(best)}(\theta) = \phi^{(u)}(\theta)$
7: **for all** active inputs $v \in A_u$ **do**
8: Add a component: $\phi^{(test)} = \phi^{(u)} \cup \theta_v^{o_v+1}$ with order $o_v + 1$
9: Regress model $g^{(test)}(\theta, \mathbf{w}^{(test)})$ to the data $\mathbf{d}^{(u)}$.
10: **if** $\frac{p(\mathbf{d}^{(u)}|g^{(test)})}{p(\mathbf{d}^{(u)}|g^{(best)})} > 1$ **then**
11: $\phi^{(best)}(\theta) = \phi^{(test)}(\theta)$
12: **end if**
13: **end for**
14: **if** $\frac{p(g^{(best)}|\mathbf{d}^{(u)})}{p(g^{(u)}|\mathbf{d}^{(u)})} > 1$ **then**
15: $\phi^{(u)}(\theta) = \phi^{(best)}(\theta)$
16: $stop = false$
17: **else**
18: $stop = true$
19: **end if**
20: **until** $stop$

$$p(\mathbf{d}|\mathbf{w}, g) = \prod_k \mathcal{N}(d_k|g(\theta_k, \mathbf{w}), \sigma),$$

where $\mathcal{N}$ is a Gaussian with mean $g(\theta_k, \mathbf{w})$ and variance σ evaluated in d_k. The definition of the prior $p(\mathbf{w}|g)$ requires some preliminary considerations. The model selection step does not recalibrate the subject for each marker motion model, but compares the models on a fixed residual obtained assuming static markers. Therefore we can expect a bias between this approximated residual and the actual one. The bias magnitude is unknown a priori and should not affect the model selection result. Consequently we use a uniform prior for the zero order parameter $w_0 \in \mathbf{w}$ while for all the other parameters we use a standard Gaussian regulariser, that is

$$p(\mathbf{w}|g) \propto \mathcal{N}(\hat{\mathbf{w}}|0, \Sigma_{\hat{\mathbf{w}}}), \tag{3}$$

where $\hat{\mathbf{w}}$ is the vector containing all the parameters but w_0 (i.e., $\mathbf{w} = [w_0\ \hat{\mathbf{w}}]$), and $\Sigma_{\hat{\mathbf{w}}}$ is a diagonal prior covariance. Also, note that the regression steps 4 and 9 in Algorithm 2 use the same regulariser.

We evaluate the regression model on capture data acquired with a rig of nine 4 megapixel Vicon MX cameras. As a proof of concept we limited the capture to two fingers: the right thumb and index of one healthy subject. 31 markers with 3mm diameter were glued to the latex glove wore by the subject as showed in Fig. 15. Also, to ensure accuracy we of the global position we glued one larger (7mm) marker over the wrist and limited the capture volume to about 1m. Finally, to reduce the

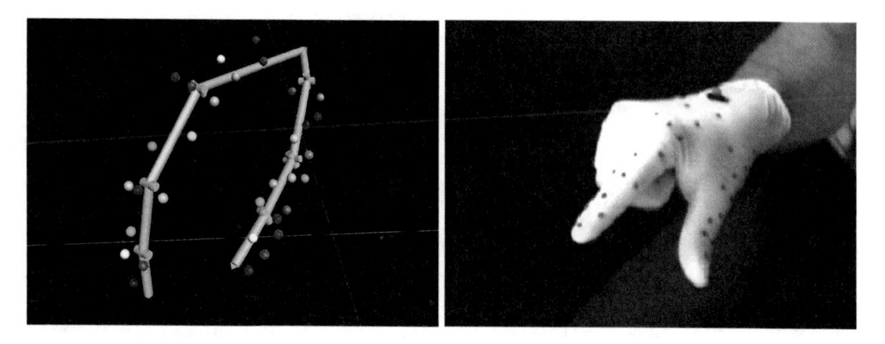

Fig. 15 High density markerset. The thumb and the index are sensorised with 32 markers. The markers are glued to a latex glove.

Table 2 Performance comparison between the standard model with static markers and the proposed moving marker model. The RMSE are in millimeters.

	RMSE		
	Static Markers	**Moving Markers**	**Perc. difference**
Trial 1	0.91	0.66	27.8%
Trial 2	1.02	0.79	23.0%
Trial 3	0.97	0.80	18.2%
Trial 4	0.92	0.74	19.7%

occurrence of marker occlusions under wrist rotations we pointed two of the nine cameras upwards.

We compared the results of the standard Static Marker (SM) model and the enhanced model with Moving Markers (MM) (refCB:eq:modelSkin) on three capture trials. *Trial 1*, *2* and *3* are three ROM trials. In *Trial 4* the subject repeatedly picks a piece of plastic cutlery (a knife) from a small container that he holds with the other hand. For our experiments we used 100 frames from *Trial 1* to calibrate the two subjects. Then, for the remaining frames in *Trial 1* and for the other two trials, we computed the joint angles and the Root Mean Square Error (RMSE) of the unnormalised marker residuals. Also, we set the maximal degree of the polynomials in the marker motion model to three. Figure 16 and Table 2 summarise the results. For all four trials MM has a significant lower RMSE than SM. The reduction was expected on *Trial 1* as this is the trial used to calibrate the subject and the proposed model has a larger number of free parameters than the standard one. However the improvement on the other three trial shows that MM generalises well on unseen data. This result is consistent for motions similar to the training ones (i.e., *Trial 2* and *Trial 3*) as well as for fairly dissimilar movements as in *Trial 4*. Also, Figure 16 shows that the performance improvement is consistent over time. The marker motion model outperforms the standard model both on extreme poses, when the fingers are fully flexed (see RMSE peaks in Fig. 16), and near the mean pose.

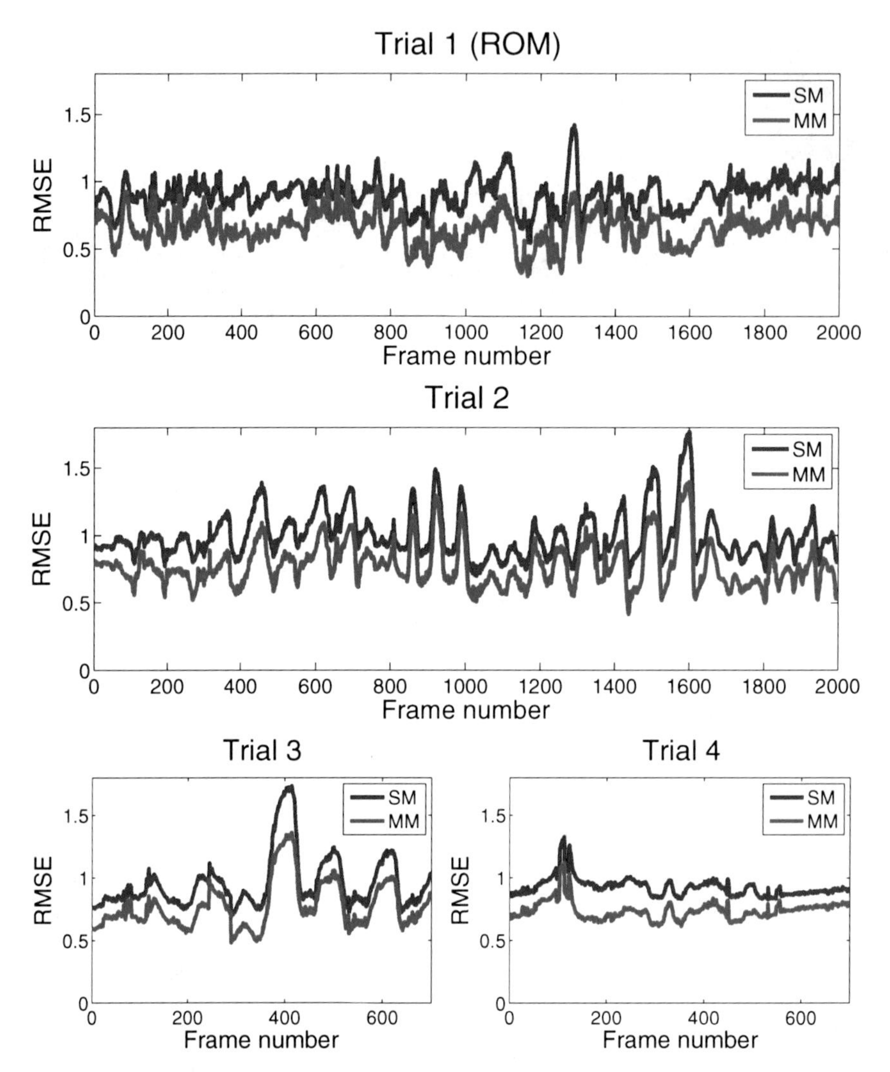

Fig. 16 RMSE comparison between the standard calibration model with static markers (SM) and the proposed model with polynomial moving markers (MM). For all test trials MM better predicts the marker positions.

5 Low-Level Sensor Fusion Algorithm

Since the motion capture system presents the problem of marker occlusion and the angular sensors are less accurate and can not be applied to all the hand DOFs, the sensor fusion seems to be the best approach to tackle the problem of the real-time observation of human manipulation. The first step to deal with a Bayesian sensor fusion problem is defining a stochastic model of the system. Then, with the aid of computer simulations, an EKF-based sensor fusion algorithm has been designed.

5.1 System Modelling

To limit the notation complexity, the algorithm will be presented for a single finger. Firstly, define a state-space model of the system with state vector

$$\mathbf{x} = \left(\vartheta \ \dot{\vartheta} \ \mathbf{l} \ \mathbf{h}\right)^T \tag{4}$$

being

$$\vartheta = \left(\vartheta_1 \ \vartheta_2 \ \vartheta_3 \ \vartheta_4\right)^T \tag{5}$$

$$\dot{\vartheta} = \left(\dot{\vartheta}_1 \ \dot{\vartheta}_2 \ \dot{\vartheta}_3 \ \dot{\vartheta}_4\right)^T \tag{6}$$

$$\mathbf{l} = \left(l_1 \ l_2\right)^T \tag{7}$$

$$\mathbf{h} = \left(h_1 \ h_2\right)^T \tag{8}$$

where (l_1, h_1) and (l_2, h_2) are the (x, y) coordinates of markers m_1 and m_2 in the frames $c_2 - x_2y_2z_2$ and $c_3 - x_3y_3z_3$ respectively and have been considered as state variables to be estimated, instead of system parameters, to take into account the sliding of the markers with respect to the finger bones.

As usual in tracking problems tackled via state estimation techniques, in the state update model, the joint variables to be tracked are assumed to vary with a constant velocity, while the unknown parameters are assumed to be constant. Hence the state update equations are

$$\begin{aligned}
\vartheta_{k+1} &= \vartheta_k + \dot{\vartheta}_k \Delta t + w_k^{\vartheta} \\
\dot{\vartheta}_{k+1} &= \dot{\vartheta}_k + w_k^{\dot{\vartheta}} \\
\mathbf{l}_{k+1} &= \mathbf{l}_k + w_k^{l} \\
\mathbf{h}_{k+1} &= \mathbf{h}_k + w_k^{h}
\end{aligned} \tag{9}$$

with Δt the sampling time of the filter, chosen as the minimum of the sampling times of the sensors, in case the sensors have different sampling rates.

It is easy to see that (9) is linear, then it can be written in matrix form as

$$\mathbf{x}_{k+1} = \mathbf{F} x_k + \mathbf{w}_k \tag{10}$$

where

$$\mathbf{F} = \begin{bmatrix} \mathbf{I}_4 & \Delta t \mathbf{I}_4 & \mathbf{O}_4 \\ \mathbf{O}_4 & \mathbf{I}_4 & \mathbf{O}_4 \\ \mathbf{O}_4 & \mathbf{O}_4 & \mathbf{I}_4 \end{bmatrix}, \tag{11}$$

and

$$\mathbf{w}_k = \left[(\mathbf{w}_k^{\vartheta})^T \ (\mathbf{w}_k^{\dot{\vartheta}})^T \ (\mathbf{w}_k^{l})^T \ (\mathbf{w}_k^{h})^T \right]^T \tag{12}$$

is a (12×1) vector of stochastic processes. The hypothesis under the Kalman-like filters is that n_k and w_k are additive Gaussian white noise (AGWN). The measurement equation is

$$\mathbf{y}_k = \begin{bmatrix} \mathbf{m}_1^b(\mathbf{x}_k) \\ \mathbf{m}_2^b(\mathbf{x}_k) \\ \mathbf{m}_3^b(\mathbf{x}_k) \\ v_2(\mathbf{x}_k) \\ v_3(\mathbf{x}_k) \\ v_4(\mathbf{x}_k) \end{bmatrix} + \mathbf{n}_k = \mathbf{h}(\mathbf{x}_k) + \mathbf{n}_k \tag{13}$$

where $v_i(x)$ is the voltage measured by the angular sensor applied to i-th joint and its analytic expression has been obtained as a result of the data glove calibration phase. The marker positions with respect to the D-H frame 0 are computed as

$$\tilde{\mathbf{m}}_1^b = \mathbf{T}_w^b(k)\mathbf{T}_p^w\mathbf{T}_0^p\mathbf{T}_2^0(k)\tilde{\mathbf{m}}_1^2 \tag{14}$$

$$\tilde{\mathbf{m}}_2^b = \mathbf{T}_w^b(k)\mathbf{T}_p^w\mathbf{T}_0^p\mathbf{T}_3^0(k)\tilde{\mathbf{m}}_2^3 \tag{15}$$

$$\tilde{\mathbf{m}}_3^b = \mathbf{T}_w^b(k)\mathbf{T}_p^w\mathbf{T}_0^p\mathbf{T}_4^0(k)\tilde{\mathbf{m}}_3^4 \tag{16}$$

where $\mathbf{T}_w^b(k)$ and $\mathbf{T}_p^w$ result from the glove calibration phase, $\mathbf{T}_i^{i-1}(k)$, $i = 0, 1, 2, 3, 4$ are the D-H transformations, $\mathbf{T}_0^p$ can be computed as

$$\mathbf{T}_o^p = \mathbf{T}_1^p(1) \begin{bmatrix} \mathbf{R}_x(\pi/2) & 0 \\ \mathbf{0}^T & 1 \end{bmatrix},$$

being $\mathbf{T}_1^p(1)$ the constant matrix calculated again during the calibration phase, and, by definition,

$$\tilde{\mathbf{m}}_1^2 = \begin{pmatrix} l_1 & h_1 & 0 & 1 \end{pmatrix}$$

$$\tilde{\mathbf{m}}_2^3 = \begin{pmatrix} l_2 & h_2 & 0 & 1 \end{pmatrix}$$

$$\tilde{\mathbf{m}}_3^4 = \begin{pmatrix} 0 & 0 & 0 & 1 \end{pmatrix}.$$

According to the Kalman filter framework, state variables are assumed to be Gaussian stochastic processes, with zero mean and diagonal covariance matrix, set by aid of computer simulations. The hypothesis of Gaussian pdf measurement error, on which the Kalman-like filters are based, is a restrictive hypothesis, since the measurement model is strongly nonlinear and the angular sensor characteristics may be not accurate. An improvement of filter performance can be obtained through more complex filtering techniques like particle filters [41], which allow estimating effectively the state of nonlinear systems even if the noise pdf is not Gaussian. The aim of the Extended Kalman Filter (EKF) is to track joint angular positions and velocities in a way robust to occlusions and marker slipping phenomena. Occlusion marker problem is handled by fusion of camera measurements and angular sensors measurements; the sensor fusion algorithms improve the measurement system robustness, since one sensor can contribute information while others are unavailable, jammed, or lack coverage of a target or event. The marker slipping phenomenon is handled by inserting in the state vector the marker positions expressed in the reference frames fixed to the finger links, hence they are estimated by the EKF as well. To model a system for filtering or data fusion purposes it is very important to evaluate the variance of measurement errors. If the error variance of the sensors is badly evaluated, the sensor fusion algorithm may become ineffective. It is assumed that the measurement noises are independent from each other, hence the covariance matrix is assumed diagonal. The experimentally estimated variances of camera measurement noises are

$$\begin{aligned} \sigma_{m_1}^2 &= 9 \cdot 10^{-6}\,\mathrm{m}^2 \\ \sigma_{m_2}^2 &= 9 \cdot 10^{-6}\,\mathrm{m}^2 \\ \sigma_{m_3}^2 &= 9 \cdot 10^{-6}\,\mathrm{m}^2, \end{aligned} \tag{17}$$

assumed equal in all the directions. Whereas, the experimentally estimated variances of angular sensor noises are

$$\begin{aligned} \sigma_{v_2}^2 &= 3.5 \cdot 10^{-5}\,\mathrm{V}^2 \\ \sigma_{v_3}^2 &= 4.7 \cdot 10^{-5}\,\mathrm{V}^2 \\ \sigma_{v_4}^2 &= 2.3 \cdot 10^{-5}\,\mathrm{V}^2. \end{aligned} \tag{18}$$

5.2 *Filter Design*

Whenever a marker is occluded, the model of the system changes, then also the sensor fusion algorithm parameters have to change in order to estimate effectively the joint angles. The whole system is then modelled as a switching nonlinear system, in which each state of a finite state machine matches a nonlinear set of differential equations describing the system. The state of the finite state machine is represented by the variable S_{ijk}, where i, j, k are equal to zero if the marker m_1, m_2, m_3, respectively, is not occluded and they are equal to one otherwise. As a consequence, to cope with the problem of marker occlusion, a switching EKF (SEKF) has been proposed. When no marker occlusion occurs, i.e. when the system is in the state S_{000}, we can design the EKF by setting the covariance measurement matrix Q as a diagonal (12×12) matrix, having measurement noise covariance as diagonal element

$$\mathbf{Q} = \mathrm{diag}\{\sigma_{m_1}^2 I_3, \sigma_{m_2}^2 I_3, \sigma_{m_3}^2 I_3, \sigma_{v_2}^2, \sigma_{v_3}^2, \sigma_{v_4}^2\}. \tag{19}$$

Instead, it is more difficult to choose the model covariance matrix $\mathbf{V}$; the simplest method, adopted here, is to set the model covariance matrix as a diagonal matrix, i.e. to assume that the process modelling errors are mutually independent. In [41] a set of empirical and semi-empirical methods for selecting the $\mathbf{V}$ matrix entries are described, while [31] proposes a method which ensures convergence of the filter.

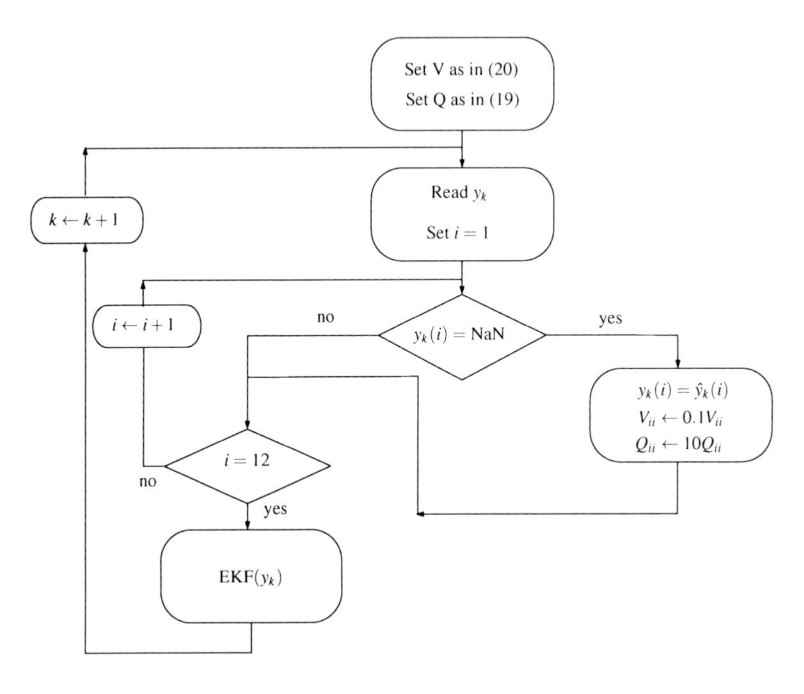

Fig. 17 Flowchart of the switching EKF used as sensor fusion algorithm for estimation of the hand joint angles.

The entries of **V** have been set through the aid of computer simulations; the rule of thumb most frequently adopted for tuning the **V** matrix is that decreasing the value of elements of **V**, implies that the filter "bandwidth" decreases and the measurement noise is attenuated; increasing the values of **V** entries process modelling error is attenuated and the filter "bandwidth" increases. The performed simulations show that the filter presents good performance when **V** is set as

$$\begin{aligned}\mathbf{V} = \text{diag}\{&5 \cdot 10^{-8}, 5 \cdot 10^{-8}, 1 \cdot 10^{-7}, 1 \cdot 10^{-6}, 5 \cdot 10^{-8}, 5 \cdot 10^{-8},\\ &10^{-6}, 10^{-7}, 5 \cdot 10^{-11}, 10^{-11}, 10^{-11}, 10^{-11}\}.\end{aligned} \quad (20)$$

In presence of marker occlusions (i.e. when the finite state machine is not in state S_{000}), the filter receives in input the value NaN. Computer simulation has been used also to tune the filter in presence of occlusions, by using a circular movement in 2D space. It is important to specify that in the general case of a movement in 3D space, since no abduction angular sensor is present, measurement of ϑ_1 is not possible and thus accurate tracking of the finger abduction movement is not possible in presence of permanent occlusion. The algorithm to switch from an EKF model to another is based on camera measurement observations. In detail, when $\mathbf{y}_k^{m_i} = \text{NaN}$, the filter sets

$$\mathbf{y}_k^{m_i} = \hat{\mathbf{y}}_k^{m_i},$$

where $\mathbf{y}_k^{m_i}$ is the vector containing the components of $\mathbf{y}_k$ corresponding to marker m_i and $\hat{\mathbf{y}}_k^{m_i}$ is its estimate computed by taking the same components of the vector $\mathbf{h}(\hat{x}_{k-1})$ in (13), with $\hat{\mathbf{x}}_{k-1}$ the estimated state at step $k-1$. Moreover, since the angular sensor measurements are less reliable than the camera measurements, the model covariance matrix is modified as

$$\mathbf{V}^{m_i}_{\text{occlusion}} = 0.1 \cdot \mathbf{V}^{m_i} \tag{21}$$

$$\mathbf{Q}^{m_i}_{\text{occlusion}} = 10 \cdot \mathbf{Q}^{m_i}, \tag{22}$$

where $\mathbf{V}^{m_i}_{\text{occlusion}}$ and $\mathbf{Q}^{m_i}_{\text{occlusion}}$ are the sub-matrices of $\mathbf{V}_{\text{occlusion}}$ and $\mathbf{Q}_{\text{occlusion}}$ containing the elements corresponding to the occluded marker m_i, being $\mathbf{V}_{\text{occlusion}}$ and $\mathbf{Q}_{\text{occlusion}}$ the model and the measurement covariance matrices, respectively, when an occlusion happens. Whereas, $\mathbf{V}^{m_i}$ and $\mathbf{Q}^{m_i}$ are the sub-matrices of $\mathbf{V}$ and $\mathbf{Q}$ which contain the elements corresponding to the marker m_i when the marker is not occluded. The algorithm flow chart is sketched in Fig. 17.

6 High-Level Sensor Fusion Algorithm

The high-level sensor fusion module aims at improving the observation of the human hand motion, exploiting the measurements of fingertip contact forces and a virtual environment. The main idea of the proposed algorithm is to compare the fingertip contact information, obtained by commercial tactile sensors, with the contact information computed in a virtual environment, that reproduces the real one. In case the estimation of the joint angles and the relative pose between the hand and the object are accurate, the contact information in the virtual and in the real environment are fitting, i.e the contact-consistence condition is satisfied. On the other hand, when the two sources of information are not consistent, a correction of the hand posture is carried out. The high-level module is completely independent from the particular technology adopted by the lower level, then can be integrated with any other low-level measurement system that provides estimation of the hand posture and of fingertip contact forces.

6.1 Correction Method

The observation phase including the proposed correction algorithm is constituted by three steps:

1. data acquisition and set up of the virtual environment that reproduce the "real" one;
2. use of the geometric information and contact information to define the correction of the fingertip positions and orientations;

3. use of an inverse kinematic algorithm, that finds the pose of the hand and the joint angles required to implement the correction.

The virtual environment is constituted by the 3D models of the hand and of the objects involved in the observed task. The hand is animated with the estimated pose and joint angles during the observation phase and the position of the object is obtained by the stereo vision system. Nevertheless, the proposed algorithm is absolutely general and can be adopted with every measurement system able to estimate the hand posture and the position of the object. With no measurement error and no 3D modeling error, the virtual environment reproduces exactly the real one. In this work, the errors in object 3D models are supposed to be much smaller than the measurement errors. These often cause inconsistencies between measured contact forces and hand posture relative to the object. In order to improve the accuracy of the observed data and specifically to make the motion data consistent with the contact force data, corrections of the hand pose and finger configurations have to be computed and implemented.

The pseudo-code of the correction method is described in Algorithm 3. When a measured fingertip contact force exceeds the empirical defined threshold, the finger is considered in contact with the object (line 5). In this work, the threshold has been fixed at 0.5 N. After the contact checking in the "real environment", in order to determine if a collision in the virtual environment occurs, the collision checker described in Sect. 6.2 is called for each finger (line 10). When the collision checker is called, for each finger f, it requires as input two 3D models (see Algorithm 4). The first 3D model, $\mathbf{CA_f}$, is the tactile sensor pad model, which is a thin plate modeling the expected contact area of the finger f. The second 3D model, $\mathbf{O}$ is the model of the grasped object.

If in the real world, according to the force measurement, a finger is considered in contact with the object and in the virtual environment no collision is detected or vice-versa (line 11), the finger is considered not-consistent and the correction is computed on the basis of the information provided by the collision checker, i.e. the unit vector $\hat{\mathbf{n}}$, the point $\mathbf{p}_0$ and the scalar d, defined in Sect. 6.2.

As explained in Algorithm 4, these data allow finding the point $\mathbf{p}_f$ on the object $\mathbf{O}$, such that the distance between the pad model of the finger f and the object is minimum.

The object point $\mathbf{p}$ and two of three components of the unit vector $\hat{\mathbf{n}}$ are included in the vector $\mathbf{x}_{goal}$. Its elements specify the desired poses of the fingertips for the correction step.

In line 15, the vector $\mathbf{x}_{goal}$ is given as input to the CLIK algorithm (see Algorithm 5 in Sect. 6.3), which is in charge of finding the hand configuration $\mathbf{q}_{goal}$ such that, for each non-consistent finger, the fingertip position is $\mathbf{p}_f$ and the unit vector normal to the pad $\mathbf{CA}_f$ is aligned with the unit vector $\hat{\mathbf{n}}_f$. The starting configuration $\mathbf{q}_0$ of the CLIK algorithm is taken as the hand configuration measured by the sensor setup.

Algorithm 3. Correction algorithm

```
1: for each frame k do
2:     (q_0, force) = readSensorData()
3:     x_goal = []
4:     for each finger f do
5:         if force_f > threshold then
6:             realContact_f = true
7:         else
8:             realContact_f = false
9:         end if
10:        (n̂_f, coll_f, d_f, p_0f) = checkCollision(CA_f, O)
11:        if coll_f ≠ realContact_f then
12:            p_f = p_0f + d_f · n̂_f {correction required}
13:            x_goal = append(x_goal, [p_f, n̂_f(1 : 2)])
14:        end if
15:        q_goal = CLIK (x_goal, q_0, W, α)
16:    end for
17:    moveHand(q_goal)
18: end for
```

6.2 *Geometric Information from the Scene*

To obtain the geometric information necessary to define the correction of the fingertip poses, the collision checker Proximity Query Package (PQP) [27] has been used in combination with Openrave [13]. Algorithm 4 describes the high level behavior of the collision checker, based on PQP. The input data of the collision checker module are two 3D models: $model_1$ and $model_2$.

Algorithm 4. Collision checker behaviour

```
1: (n̂, coll, d, p_0)=checkCollision(model_1, model_2) :
2: collision = isColliding(model_1, model_2)
3: if collision == false then
4:     p_0 = computePoint(model_1, model_2)
5:     d = computeDistance(model_1, model_2)
6:     n̂ = computeNormal(model_1, model_2)
7:     return (n̂, collision, d, p_0)
8: else
9:     p_0 = computePenetrationPoint(model_1, model_2)
10:    d = computePenetrationDepth(model_1, model_2)
11:    n̂ = computeNormal(model_1, model_2)
12:    return (n̂, collision, d, p_0)
13: end if
```

The output variables $\mathbf{p}_0$, $\hat{\mathbf{n}}$, d are such that:

- When $model_1$ and $model_2$ are not in collision, the point $\mathbf{p} = \mathbf{p}_0 + d \cdot \hat{\mathbf{n}}$ is the point on $model_2$ closest to $model_1$, d is the minimum distance between $model_1$ and $model_2$, $\mathbf{p}_0$ is the point on $model_1$ closest to $model_2$, $\hat{\mathbf{n}}$ is the unit vector of the line connecting $\mathbf{p}$ and $\mathbf{p}_0$.
- When $model_1$ and $model_2$ are in collision, the translation $\mathbf{p} - \mathbf{p}_0$ brings $model_1$ outside $model_2$ and it is $d = 0$.

Since, for each non-consistent finger f, the point $\mathbf{p}_f$ and the unit vector $\hat{\mathbf{n}}_f$ will be set as corrected positions and orientations in the task space, the correction is computed according to a minimum distance criterion.

6.3 Inverse Kinematics Method

Algorithm 5 explains the pseudo-code for the inverse kinematics. Line 1 is the definition of the function, which specifies the input and output variables.

The task space goal $\mathbf{x}_{goal}$ contains, for each not-consistent finger, the desired position and orientation, computed by calling the collision checker. For a finger, the position is the point $\mathbf{p}$ and the orientation is represented by the first two components of the vector $\hat{\mathbf{n}}$, defined in Sect. 6.2.

The authors have empirically observed that the performance in terms of velocity and stability is often better, if the orientation is specified only for one or two fingers. In fact, when the size of Jacobian matrix increases too much, a significant decreasing in the performance may occur. Hence, in all the experiments presented in Sect. 7, the orientation has been fixed only for the first two not-consistent fingers.

The vector $\mathbf{q}_0$ contains the measured hand posture, i.e. position, orientation and joint angles. It is the starting point of the IK algorithm. The more accurate the measurement system is, the closer $\mathbf{q}_0$ is to the correct minimum and then the faster and more effective the search is.

To take into account the differences in variances of the sensors and in magnitude order, the weighted pseudo-inverse can be computed instead of the simple right pseudo-inverse:

$$\mathbf{J}^{\dagger} = \mathbf{W}^{-1}\mathbf{J}^{\mathrm{T}}(\mathbf{J}\mathbf{W}^{-1}\mathbf{J}^{\mathrm{T}})^{-1} \tag{23}$$

where the weight matrix $\mathbf{W}$ can be chosen in this form:

$$\mathbf{W} = f(\Sigma) \tag{24}$$

with f monotonically increasing function of the sensor standard deviations $\sigma_1, \ldots, \sigma_n$ and

$$\Sigma = \mathrm{diag}\,(\sigma_1, \sigma_2, \ldots, \sigma_n)\,. \tag{25}$$

The simplest choice for $\mathbf{W}$ is the following:

$$\mathbf{W} = w\Sigma \tag{26}$$

where w is a positive scalar.

Fig. 18 Pictures of the manipulated objects: can (top left); big bottle (top right); small bottle (middle left); box (middle right); plastic bottle (bottom left); pencil (bottom right).

With this choice, the algorithm will tend to modify more the variables whose measurement are less reliable and less the variable whose measurements are more reliable. On the other hand, when the magnitude order of the sensor variance is too different, a possible choice is

$$\mathbf{W} = \mathbf{K}\Sigma \tag{27}$$

$$\mathbf{K} = \mathrm{diag}\left(k_1, k_2, ..., k_n\right). \tag{28}$$

With this choice, the parameters $k_1, k_2, ..., k_n$ have usually to be fixed empirically.

The scalar α is the so-called CLIK gain. The gain strongly influences the speed of convergence and the stability of the algorithm. In [17], a theoretical study on the stability of closed loop inverse kinematics algorithms is proposed and it can be an useful tool to correctly tune the parameters of the IK algorithm and, in particular, the gain α. The last input parameter is $\mathbf{k}(.)$, the direct kinematics function. In this application, k is a function of the position of the hand $\mathbf{p}_{hand}$, the

Algorithm 5. Inverse Kinematics Algorithm

1: $\mathbf{q}_{goal} = \mathrm{CLIK}\left(\mathbf{x}_{\mathrm{goal}}, \mathbf{q}_0, \mathbf{W}, \alpha, \mathbf{k}(.)\right)$:
2: $\mathbf{e} = \mathbf{x}_{goal} - \mathbf{k}(\mathbf{q}_0)$
3: **while** $\|\mathbf{e}\| > \varepsilon$ **do**
4: $\quad \mathbf{J}(\mathbf{q}) = \mathrm{computeJacobian}\left(\mathbf{k}(.), \mathbf{q}\right)$
5: $\quad \mathbf{q} = \mathbf{q} + \alpha \mathbf{J}^{\dagger}(\mathbf{q})(\mathbf{x}_{goal} - \mathbf{k}(\mathbf{q}))$
6: $\quad \mathbf{e} = \mathbf{x}_{goal} - \mathbf{k}(\mathbf{q})$
7: **end while**
8: **return** $\mathbf{q}_h$

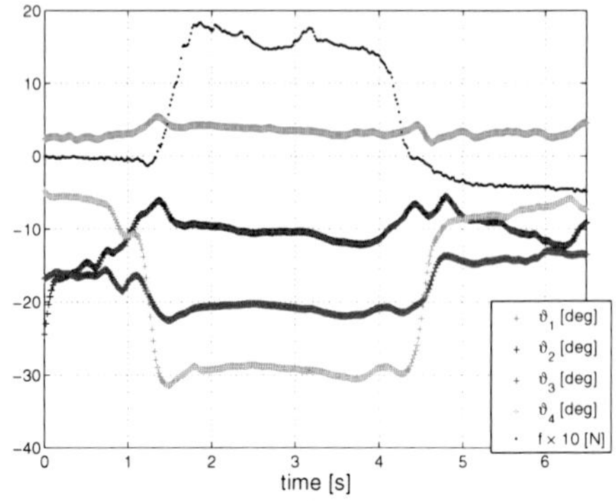

Fig. 19 Index joint angles and estimated tactile forces for the box shown in Fig. 18.

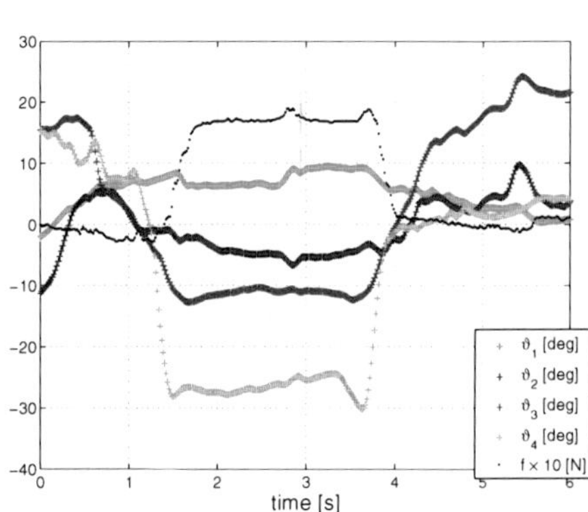

Fig. 20 Index joint angles and estimated tactile forces for the empty plastic bottle in Fig. 18.

orientation of the hand Φ_{hand} and the finger joint angles $\mathbf{q}_{joint}$. It is defined as: $\mathbf{x} = \mathbf{k}(\mathbf{q})$, $\mathbf{q} = \left[\mathbf{p}_{hand}\ \Phi_{hand}\ \mathbf{q}_{joint}\right]^T$ and $\mathbf{x} = \left[\mathbf{x}_{pos}\ \mathbf{x}_{orientation}\right]^T$. The main advantage to have as input the direct kinematics function is that the algorithm does not require any changes if a different hand kinematic model is adopted. For this reason, the Jacobian is also numerically evaluated (line 4). Since the Jacobian computation is much faster than the pseudo-inverse computation, the Jacobian numerical evaluation does not cause a significant delay in finding the IK solution.

7 Experimental Results

After the encouraging results of the preliminary experiments, a repository of observed grasp and manipulation tasks has been set up. For each trial, it contains estimations of the hand pose in the space, joint angles and normal components of fingertip contact forces. The grasp and manipulation tasks have been performed on a set of elementary objects, which includes a cup, a pencil, a soft empty bottle, a soft

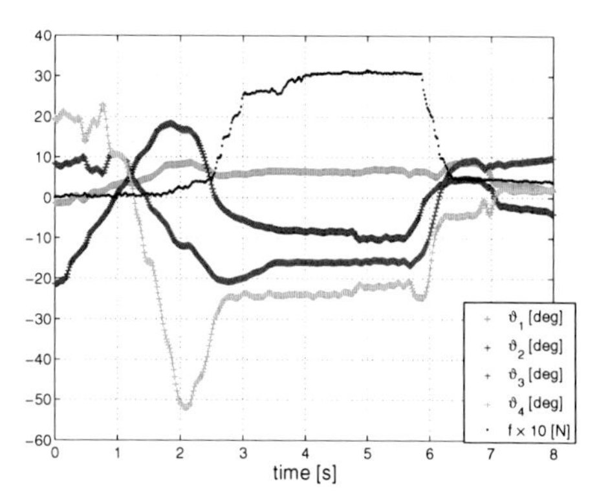

Fig. 21 Index joint angles and estimated tactile forces for the half-filled plastic bottle shown in Fig. 18.

half-filled bottle, a small empty rigid bottle and a large empty rigid bottle. Some objects can be distinguished only through contact force measurements, e.g. soft empty bottle and soft half-filled bottle. This choice has been made in order to empathize the importance of contact force knowledge in the observation of human grasping and manipulation.

The gathered data has been used within DEXMART to tackle several problems such as PbD, object recognition, segmentation of elementary actions, activity recognition, low-level trajectory generation and other learning issues, by using not only the kinematic data, but also contact force information.

The number of the objects has been kept intentionally low, since each objects is representative of an entire class. For example a can represents all the cylindric-shaped object involved in typical manipulation tasks. Some of the objects involved are shown in Fig. 18. Figures 19, 20 and 21, show examples of the estimated joint angles and the measured contact force for the index finger.

8 Robotic Execution

As long as the robotic hand has a very similar kinematic and very similar strength to apply the recognized contact forces during the manipulation action, the previously described very direct application of observed manipulation trajectories and forces works quite well. If the kinematic and force constraints posed by the robotic device are different form those of the human beeing an additional uncertainty through the direct mapping to the robotic device occurs. In such cases it is more useful to segment the observed action and try to extract the human and robot independent (geometric) task constraints and try to use them in the execution phase.

This of course makes it necessary to plan and generate the grasps and motion trajectories of the robot from scratch considering the information stored from the observation. For some special cases where the start and goal configurations are clearly defined in the task constraints this can be solved by state of the art geometric motion planners. If the manipulation plan just implicitly describes a continuous set of configurations implementing the given task constraints the robotic system needs a software module to choose a certain configuration which is promising to be successfully reached and executed. A certain manipulation action, filling a glass of water from a bottle for example might be simple to plan and execute in some situations, while in other cases the scene might be blocked through obstacles. The task constraints given by the observation are only that the bottle has to be moved in a certain way above the glass to avoid spilling. Where in the workspace of the robot this is executed or if the table is cleared before the action might be decided by the smart robotic system. Figure 22 shows variations of the pouring task execution.

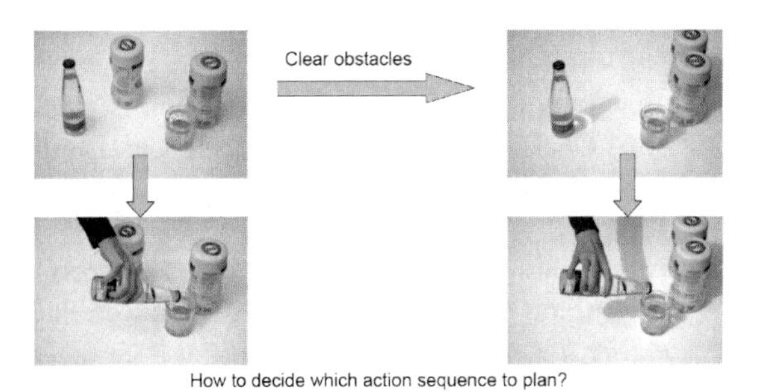

Fig. 22 The task of pouring water into a glass might result in different action plans according to the found scene.

One way to solve this mixed logic-geometric problem is to use a hybrid planning approach. The logical planner queries a motion planning system, an RRT planner for example to verify if the plan is executable by the system. This however drastically increases the time used to generate the plan. The time to find a motion plan might be bound to make sure the system does not get locked but this poses the risk that for complicated planning problems no solution is found even if there exists one. Tuning the parameters is then a difficult procedure for robotic experts.

In DEXMART we tried to solve the problem with implementing an intermediate layer that allows reasoning on the task plan using compact and pre-computed representations. These representations allow matching task and robot constraints in a simple way to speed up the generation of a valid manipulation plan. In the following a representation to cover the reachable workspace of a robotic system is presented, ways are shown to search for task constrained trajectories in such workspaces and the treatment of obstacles and grasp capabilities are discussed.

9 Workspace Representations to Support Planning

9.1 Criteria Used in Robot Design Optimization

In the design stage, a robot manipulator is optimized with respect to its kinematic and dynamic properties. As the envisioned application area in DEXMART is the dexterous manipulation of light objects compared to the robot the focus is only on kinematic aspects here. This is also aligned with the intention to exploit geometric planning methods to enhance flexibility of the system in different scenarios. The kinematic design process of a robot can be furthermore divided into task-oriented robot design (e.g.[36]) and the design of general purpose manipulators to accommodate a large variety of tasks.

In general, the robot kinematics can be optimized to maximize its workspace or to maximize various dexterity indices with respect to specific positions or with respect to the entire workspace. Park *et al.* [35] introduce general performance criteria for workspace volume and dexterity using differential geometry. Global indices are obtained through the integration of local criteria. Sturges *et al.* [49] define a dexterity measure that relates the difficulty of an assembly task to the capabilities of a planar robot arm. Some distinction about from which direction to execute the considered tasks is included in this task-dependent difficulty measure, especially concerning the accuracy of TCP movements at specific positions. However, these indices are hard to extend to redundant spatial manipulators for service tasks.

A popular means used in robot design is the analysis of the Jacobian matrix of a manipulator. Several indices describing the dexterity of the manipulator are based on this method. Klein *et al.* [25] examined the relationship of the determinant, the condition number and the smallest singular value as a dexterity measure. With the goal to obtain a global isotropy design parameter, Stocco *et al.* [48] optimized the ratio of the maximum and the minimum singular value of the Jacobian in the entire workspace to obtain a global version of the condition number.
For representing the robot workspace according to grasping or manipulation task, the positions and directions where the manipulator can carry out the task best are of special interest. This is in principle addressed by the manipulability ellipsoid introduced by Yoshikawa [53]. In the following this will be given a closer look.

9.1.1 The Manipulability Measure

The manipulability ellipsoid in the m-dimensional Euclidean space is intended to quantify how easily, by means of joint movements necessary, the end-effector of the robot can change its position and orientation. The measure is also based on the analysis of the Jacobian matrix of the manipulator (29). The Jacobian matrix relates the joint velocities $\dot{\mathbf{q}}$ with the total end-effector velocity in the Cartesian space (the angular velocity ω_E and the translational velocity $\dot{\mathbf{p}}_E$) at a single configuration $\mathbf{q}$ of the manipulator.

$$\begin{pmatrix} \dot{\mathbf{p}}_E \\ \omega_E \end{pmatrix} = \mathbf{J}(\mathbf{q})\dot{\mathbf{q}} \tag{29}$$

$$\mathbf{J} = \mathbf{U}\Sigma\mathbf{V}^T . \tag{30}$$

The principal axes and singular values σ_i of the Jacobian define the orientation and the shape of the so-called manipulability ellipsoid. They can be obtained by singular-value decomposition of the Jacobian matrix $\mathbf{J}$ (30). Here $\mathbf{U}$ and $\mathbf{V}^T$ are orthogonal matrices, and Σ is a diagonal matrix containing the singular values of $\mathbf{J}$.

The extension of the ellipsoid and its major and minor axes are assumed to represent an ability of manipulation at a certain configuration. The singular values are interpreted as the radius of the ellipsoid in the direction of the corresponding principal axis. The ratio of the minimum and maximum singular value can be used to describe the directional uniformity of the ellipsoid and thus the uniform ability of the robots end-effector to move in any desired direction given a certain joint configuration. The volume of the ellipsoid is known as the *manipulability measure* and can be interpreted as a distance of the manipulator from a singular configuration. One major drawback of the original manipulability measure is that joint limits of the manipulator are neglected, so the capabilities of the system are somehow overestimated. To overcome this Abdel-Malek *et al.* [1] augment the Jacobian matrix with joint limit criteria and redefine the manipulability measure. The resulting measure is then used in a way also intended here to evaluate a good or optimal robot placement with respect to certain manipulation goal.

9.1.2 Evaluation of Jacobian-Based Approaches as Workspace Representation for Planning

Criteria used for manipulator design often aim either at reaching an isotropic performance of the manipulator in its whole workspace or maximizing some global performance index derived from local ones. These criteria are naturally not aimed for representing directional structure in the workspace, as they are used to optimize the uniformity of the workspace. The manipulability measure (volume) itself is directionless making it impossible to discern directional preferences at certain positions in the workspaces

Figure 23 shows the volume of the manipulation ellipsoid for the right arm of the DLR Humanodd Justin [32]. To compute the measure for the whole arm workspace a large number of joint configurations is sampled and the TCP position is calculated using the forward kinematics. For such configuration the value of the manipulability ellipsoid is stored in a map evenly discretising the arm workspace. The figure shows the minimal manipulability measure across the workspace. This way it is a conservative approximation of the capabilities of the robot.

Areas colored red or close to red, mark regions where the manipulability of the arm is low. They are coincident with the singularities of the arm. However, when evaluating the manipulability measure, the choice of the TCP influences the areas

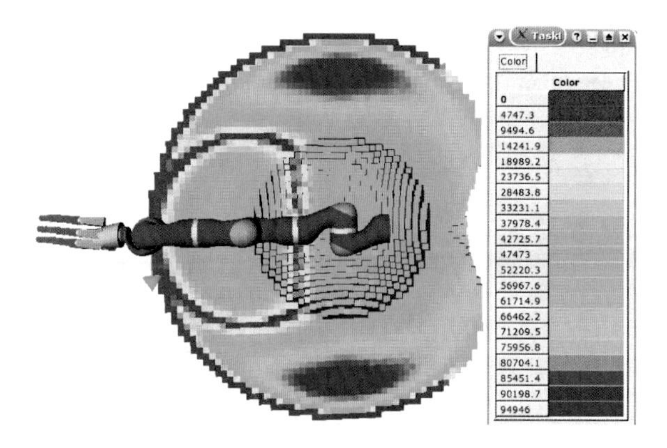

Fig. 23 Manipulability measure for the right arm of DLR's Justin with tool center point (TCP) placed at the wrist. The red areas show the singularities in the workspace of the arm.

and structure of the singularities. In the figure here the TCP is placed in the wrist, therefore only the translational part of the Jacobian get's rank deficient. Choosing a different TCP changes the kinematics and therefore also the manipulability measure through the workspace.

The main question that appears now, is how suitable is the manipulability measure for instructing geometric and logic planners where to plan and execute manipulation actions?

The manipulability measure outlines only areas where the overall manipulability is high, there are no preferred directions given. If the principal axes provide such information but the interpretation of link velocities related to translational and rotational velocities in the Cartesian space is not very intuitive. Furthermore the measure is local and bound to a certain link configuration. The measure is only valid in a small for a small ε-neighborhood. For different configurations resulting in the same TCP position the measure is different.

To overcome these drawbacks we try to specify the needs on the workspace representation more detailed and define a new measure for the capabilities of a robotic manipulator in its workspace.

9.1.3 Specifying the Needs for a Workspace Representation

For supporting logical or geometric planners the kinematic design of the robot is already given and cannot be optimized any more. What has to be accounted during the manipulation planning and execution are the kinematic constraints of the robot. Figure 24 illustrates the problem faced while planning a simple grasping action. Take a rotational symmetric object like the bottle. Given a relative position of the arm there are grasp directions which can be more easily reached than others. Such information should be reflected in the chosen representaion.

This reflects that an arm's workspace is not uniform with respect to reachability. There are regions that can only be reached from specific directions.This directional information needs to be captured.

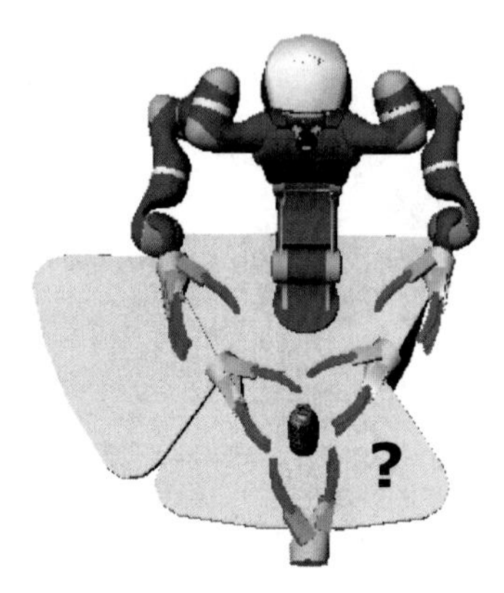

Fig. 24 A simple grasping action on a bottle, illustrates the choices to be made by the robotic system concerning arm usage and approach direction.

As the robot has two redundant arms with 7 DOF each the decision which arm to use for the manipulation action should also be supported. Moreover, due to the redundancy of the robot arms it has to choose among an infinite number of alternative configurations that can be used to approach and grasp an object. Considering a mobile manipulator the question arises how best to position the mobile platform to have optimal manipulation capabilities with respect to the operating area, e.g. a table.

In general, we need a representation of manipulator capabilities that can be used to characterize which places in the workspace can be easily reached. If only a specific direction is of interest, this direction should be applicable to the map, resulting in a filtered representation that masks all information but that lying in the requested direction. Structure inherent to the robot arm's capabilities inside its workspace should be easily recognizes. Using this representation the robotic system will be able to choose good approach directions for objects to grasp and manipulate.

Representing the reachable workspace has already received attention from other research groups. A Monte Carlo approach to represent the reachable workspace by randomized sampling was introduced by Guan *et al.* [22]. However, this approach only provides true/false information concerning the reachability of regions. No directional structure can be discerned from the representation.

9.2 *The Capability Map Approach*

The ability of humans to manipulate objects depends on the position of their arm in the workspace. Two-handed manipulation is limited to a region where the workspaces of both arms overlap. The best performance is achieved in an even smaller subspace. The same is true for two armed robots like Justin, whose kinematic design is oriented at the human example. While humans seem to have internal models of their arm's capabilities in the workspace, such models are still missing for humanoid robots.

In general, every robot arm is designed differently, and therefore has different capabilities. We show that these capabilities result in directional structures specific to workspace regions, and that these structures can be captured and represented in the form of a directional map. As a first step, we propose an algorithm to calculate such a workspace map and a visualization scheme to show the inherent structure. The approach will be illustrated using the right arm of the DLR humanoid robot Justin.

In a nutshell, the workspace structure is extracted through discretisation, randomized sampling, analysis and optimization processes. The robot arm reachability in a certain region of the workspace is examined using inverse kinematics.

9.2.1 Discretisation

The workspace of the robot arm can be encapsulated by a cube with a sidelength of two arm lengths centered at the robot arm base (Fig. 25). The maximum workspace of the arm is thereby overestimated. The envelopping cube is then subdivided into equally sized smaller cubes (voxels). Using this discretisation, we make the computation and visualization of the structure within the workspace possible and clear the way for the analysis of specific workspace regions in task planning processes.

9.2.2 Randomized Sampling to Fill the Workspace Representation

The *configuration space* is randomly sampled according to a uniform distribution. For each configuration the position of the TCP is computed via the direct kinematics. In Fig. 25 the TCP position is indicated by the coordinate frame in the hand. The TCP position is then mapped to the subcube that contains this position.

It could be considered to use the number of randomly sampled configurations mapped to a subcube as a measure of reachability for a region. The hope would be that the number of sampled configurations assigned to a specific subcube correlates with that region being easily reachable especially with respect to exploiting redundancy and versatile manipulation. However this is a false conclusion. When a robot is in a singular configuration, large steps in the configuration space for the links, causing the singularity, result only in small motions in the Cartesian workspace. Thus especially in regions containing singularities a large amount of sampled

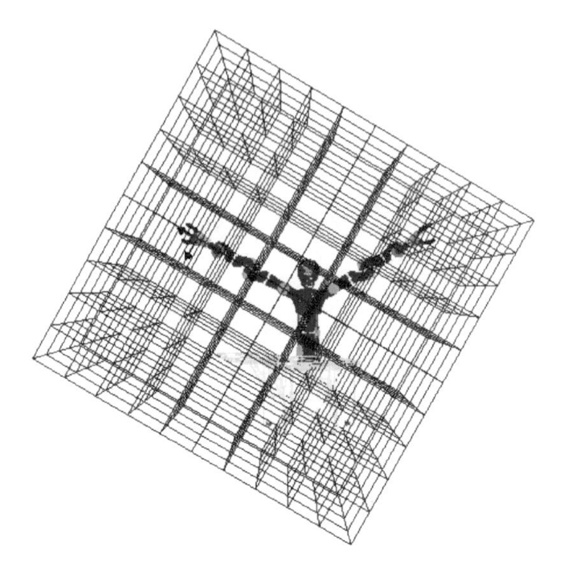

Fig. 25 The maximum workspace of the right arm is enveloped by a cube which is further subdivided into smaller cubes of 300 mm sidelength.

configurations are mapped to subcubes. Therefore the number of configurations mapped to a cube cannot be used to distinguish regions where versatile manipulation is possible. Furthermore such a mapping does not allow discerning the searched directional structure of the workspace. As a consequence, instep we use inverse kinematics to examine the workspace.

9.2.3 Using the Inverse Kinematics to Examine the Workspace

To get an evenly mapping of the directional information of the workspace we use another approach. We inscribe a sphere into each cube with a diameter equal to the width of the cube (Fig. 26(a)). Using the spiral point algorithm proposed by Saff *et al.* [43] we generate N equally distributed points on the sphere. For each point thus obtained, we generate a frame. In Fig 26(b) the frame is shown with the x-axis (red) and the y-axis (green) tangential to the sphere and the z-axis (blue) pointing towards its center. The frame is then turned around its z-axis according to a fixed step size. Each resulting frame is considered to constitute a TCP frame to be reached by the arm in question and an inverse kinematics solution is computed. If for one of the rotated frames at a specific point p on the sphere an inverse kinematics solution is available, that point p is marked in the underlying data structure. It is important to mention that the z-orientation of the TCP with respect to the sphere center is stored in the map but cannot be visualized in the graphical representations of the map.

The inverse kinematics for our 7-DOF redundant robot arm is computed by combining an analytical solution as proposed by Craig [10] with procedures to optimize

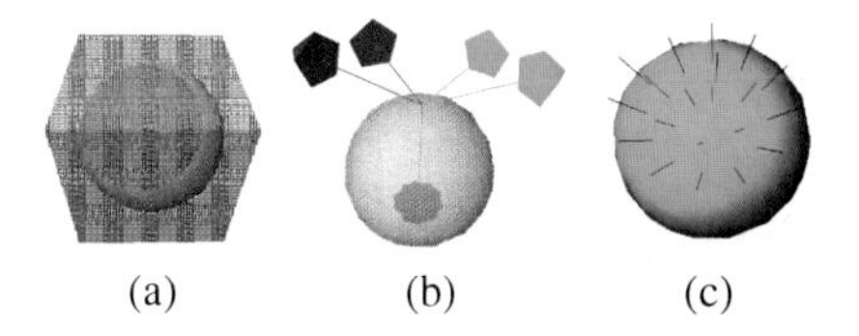

Fig. 26 A sphere is inscribed into the cube (a), exemplary frames for a point on the sphere (b), valid inverse kinematics solutions on a sphere (c).

the redundant DOF [26]. The randomly sampled configuration is taken as the initial configuration supplied to the inverse kinematics. Since an inverse kinematics for redundant robots does not have a single unique solution and involves some iterative optimization, a starting solution, that is already near the desired solution is beneficial for the computation.

9.2.4 Reachability Spheres to Characterize the Workspace

In the visualization, for each valid inverse kinematics solution on a sphere, a line is drawn originating in the sphere center (Fig. 26(c)). The spheres visualize the reachability for a region. We therefore call them reachability spheres. We assign a measure called *the reachability index* D as in (31) to each sphere characterizing the reachability of the region enclosed by the sphere. In equation 31, N is the total number of points on a sphere and R is the number of valid inverse kinematics solutions recorded. The resulting value informs about the percentage of points on the sphere, having an inverse kinematics solution:

$$D = \frac{R}{N} \cdot 100 \text{ with } R \leq N. \tag{31}$$

Using the reachability index, already some structure inherent to the robot workspace can be visualized. To achieve this, the spheres are colored with respect to their reachability index D.

Figure 27 presents the change of the reachability index across the robot arm workspace. As expected, as we move into the interior of the workspace the index gets better reaching its optimum in the blue region. Near to the robots arm base the more the index gets again smaller. The reachability index D for our robot arm ranges from 0 to 76. For better visibility, the full workspace is again cut in half along the arm. Figure 28 (left) shows all spheres with an index D in the lowest 10 percent of the reachability index ($D \in [0,8]$) across the workspace. As expected the spheres with the lowest index are on the border of the workspace. Figure 28 (right) shows spheres with an index D in the top 10 percent of the reachability index ($D \in [68,76]$) across the workspace. It can be seen that spheres with a good index D lie on somewhat more than a half a sphere shell around the robot arm base (also compare Fig. 27) with a diameter of approximately half the robot arm length. We

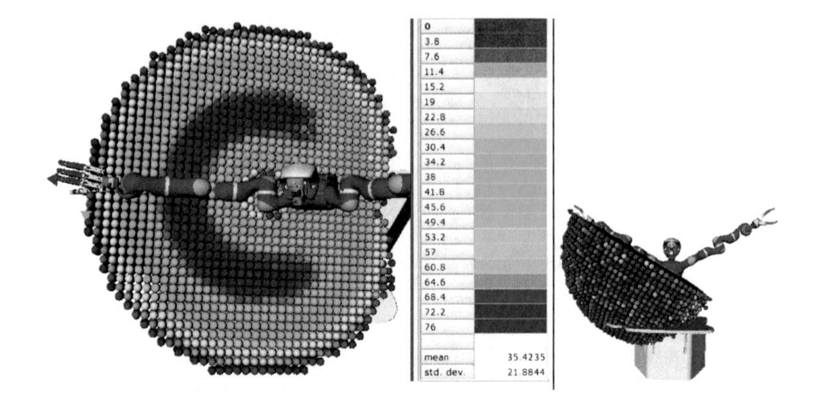

Fig. 27 Shows the reachability spheres across the workspace. The workspace representation was cut as shown on the right for better visibility of the structure.

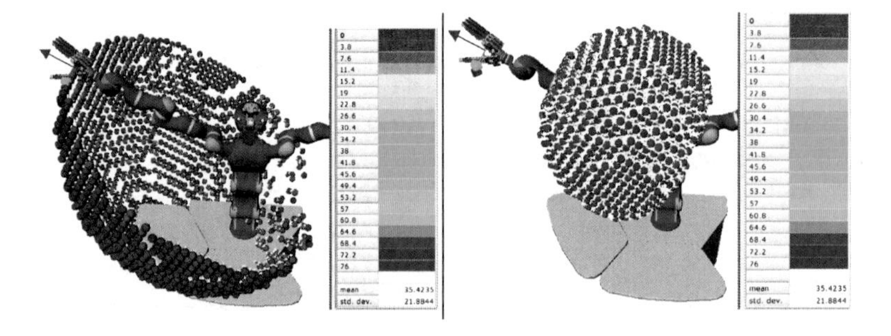

Fig. 28 Shows the spheres with an index D in the lowest 10% of the reachability index (left) and in the upper 10% of the reachability index (right).

would get a complete sphere shell when disregarding the link limits. For Figs. 27 and 28, 10^6 random samples were drawn. The spheres have a radius of 25 mm and 200 points are distributed on a sphere. The stepsize for turning the frame around its z-axis is 30 degrees.

9.3 Remarks on the Workspace Representation

The above described representation gives some insight in the structure of the kinematic capabilities of a robotic manipulator. However the detailed information can be also easily obtained by state of the art inverse kinematic solvers if the interest is only local. The nice property of the representation is that it can be pre-computed once for the manipulator and then can be quickly used for subsequent queries of logical or geometric planners. It also allows separating low level geometric algorithms and inverse kinematics from higher level planning. The following sections will illustrate how the structure can be exploited for task plans more complex than just a simple pick and place operation.

10 Representing and Searching of Task Constrained Trajectories

Tasks to be executed by bimanual robots should include more complex operations than just pick and place with a multifingered hand. Kitchen tasks for instance require beyond fetching and carrying things also the manipulation and interaction with the environment, like doors, drawers, can or bottle fasteners. To accomplish these tasks, the robot has to use knowledge about the specific environment, which we assume to come from human demonstration and knowledge about its own capabilities, which we want to retrieve from the above described capability map.

Another interesting question is which part of the kinematic chain of a more complex bimanual robot like Justin must participate in the execution of a certain task. One aspect is here the potential speedup of the geometric planning action, which has been already shown for walking robots and humanoid upper bodies by Diankov [14] and Pettre [37].

We argue that not only the question when to include the upper body is important to accomplish tasks but also when to use the mobility of the base. To execute simple trajectories it is not always necessary to use the mobile base. On the contrary, if the mobile base is unnecessarily used, e.g., while opening a kitchen closet, additional forces have to be compensated. These forces arise if the handle is grasped in a form closure grasp and the potential navigation or localization errors of the mobile base have to be compensated by a compliant arm.

In the following we propose an online method to determine the optimal base position to execute a constrained trajectory as well as methods to search for bimanual trajectories in the overlapping workspace. Therefore the method enables a planner to reason whether the mobile base is needed in a given task.

In a previous work [54], a mobile manipulator is positioned to execute linear constrained trajectories. However, more general types of trajectories, are needed in environments like a kitchen for opening doors and cupboards. The approach might also help in order to imitate prototypical movements demonstrated by humans to reach an object [51]. In such cases the task planner has also to reason about where to place the robot.

We propose an algorithm that uses a model of the reachable workspace of a robot arm to determine where the robot can be placed or if a given task is solvable at all. We present results for 3D trajectories using the example of opening a closet. Once the mobile manipulator is positioned, the trajectory is executed without using the mobile base.

10.1 Related Work

The use of models encapsulating robot specific knowledge was recently taken up by several research groups. Pettré *et al.* [37] make the animation of a digital actor more efficient by dividing the large number of DOFs of a humanoid into functional units providing the locomotion and the manipulation capabilities. Diankov *et al.* [14] use a similar functional structure for their humanoid robot to plan a path from a given start position to an object to be manipulated. In the process they furthermore consider a model of the reachable workspace of the robot arm to decide where the robot may stand to grasp an object and thus focus the search. Gienger *et al.* [20] use an object-specific model of the grasping capabilities of their humanoid robot to optimize the whole body motion to reach and grasp an object.

Most approaches to constrained trajectory planning for mobile manipulators combine the positioning of the robot with the search for feasible trajectories for the robot arm in the configuration space (C-space). Optimization and path planning techniques are used. Optimization techniques are applied to the whole kinematic chain. Multi-criteria optimization can also be used for positioning a mobile manipulator to reach a point. However choosing criteria weights, competing criteria, and the resulting local minima pose a great challenge [38]. When planning constraint motions for a mobile manipulator Stilman [47] uses the Jacobian transpose to project a given sample configuration into the subspace of configurations valid for the motion. However, the system always moves the mobile base to accomplish a task. For simple tasks this may not always be necessary. This one the one hand slows down the planning process as more DOFs have to be planned. On the otherhand the potential positioning inaccuracy of the mobile base might pose large forces on the manipulator when opening geometric constraint drawers for example. Diankov *et al.* [15] claim that fixed grasps on objects limit manipulation capabilities. They propose to use a set of caging grasps during path planning and execution to extend the possibilities of a mobile manipulator to fulfill the task. Whether or not a trajectory exists is determined by a timeout for the path search.

In contrast, we aim at providing a decision module that can predict with high probability whether or not and where a trajectory can be conducted in the workspace of the manipulators. The module is based on a reachability model for a robot arm and not on traditional path planning techniques.

10.2 A Closer Look on the Execution of Task Constrained Trajectories

In contrast to a pick and place operation, where the path the object is moved between the fixed fetching and parking position has only to provide no collisions, the opening of a drawer or cupboard is subject to constraints that come from the nature of the

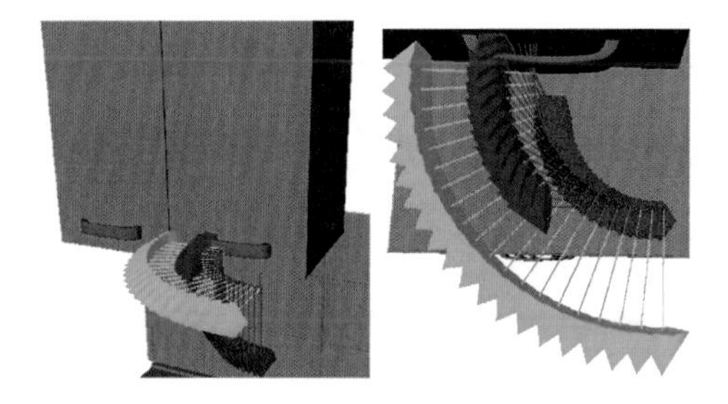

Fig. 29 (Left) Trajectory for opening a closet. (Right) A zoomed view.

object or task. For the first case a standard path planning algorithms can be queried for a suitable path. What are the specific properties for the second case?

For opening a closet, the tool center point (TCP) attached to the last link of a robot arm is constrained to move on a circular path (Fig. 29). For a frame attached to the handle of the closet, the orientation of the z-axis (blue arrow) constantly changes. The radius and orientation of this path are connected with the design of the closet. Due to the robot arm kinematics and link limits, executing such a constrained trajectory might not be possible at arbitrary positions in a robot arm's workspace. Depending on a robot arm's capabilities or the arm's attachment to an upper body, some mobile manipulators may not be able to perform certain tasks at all, like opening a closet at a certain height. A method is needed to analyze the capabilities of a robot given the environment and typical tasks performed therein.

Also for the execution of bimanual tasks like the unscrewing of a bottle fastener or the pouring action to fill water from a bottle into a grasped glass, the task determines the trajectory to be executed by the robot. The place in the overlapping workspace as well as the directions and contacts where to grasp the objects are to be chosen by the task planning module.

In the following a way to represent such task constraint trajectories and a way to match them with the capabilities of the robot, extending previous work [54], will be presented. Section 10.5 describes how regions are extracted where the given 3D Cartesian space trajectory is possible. These are used to infer placements for the mobile manipulator. In the second stage, the placements are checked for collision-free execution of the trajectory. Results are reported in Sect. 10.8.

10.3 Definition of the Search Pattern

We will use the example of opening a closet introduced in Sect. 10.2 to illustrate our method. If a closet has to be opened, the end-effector grasps the handle and moves on an arc (Fig. 29). We assume the trajectory template followed by the robot arm TCP to be given as a sampled sequence of frames F_l, $l > 0$ with respect to a local reference system. A frame F_l is represented as a homogenous matrix $F_l = \begin{pmatrix} R_l & \mathbf{t_l} \\ \mathbf{0}^T & 1 \end{pmatrix}$ with $R_l = (\mathbf{x},\ \mathbf{y},\ \mathbf{z}) \in SO(3)$ and $\mathbf{t_l} \in \mathrm{IR}^3$ describing rigid body rotation and translation. The frames are mapped to their discrete representations in the model. The set of accordingly mapped frames F_l is called search pattern p hereafter. The mapping works in the following manner. For each position of the Cartesian trajectory we first determine the sphere it maps to. Fig. 30 (left) exemplarily shows a trajectory superimposed on the workspace discretisation underlying the reachability sphere maps. The 2-d projection was chosen for illustration. The filled spheres symbolize those spheres the trajectory is mapped to.

$$f(\mathbf{t_l}) : \mathrm{IR}^3 -> \mathrm{IN}^3 \ with \ f(\mathbf{t_l}) = (p[l].x, p[l].y, p[l].z) \tag{32}$$

Let f be the function that maps $\mathbf{t_l}$ to a sphere in the pattern given the discretisation, i.e. the sphere diameter, also underlying the reachability sphere map.

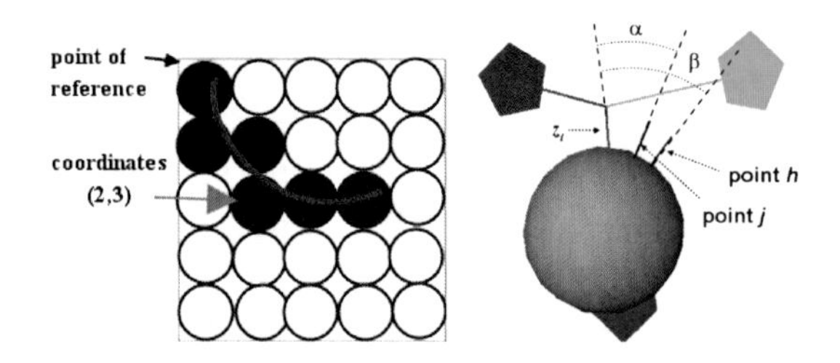

Fig. 30 Discretisation of a trajectory. (Left) 2-d view of mapping translations on the sphere map grid. (Right) Mapping of the frame orientation to a point on the sphere.

Each sphere is represented by an offset in $p[l].x, p[l].y, p[l].z$ of the sphere space with respect to the point of reference of the pattern.

The frame F can equally well be interpreted as a coordinate system base $F = \begin{pmatrix} \mathbf{x}\ \mathbf{y}\ \mathbf{z}\ \mathbf{t} \\ 0\ 0\ 0\ 1 \end{pmatrix}$ with $\mathbf{x},\ \mathbf{y},\ \mathbf{z},\ \mathbf{t} \in \mathrm{IR}^3$. In a second step, the z-axis $\mathbf{z_l}$ of frame F_l is mapped to the best fitting point on the sphere and determines the pattern element $p[l].k \in [1,n]$. Let $F_{i,0}$ be the coordinate system attributed to point i and nominal orientation 0 around $\mathbf{z}_{i,0}$. Note that $\mathbf{z}_{i,0} = \mathbf{z}_{i,m} = \mathbf{z}_i$ and $|\mathbf{z}_i| = |\mathbf{z}_l| = 1$. Equation 33 describes the mapping which is illustrated in Fig. 30 (right)

$$p[l].k = \mathrm{argmin}_{i \in [1,n]}(\mathrm{acos}(\mathbf{z_i}^T \cdot \mathbf{z_l})) \tag{33}$$

For the computation of the reachability sphere map the orientation around the z-axis of a frame (Fig. 26(b)) was discretised into m steps. In the last part of the mapping process the orientation around the z-axis of frame F_l has to be mapped to the sphere data structure, i.e. to one of these m orientations. Let $F_{k,0}$ be the frame belonging to the previously computed sphere point k and nominal orientation 0. The x-axis $\mathbf{x}_l$

of frame F_l is projected onto the xy-plane of the coordinate system defined by the frame $F_{k,0}$ in (34). Let P_{xy} be the projection matrix for projection onto the xy-plane.

$$\mathbf{x}_l^{new} = R_{k,0} \cdot P_{xy} \cdot R_{k,0}^{-1} \cdot \mathbf{x}_l \tag{34}$$

$$\alpha = \operatorname{acos}\left(\frac{\mathbf{x}_l^{new^T} \cdot \mathbf{x}_{k,0}}{|\mathbf{x}_l^{new}| \cdot |\mathbf{x}_{k,0}|} \right). \tag{35}$$

The angle between the projected axis and the x-axis of frame $F_{k,0}$ is then computed as in (35) and discretised. It determines the pattern element $p[l].o \in [1..m]$ with $p[l].o = \lfloor \frac{\alpha}{\Delta o} + 0.5 \rfloor$. Δo denotes the discretisation step width of the orientation around $\mathbf{z}$.

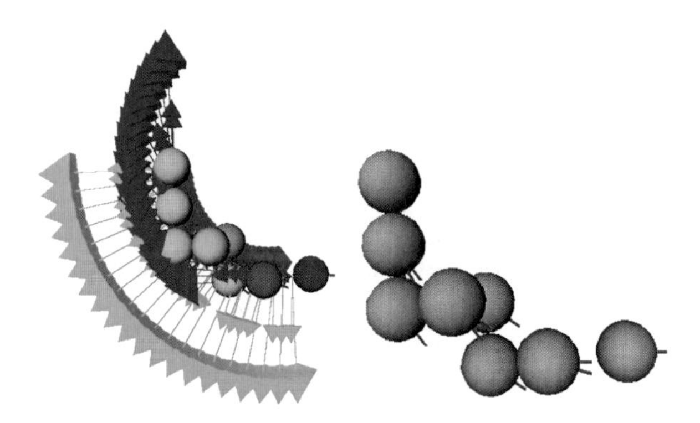

Fig. 31 (Left) The trajectory mapped to spheres. (Right) A zoomed view showing the sphere points mapped.

Figure 31 shows an example trajectory for opening a cupboard and the search pattern in the space of spheres. In Fig. 31 (left) the large coordinate frames represent the original trajectory frames. The smaller frames represent the mapped frames in the sphere data structure. In Fig. 31 (right) the red lines show to which lines on the sphere the frames were mapped.

10.4 The Search for the Trajectory in the Workspace Representation

Cross-correlation is a standard technique in signal processing to determine the shift between two signals. The signals are specified over the same domain, e.g. $\mathbb{R} \rightarrow \mathbb{R}$ for audio signals over time or $\mathbb{R}^2 \rightarrow \mathbb{R}$ for static grey images. The result of a correlation is a signal in the same domain, showing peaks at those locations where the two signals best match each other. We use this idea to find the search pattern obtained in Sect. 10.3 in the reachability sphere map. The search is done by correlating the sphere data structure with the given search pattern. Figuratively speaking, the pattern is moved across the 3D data structure and compared with the data present. Equation (36) implements the correlation between the two signals. Let D be the 3D data structure which represents the reachability sphere map. The search pattern p is obtained as described in the last section:

$$(D*p) = \sum_{ix}\sum_{iy}\sum_{iz}\sum_{l=0}^{p.length} S(ix_l, iy_l, iz_l)[p[l].k][p[l].o] \tag{36}$$

$$S(ix, iy, iz) = D[ix]D[iy]D[iz] \tag{37}$$

$$ix_l = ix + p[l].x,\ iy_l = iy + p[l].y,\ iz_l = iz + p[l].z \tag{38}$$

$S(ix, iy, iz)$ in (37) describes the location of a sphere in the 3D reachability sphere map. ix, iy, iz iterate over the whole workspace D minus the dimension of the 3D search pattern. In (38) the sphere offset of the pattern element l is added to the current starting point of the pattern in the 3D sphere space. Given a number of discretised orientations around the z-axis (Fig. 26(b)) the value of $S(ix, iy, iz)[p[l].k][p[l].o]$ encodes for point $p[l].k$ on the sphere whether the orientation $p[l].o$ around the z-axis is reachable. The variable is 1 if the orientation is reachable and 0 if it is not reachable. As a result $(D*p)$ is a representation of how well the trajectory fits across the map. We search those places in the robot arm workspace where the pattern fits completely. Figure 32 exemplifies the correlation result for opening a closet at a certain height. Justin's torso is in its zero position and the trajectory is at about shoulder height (Fig. 32 (top)). Since the trajectory is composed of 20 frames, the correlation result ranges from 0 to 20 (Fig. 32 (middle), (bottom)). Note that the positive x-axis indicates the front of the robot. A value of 20 (dark red) means all frames of the trajectory are predicted to be reachable if the trajectory is started at the corresponding point in the robot arm workspace. It can be seen that the region in which the trajectory can be performed completely is quite small in this example.

Note that with this method, the pattern will not be found if it occurs in the image in a different orientation. The standard solution to this is to rotate the pattern using a fixed step size. Accordingly, the original Cartesian space trajectory can be rotated before being mapped into the sphere space.

10.5 Computing the Robot Base Position for Optimal Execution

The reachability map is computed with respect to the robot arm base. Thus if the robot arm base is moved, the map moves accordingly. Once we have the position of the trajectory in the robot arm workspace with respect to the robot arm base, the position of the mobile manipulator with respect to the world can be determined easily.

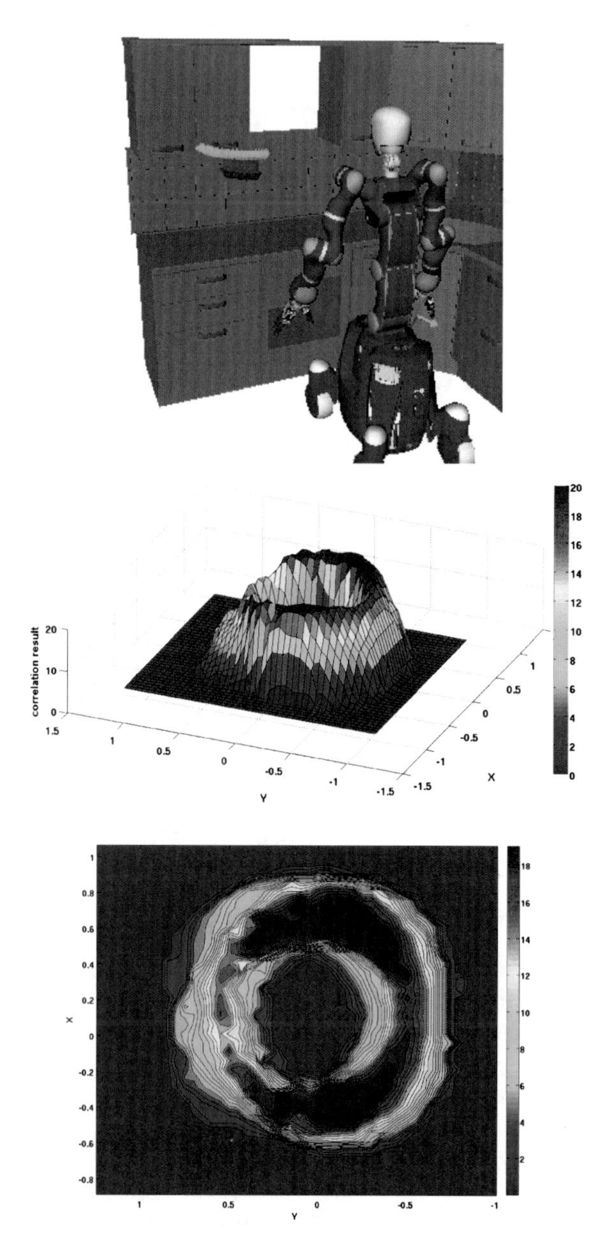

Fig. 32 (top) Justin in the Kitchen. (middle) Correlation result for the trajectory. (bottom) Contour view of correlation result.

The z-axis of the world system is assumed to point upwards. The transformation from the old to the new robot base position involves only a rotation R around the z-axis of the world system and a translation t in the xy plane. Let $\mathbf{t} = (x,y,z)^T$ be the translation and R be the rotation of the search pattern with respect to the map space for which the correlation of (36) reaches the maximum. Equation (39) gives the target position of the arm base in a reference frame F_0, e.g. the world base. The placement of the mobile base follows directly

$$T^0_{arm} = T^0_{object} \cdot (T^{arm}_{object})^{-1} = T^0_{object} \begin{pmatrix} R & \mathbf{t} \\ \mathbf{0}^T & 1 \end{pmatrix}^{-1} \tag{39}$$

10.6 Computational Complexity

In this section we present the examination of complexity that led us to perform the search in position space as opposed to in frequency domain. Cross-correlation in image processing customarily transforms two images into frequency domain using fourier transformation. The reachability sphere map and the search pattern are the correspondences of the image. In frequency domain the spectra are multiplied and the result is transformed back using inverse Fourier transformation.

It is known that the fast discrete fourier transform (FFT) using Cooley-Tuckey's radix-2 algorithm has a time complexity of $O(N\log(N))$, where N is a power of factor 2. Let N_D^3 denote the volume of the discretised robot workspace with side length N_D constraint to be a power of factor 2. The complexity is highest if the search pattern has the dimension of the workspace, i.e. $N_P = N_D$. According to the number of multiplications for a single FFT [3], the total cost for two fourier transformations and for the multiplication in frequency domain involved are

$$\mathrm{Cost}_\mathrm{freq} = |O| N_\mathrm{D}^3 \log_2(N_\mathrm{D}^3) + |O| N_\mathrm{D}^3 \,. \tag{40}$$

where $|O| = n \cdot m$. n is the number of discretised orientations, i.e. points on the sphere and m the number of discretised orientations around $\mathbf{z}$. Note that the cost of the fourier transformation of the reachability map is neglected because it can be computed once and used for different planning tasks. In the case of the discretisation of Justin's arm workspace with side length $N_\mathrm{D} = 40$ spheres and $|O| = 200 \cdot 12$ orientations, the total number of multiplications $\mathrm{Cost}_\mathrm{freq}$ amount to $2.6 \cdot 10^9$.

In contrast, the number of multiplications for general cross-correlation in the space domain amount to $|O| \cdot N_\mathrm{D}^3 \cdot N_\mathrm{P}^3 = 1.5 \cdot 10^{11}$, whereas a side length $N_\mathrm{P} = 10$ of the trajectory map is assumed. Contrary to the general case, significant optimizations can be applied here because only a few entries in the trajectory volume are non-zero. Let $|p|$ denote the length of the discretised trajectory, then the multiplications amount to

$$\mathrm{Cost}_\mathrm{space} = (N_\mathrm{D} - N_\mathrm{P})^3 \cdot |p| \,. \tag{41}$$

Only a subvolume of the robots reachability map is considered, because the trajectory is requested to be completely within the workspace of the robot. Whenever $|p| = N_P$, the costs reach its maximum at $N_P = N_D/4$. In the case of Justin's arm, this corresponds to only $2.7 \cdot 10^5$ multiplications[1], which are four orders of magnitude fewer then assessed for the correlation with the fourier transformations.

10.7 Discretisation Issues

In this section we describe the requirements for the data representation and the trajectory representation that have to be met to ensure the success of our approach. The requirements for the reachability map concern the workspace discretisation step width, the number of points on the spheres and the orientations around the z-axis. In general it can be stated the greater the workspace discretisation step width, i.e. the sphere diameter, the worse the prediction performance. A corner stone of our approach is that the spheres inscribed in a subcube describe the reachability for this region. Due to memory consumption the spheres cannot be arbitrarily small. Therefore a sphere diameter of 0.05 m was empirically chosen. If too few points are distributed on the sphere, then direction-specific reachability is not represented well anymore. Here the orientation of the z-axis is captured by uniformly distributing 200 points across a sphere. In this case the minimum angle between two points is 8.12°. The orientation around the z axis was discretised into 12 steps.

To unambiguously represent the task-specific trajectory template, the trajectory has to be sampled according to the Nyquist-Shannon sampling theorem [45]. If this theorem is violated and too few frames characterize the trajectory, the trajectory cannot be correctly reproduced, i.e. the pattern does not correctly represent the trajectory. In this case the trajectory is aliased with a less frequent one (Fig. 33 left). The search results are not valid for the original trajectory. In this case the ratio of predicted to actually reachable trajectories is low.

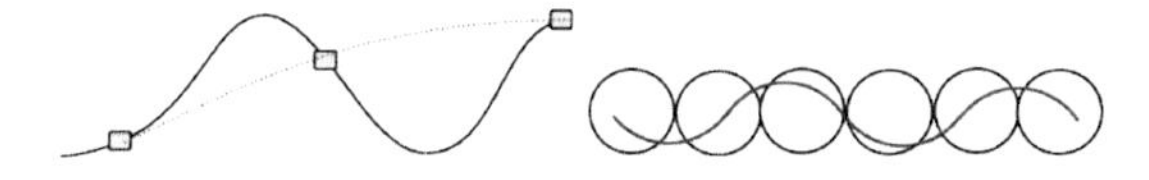

Fig. 33 (Left) The aliasing effect when sampling a 2-d trajectory. (Right) A trajectory with a low amplitude.

For trajectories with low amplitude i.e. $amplitude < sphere\ radius$ (Fig. 33 right) we assume that the interpolation assumption holds which underlies the discretisation of the robot arm workspace. It is assumed that at each point of a subcube the same orientations are reachable as on the sphere located at its center.

[1] Since both volumes are binary, the multiplication can be replaced with a simple binary AND operation.

10.8 Evaluation

In this section we evaluate the presented approach for different types of tasks and corresponding trajectories on the mobile Humanoid Rollin' Justin [18]. For reference the approach is summarized in Fig. 34.

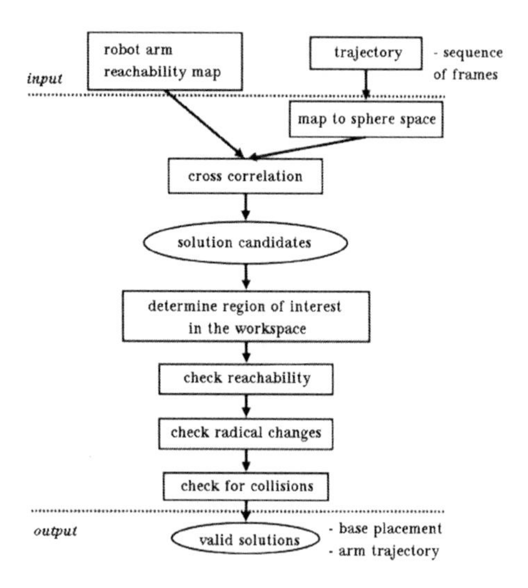

Fig. 34 The algorithm at a glance.

10.8.1 Naive Search vs. Model-Based Search

Without a reachability map, a brute force search could be thought of, to determine whether a given trajectory can be performed by the robot. The start point is randomly sampled using a uniform distribution. The trajectory is attached to this start point. However the sample only contributes to the computed success rate if the position of the first and last frame of the trajectory lies within the hull of the reachable workspace of the robot arm. The trajectory is checked for reachability using inverse kinematics. A relative success measure is computed from the number of successes. In Tab. 3 we report the likelihood to find the arc for opening a closet at a certain height in the workspace. Results are reported for the brute force search and the reachability model based search for the trajectory in one orientation.The results for the brute force approach show that a lot of effort is wasted to find valid trajectories. While our efficient approach needs 1.6 s to find and verify the 96 trajectories, the naive approach finds the same number of valid trajectories in 20 s. Considering that the trajectories may still be inconsistent or colliding, the advantages of our model-based approach are emphasized.

Table 3 Results of the brute force search are compared with the reachability model-based approach.

brute force search			reachability model		
samples	valid	%	predicted	valid	%
$1 \cdot 10^6$	39807	3.98	119	96	80.7

10.8.2 Validation of Solutions

The correlation process itself is very efficient. It needs 25 ms to find all occurrences of an arc trajectory (Fig. 29) with 20 frames in one orientation within the discretised workspace. Only those correlation results are considered that lie in an area of the workspace that is of interest for the given task. The solutions have to be validated because of the following:

10.8.3 Generalization Assumption

The correlation process provides a number of starting points for trajectories. These need to be checked for actual reachability since the reachability sphere map assumes that entries in the map generalize over the complete subregion, i.e. subcube. It is assumed that at each point of a subcube the same orientations are reachable as on the sphere located at its center, i.e. that the reachability structure does not change for small increments in space. Normally, this assumption will hold. However at the inner and outer border of the reachable workspace or at the border of structurally different regions, this assumption may be violated.

10.8.4 Collision-Free and Consistent Trajectories

Using the reachability sphere map, reachability of the trajectory is independently evaluated for the individual frames. It is not guaranteed that the robot is able to follow a smooth trajectory between frames, e.g., in the vicinity of singularities reconfigurations of the robot arm will occur. Therefore each trajectory is checked for consistency by setting a threshold on the allowed link-wise change of two configurations for two adjacent path steps. Additionally the trajectory is also checked for collisions for the robot with the environment and the robot arm with e.g. the head or the upper body.

10.8.5 Robots Working in the Kitchen

A robot working in a kitchen is required to be able to open cupboards to extract dishes or fill the dishwasher. Robot placements of the mobile manipulator are computed and validated for opening the door of a cupboard (Fig. 29). The trajectory

Table 4 The set of solutions is analyzed with respect to different criteria.

	nr. of solutions				
torso conf.	predicted	real	reconf.	colliding	valid
C1	119	96	70	82	7
C2	204	195	97	157	27

search is only performed for the original orientation of the trajectory. Results in Tab. 4 are reported for two different configurations of the movable upper-body of the robot *Rollin'Justin*. The results show significant differences for the two torso configurations. For configuration $C2$ more reachable solutions are found. These results indicate that the presented approach is able to decide on a beneficial torso configuration for a given task. Fig. 35 exemplarily shows a robot placement for each torso configuration. After the validation step, a set of valid solutions remain, that a task planner can choose from. In both cases, the number of collisions is striking.

Since the workspace of the arm is nearly symmetric solutions are found in front of and behind the torso of Justin (Fig. 32). In the latter case Justin has to be placed inside the kitchen closets to execute the trajectories. Furthermore often collisions of the arm and the head were encountered. The results for the check for consistency and freedom of collision should be seen as a proof of concept. Our inverse kinematics currently computes independent solutions for the individual frames and does not exploit the null space of the 7-DOF robot arm to avoid collisions. For trajectories that are currently invalid, we expect that there exist alternative arm configurations that lead to consistent and collision-free solutions. We measured the time consumption for each part of the algorithm for the torso in configuration C1. The computation

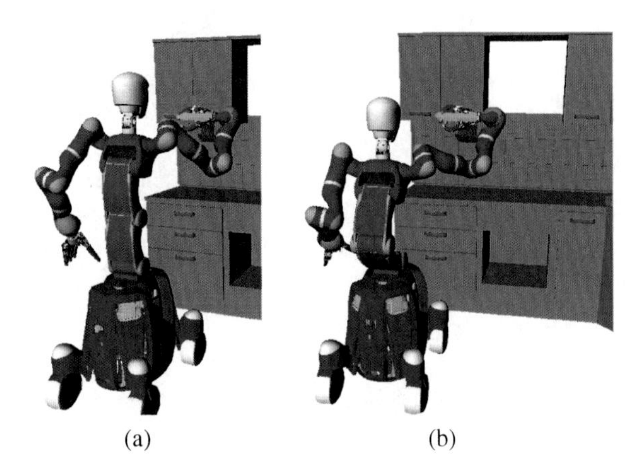

(a) (b)

Fig. 35 Rollin' Justin is placed to open a closet. (a) The torso is in configuration C1. (b) The torso is in configuration C2.

Table 5 Computation times for the steps of the algorithm.

correlation	reachability	reconfiguration	collision	total
25 ms	1564 ms	0.113 ms	31 ms	1620 ms

times are summarized in Tab. 5. The test was performed on an Intel Pentium D 3 GHz computer with 2 GB memory.

10.9 Remarks on the Trajectory Search

The presented approach allows a mobile manipulator to be optimally positioned to execute 3D trajectories determined by a certain task to be executed. The used model describes the capabilities of a robot arm and the trajectories to be executed accordingly. This allows using cross correlation to immediately find all possible positions in the workspace of the robot where the task trajectory can be executed. Once a trajectory is found, deriving the corresponding mobile manipulator position is straight forward.

The proposed algorithm is not only relevant for service robotic tasks. It can also be used for online positioning industrial robots e.g. for welding tasks or to compare the capabilities of different robot arms. The determined number of solutions for the constrained task can be assumed to correlate with the ability of the robot to cope with disturbances e.g. objects left behind by a human, or a human standing in the way. It could be argued that the mobile base can always be used for compensation. However, then the robot cannot operate in tight spaces and has to compensate forces occurring at the TCP.

The method is especially suited to decide whether or not a task e.g., opening a door, can be done without using the mobile base. This information could be used by a task planner to decide which planner or execution component to trigger.

11 The Influence of Obstacles on the Reachable Workspace

11.1 Considering Obstacles in Grasping and Manipulation of Objects

The previously described approaches add the constraints coming from the robotic system to the planning problem. Classical motion planners however deal with the problem of collision free pathes in the presence of obstacles. The above mentioned solutions are not considering any obstacles. Therefore in scenarios with targeted objects and several obstacles the solutions and implications derived from the capability map and task trajectories might no longer hold for the actual situation. A way of determining if obstacles prevent the execution of a task has to be found. This is again helpful for the simple case of pick and place where certain grasp directions might be blocked as well as for the execution of certain task trajectories. In the following the influence of obstacles to the capability map is shown and a representation is proposed to reflect the resulting restricted capabilities of robotic manipulator.

11.2 An Obstacle's Region of Influence

Obstacles have an influence on the reachability of neighboring regions. Depending on size and location of the obstacle in the workspace, different regions of influence can be expected. The capability map showed that the workspace of the right arm of the humanoid robot "Justin" has regions with different structural properties. We examine the influence of obstacles on the reachability using a simple object. We use a cube (height:0.3m, width: 0.06m, depth: 0.06m) that is positioned upright in the workspace. In rough approximation bottles or glasses can be represented by similar cubes. The influence of the obstacle is examined at the outer workspace border, in the center of the reachable workspace and near the inner border of the reachable workspace. Fig. 36 shows the position of obstacles and the positions $p1$ and $p2$ for objects to be grasped.

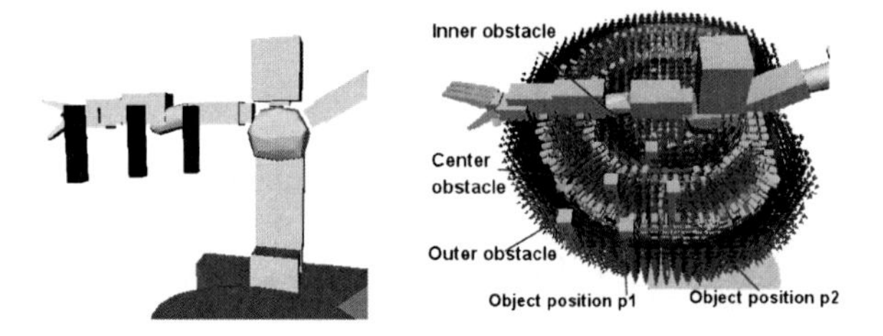

Fig. 36 Obstacles are positioned in structurally different regions of the workspace.

For reason of fitting through doors and avoiding self-collisions, the arm of the robot "Justin" is attached at an angle of 150 degree to the upper body of "Justin". Thus the workspace is oriented askew with respect to a table surface. For the analysis presented here, the arm was attached at an angle of 90 degrees to the robot to obtain more easily interpretable results. Simplified 3D models of the robot links were used in the collision tests.

The reachability sphere map will be used to visualize the influence of obstacles. All target positions in the workspace for which the inverse kinematics solution collided with the obstacle are marked in the data structure and are reflected in the visualization. By randomly sampling an initial solution for the inverse kinematics, different arm configurations for the same Cartesian position are computed by the inverse kinematics and the redundancy is explored. This is a conservative estimation of the region of influence of the object. Due to the redundancy of the robot arm, some of the target positions may be reachable without collisions using a different configuration of the robot arm. The resulting representation approximates the maximum region of influence of an obstacle. At the moment it is computed off-line. In view of our long term goal with respect to reasoning, the exploitation of these regions of influence should be seen as a proof of concept.

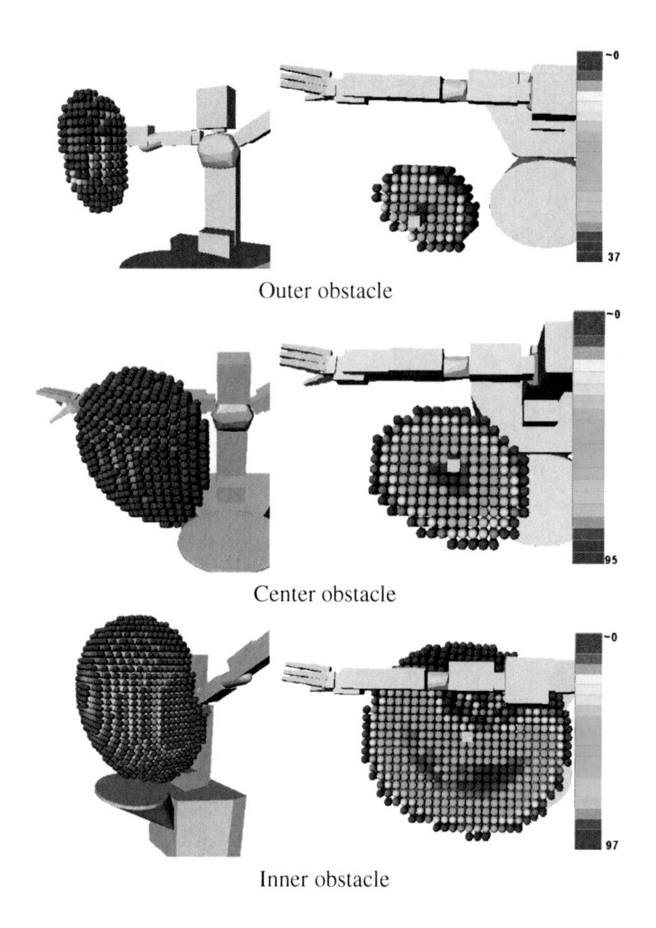

Fig. 37 The region of influence with respect to collisions is visualized for the considered obstacle (gray cube).

For each obstacle (gray cube) this maximum influence region is visualized. The right-hand side of Fig. 37 shows a cut through the representation to enable an impression of the structure. The color encodes the percentage of points on the sphere that are reachable and in collision. Thus blue spheres accumulate many reachable positions where the arm collided with the obstacle.

As expected, it can be seen that the shape of the region of influence of an obstacle changes when the obstacle is placed in the structurally different regions of the workspace. The region of influence increases as the obstacle is placed closer to the robot arm base. The region of influence of the obstacle placed near the inner workspace border is the largest. This region is often swept when the robot arm operates in the center of its workspace, therefore an obstacle placed in this region causes many collisions. If the obstacle is placed at a different position in the specific structural region of the workspace, similar structures of the region of influence can be observed. This is shown in Fig. 38 using the center obstacle and placing it at two additional positions. For this example, these regions of influence are computed off-line. If their changes could be analytically described when placed at a different position in the considered structural region, it would be possible to perform the collision checks against the environment outside of the grasp planner. Thus system modularity would be preserved. The performance that could be achieved with such a system will be examined in the following sections.

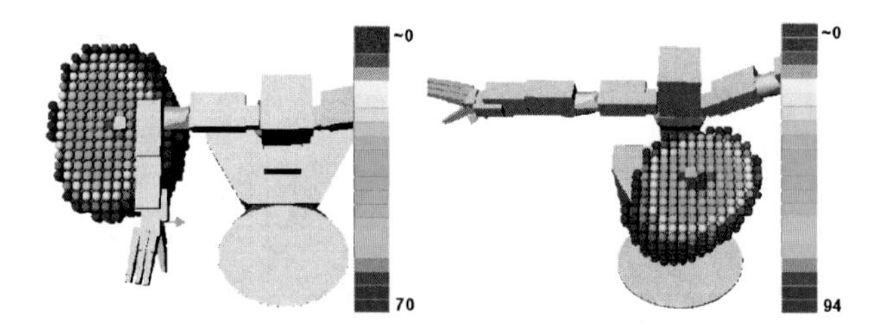

Fig. 38 The center obstacle is placed at different positions in the same structural region and the region of influence with respect to collisions is visualized.

11.2.1 Obstacles and the Influence on the Capability Map

The obstacle's region of influence is subtracted from the original reachability sphere map. Figure 39 (top) shows the original reachability sphere map. The color encodes the percentage of points reachable on the sphere, i.e. the reachability index [55]. After subtracting the region of influence for the center obstacle from the original map we get the map visualized in Fig. 39 (bottom). The influence of the center obstacle in the reachability is clearly visible. Using this modified reachability sphere map, the capability map is recomputed.

By giving this modified capability map to the grasp planner collision avoidance for the robot arm is added to the grasp planner without actually including the robot and object models into the grasp planner. However the capability map is not a perfect representation of the reachability sphere map and our robot arm is redundant. Therefore the robot arm configurations are still checked for collisions in a final step.

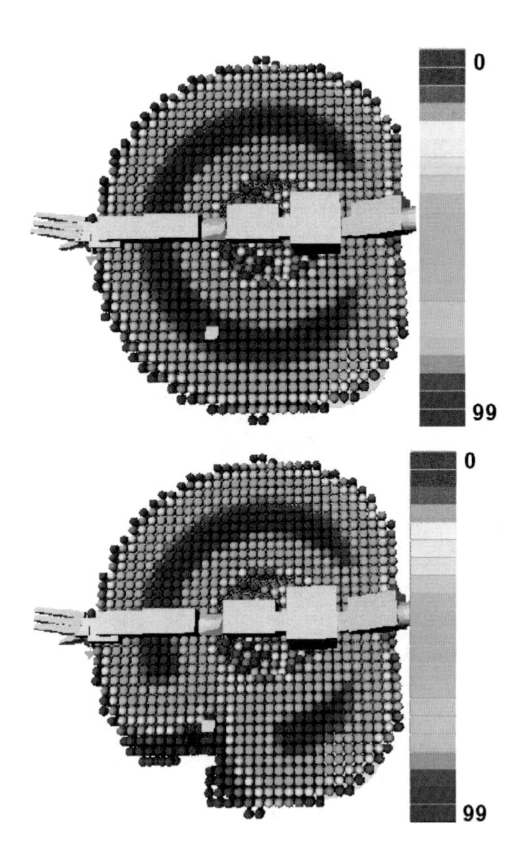

Fig. 39 (Top) The original reachability sphere map. (Bottom) The reachability sphere map minus the collision representation of the center obstacle.

Table 6 Time to generate a collision-free grasp in the presence of one obstacle.

	tight integration	loose integration	cap. map
position p1			
inner obstacle	31 ms	119 ms	60 ms
center obstacle	28 ms	91 ms	46 ms
outer obstacle	26 ms	61 ms	33 ms
position p2			
inner obstacle	33 ms	115 ms	92 ms
center obstacle	58 ms	202 ms	94 ms
outer obstacle	34 ms	120 ms	71 ms

11.3 Remarks on the Obstacle Representation

The calculation of the obstacle representation in the workspace is of high computational cost. Therefore an online computation which would be preferable for changing scenarios, seems not appropriate. If for a number of positions in the workspace certain geometric classes of obstacles are pre-computed, these generic regions of influence could be used for a geometric reasoning system to decide wether a certain manipulation action is likely to succeed or to fail. However such an operation can only give an approximative view on the real scenario. It has to be stated that the approach cannot lead to an optimal solution as the approximation of the blocked area by the obstacle is only a rather rough estimate. However, for a task planner this can be still a substantial benefit if potential penetrating objects for the execution of a task can be identified and a removal operation is instantiated. This is somewhat similar to the instruction of humans to try to keep a tidy workspace, not to unintendedly knock over spare objects.

12 Optimizing Grasp and Motion Planning

In the previous sections it has been shown how the constraints of the robotic arm can be taken into account when planning human demonstrated or otherwise specified tasks on a robotic platform. The topics of pick and place actions grasping and releasing an object have been tackled from the robotics arm perspective. The integration of an efficient grasp planner has not been intensively studied. Grasp planning poses additional problems. The geometry of the grasped object together with the kinematics and geometry of the grasping device is a well-known and complex planning problem itself. If task execution on bimanual robotic system should be addressed, the grasp and motion planning as well as the overlaying task planner must interplay well. Therefore in the following section it is shown how the concept of the capability map can be adopted and integrated efficiently.

12.1 Related Work on Grasping and Grasp Planning

If the model of the object is known beforehand, such as in applications with semi-structured environments (e.g. a home environment), information useful for speeding up the online computation of a valid grasp can be obtained and stored off-line, to be consulted when the online task planner requires it. For instance, the Columbia grasp database stores thousands of grasps for different objects and hands [21]. Using this database, a grasp planning algorithm looks for objects similar to the object to be grasped, and synthesizes new FC grasps based on the grasps stored in the database.

The object geometry has a large influence on the final FC grasps. In [56], it is considered that some regions on the object surface contribute to high quality grasps more than others. This information is represented in an object-specific *grasp map*, which can later be used to bias the generation of force closure precision grasps by concentrating on the most promising regions.

The construction of a *task map* for representing feasible power grasps on a particular object was previously proposed in [19]. The 6-dimensional space of positions and orientations for a particular hand with respect to an object frame is explored using Rapidly exploring Random Trees (RRTs). For a given initial pose, the hand moves forward towards the object until a contact is detected. Then, the fingers are closed until a sufficient enclosing force is achieved, and the object is lifted and slightly rotated to verify that the power grasp is successful. The exploration using RRTs allows the detection of contiguous regions of valid parameters in the pose space. The specification of continuous regions, defined as boxes in the pose space, has also been used for meeting task specifications despite of pose uncertainties [2].

12.2 Generation of the Graspability Map

This section presents the approach for computing the graspability map. First, the assumptions and considerations on the hand workspace are presented. Then, the algorithm for computing the graspability map is explained, and the implementation details are discussed, including the sample generation and the verification of the force closure condition.

12.2.1 Assumptions

In this work the following assumptions are considered. The object surface is represented by a set Ω of N points, specified by position vectors p_i measured with respect to a reference system located in the center of mass CM of the object, and each point p_i has an associated surface normal direction $\hat{n}_i$ pointing towards the interior of the object. N is assumed to be large enough to accurately represent the object.

The frictional contact between each finger and the object is considered punctual. Coulomb's friction model is used, which states that to avoid slipping, the force f_i applied at p_i must lie inside the friction cone defined by $f_i^t \leq \mu f_i^n$, where μ is the friction coefficient and f_i^t and f_i^n are the tangential and normal components of f_i. In the 3-dimensional space this model is nonlinear and, to simplify it, the friction cone is linearized using an m-side polyhedral convex cone. Thus, by representing the unitary vector along the j-th edge of the convex cone at the i-th contact with $\hat{n}_{ij}$, the grasping force is

$$f_i = \sum_{j=1}^{m} \alpha_{ij} \hat{n}_{ij} \, , \quad \alpha_{ij} \geq 0. \tag{42}$$

The force f_i applied on the object at p_i generates a torque $\tau_i = p_i \times f_i$ with respect to CM. f_i and τ_i are grouped together in a wrench vector given by

$\omega_i = (f_i\ \tau_i)^T$. The wrench ω_{ij} generated by a unitary force f_i along the edge j of the linearized friction cone is called a *primitive wrench*. A grasp defined by the set of contact points $C = \{p_1, \ldots, p_n\}$ is associated with the set W of primitive contact wrenches $W = \{\omega_{11}, \ldots, \omega_{1m}, \ldots, \omega_{n1}, \ldots, \omega_{nm}\}$.

12.2.2 Force Closure Workspace of the Hand

The workspace of the hand depends only on its kinematic configuration. For the computation of precision grasps, however, the whole hand workspace is not exploited, as not all the configurations of the fingers can lead to force closure precision grasps. Let $\boldsymbol{\Phi}$ be a set of spatial points located with respect to the base coordinate system of the hand, and reachable for a defined set of hand configurations. The set $\boldsymbol{\Phi}_{\text{fc}}$ is defined as the set of reachable points for the fingertips, which potentially allow a force closure grasp [42].

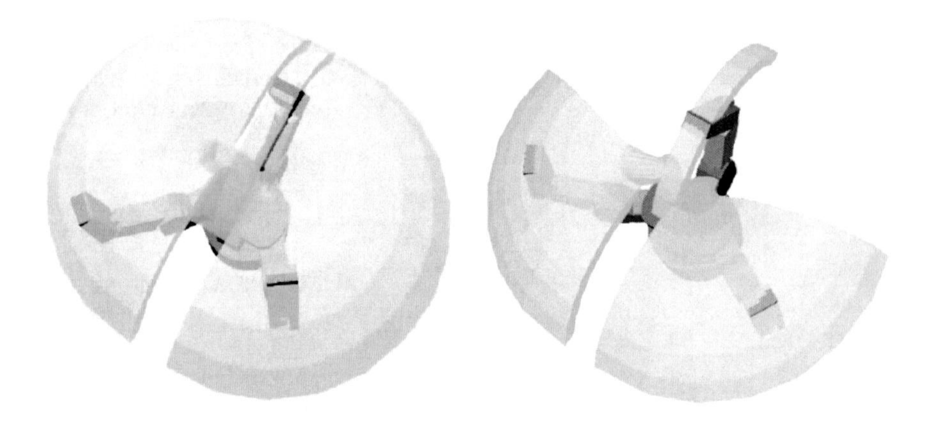

Fig. 40 Workspaces for the Barrett Hand: a) Reachable points for the fingertips; b) Set of points $\boldsymbol{\Phi}_{\text{fc}}$.

To compute $\boldsymbol{\Phi}_{\text{fc}}$, the hand configuration space is uniformly sampled, and the position of the fingertip region is computed via the direct kinematics of the hand. The FC condition can be tested using the normals to the fingertips, a predefined coefficient of friction μ_p, and a common coordinate system with its origin located at the centroid of the considered fingertip points. The artificially imposed μ_p should be the maximal value expected for the applications of the real hand-object system. As an example, Fig. 40a shows for a Barrett hand [50] the set of reachable points for a patch defined on each fingertip. Figure 40b shows for the same hand the set $\boldsymbol{\Phi}_{\text{fc}}$, computed with a friction coefficient of $\mu_p = 0.5$. Note that in this case the set $\boldsymbol{\Phi}_{\text{fc}}$ is composed by three different subsets of points, hereafter called ϕ_i, one for each fingertip ($\boldsymbol{\Phi}_{\text{fc}} = \phi_1 \cup \phi_2 \cup \phi_3$).

12.2.3 Algorithm

The computation of the graspability map requires the following data:

- A 3D object model (with the assumptions described in Sect. 12.2.1).
- A set Φ_{fc} of reachable points for the fingertips of a mechanical hand, which potentially lead to force closure precision grasps (Sect. 12.2.2).
- A friction coefficient μ_c that estimates the friction between the fingertips and the object.

The algorithm to compute the graspability map works as follows. First, the space surrounding the object that allows a contact between the robot hand and the object is enveloped by a parallelepiped (which is a scaled version of the bounding box of the object). A number of poses for the hand base frame are defined inside this parallelepiped. For each pose, a collision detection between the object and the hand is performed, with the fingers in the configuration of maximum aperture of the hand. If there is no collision, then the intersection ψ_i between the object and the workspace ϕ_i for each finger is computed. The set ψ_i includes all the points on the object reachable for each finger i. The next step creates the sets ψ_i' of intersected points whose normals are within the potential directions of force that the fingertip can apply, as shown in Fig. 41. If at least two sets ψ_i' are not empty, then it is verified whether the reachable points lead to a force closure grasp. The steps in the algorithm are as follows.

Algorithm: Computation of the graspability map

1. Voxelize the parallelepiped delimiting the possible locations of the hand base frame around the object.
2. Define a set Γ of potential locations and orientations for the hand base frame.
3. For each pose of the hand base frame in Γ:

    a. Check for collisions between the hand and the object. If there is a collision, discard the pose.
    b. For each finger i compute $\psi_i = \phi_i \cap \Omega$.
    c. Obtain the sets $\psi_i' \subset \psi_i$ of points with normals within the directions of force that each fingertip can apply.
    d. If at least two sets ψ_i' are not empty
        Verify the force closure condition
       Else
        Discard the pose.

4. Return all the poses in Γ that lead to FC grasps

Figure 42 illustrates some steps in the computation of the graspability map for a banana, using a DLR hand II [4]. Figure 42a shows different orientations for one potential location of the origin of the hand base frame. Figure 42b shows the hand in one of these potential poses. Figure 42c shows the intersection of the workspaces ϕ_i with the object, and Figure 42d shows the corresponding sets ψ_i'. A modified version of the Voxmap-Pointshell (VPS) algorithm [44] was used to compute the

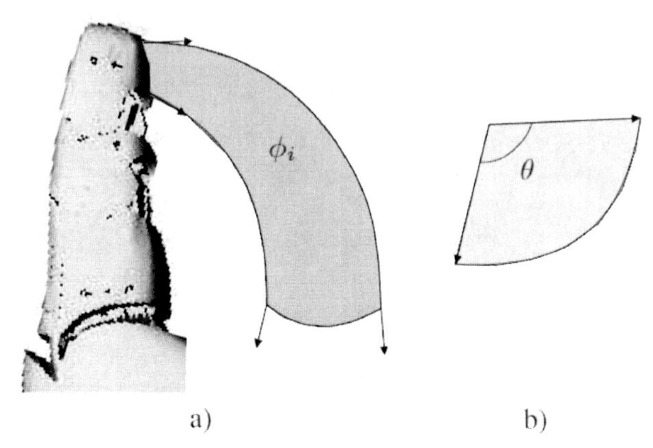

Fig. 41 Reachable directions for a fingertip i: a) Workspace ϕ_i for the fingertip; b) Directions of forces that the fingertip can apply on the object.

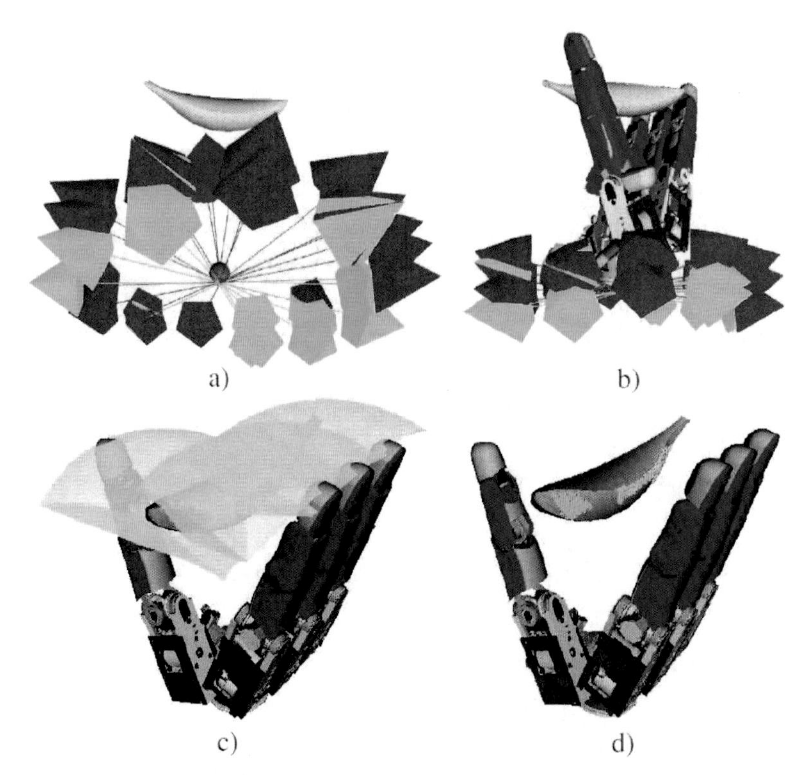

Fig. 42 Steps in the computation of the graspability map: a) Different orientations for the same location of the hand base; b) Hand in one of the potential poses; c) Intersection of the workspaces ϕ_i with the object; d) Sets of reachable points ψ_i'.

intersections, due to the fast responses (below 1 ms) to collision queries. This algorithm basically computes the intersections between *voxmaps*, voxelized volume structures for static objects, and *pointshells*, point clouds describing moving objects. For Step 3b, the computations are performed considering the workspaces ϕ_i as static objects, i.e. the global coordinate system is located in the hand base, and the coordinates of the points in Ω are transformed to that system via the transformation matrix describing the relative pose of the object with respect to the hand. The outcome of the VPS algorithm is the set ψ_i of intersection points between the workspaces and the object.

The following subsections provide more details on the generation of the set Γ, and on the force closure verification.

12.2.4 Potential Poses for the Hand Base Frame

The generation of the set Γ of potential positions and orientations for the hand base frame works similar to the discretisation of the capability map. The space surrounding the object that allows grasping the object with a robot hand is enveloped by a parallelepiped. This space is subdivided into equally-sized cubic voxels. To generate postures, a sphere is inscribed inside each voxel and on this sphere n points are uniformly distributed. Frames are generated for each point on the sphere, which describe the desired pose of the hand base frame. The normal to the sphere at a sphere point determines the z-axis (blue arrow) of the hand frame. The orientation of the x- and y-axis around the z-axis are sampled equidistantly. If a force closure grasp is found for one specific frame, this fact (but not the associated finger configurations) is saved in the data structure of the graspability map. The availability of a force closure grasp for a hand frame is visualized by a black line perforating the sphere at the corresponding point. The spheres visualize the graspability for a region of the space. The graspability map is the aggregation of all the considered spheres.

12.2.5 Force Closure Verification

Given one contact force at each fingertip, a necessary and sufficient condition for the existence of an FC grasp is that the origin O of the wrench space lies strictly inside the convex hull of the set W of primitive contact wrenches [30], from now on represented as $\mathrm{CH}(W)$. Several FC tests based on this necessary and sufficient condition have been proposed, for instance solving linear optimization problems [28], linear matrix inequalities [23], or using collision checks [58].

For the generation of the graspability map, the problem is more complex. Given the sets ψ_i' of reachable points per finger, we must find out if there exist at least one set of contact points $C = \{p_1, \ldots, p_n\}$ such that its corresponding wrenches allow an FC grasp. We use the following proposition.

Proposition: Given sets of points $\{\psi'_1, \ldots, \psi'_n\}$, a necessary condition for the existence of a force closure grasp for a contact set $C = \{p_1, \ldots, p_n\}$, where $p_i \in \psi'_i$, is that the convex hull $\mathrm{CH}(W_{\psi'})$ contains the origin of the wrench space, with $W_{\psi'}$ the set of primitive contact wrenches for all the points in $\psi'_1 \cup \ldots \cup \psi'_n$.

For the implementation of Algorithm 1, speed in testing the FC condition in Step 3d is important, as a large number of sample poses is required to generate the graspability map. To speed up the process, Proposition 1 is tested not in the 6-dimensional wrench space, but in the 3-dimensional force and torque subspaces, which leads to an empirical speed increase of about 100x. This option is chosen despite the loss of accuracy due to this approximation.

12.3 Remarks on the Graspability Map

We proposed an algorithm for the off-line computation of the graspability map, a representation of the poses for a mechanical hand that might lead to a precision force closure grasp. The map is specific for an object and hand. The algorithm is based on computing the intersection between the object and the workspaces for the fingertips. When a valid set of reachable points is obtained for each finger, a necessary condition for obtaining FC grasps is applied. The map can also be obtained by representing a database of FC grasps on the same object using the specified hand. Although that representation is not exhaustive, it allows the visualization of other type of information, such as the maximum grasp quality obtainable from a given position of the hand.

This approach can be used to do cross-correlation with the capability map of a manipulator to get clear distinction where to place the robot to most likely succeed in pick and place operations. This shows that the storing of discretised workspace and graspability information might be a beneficial way to efficiently plan manipulation tasks for bimanual robotic systems.

13 Conclusion

One aim of the DEXMART project was to make the observation of human manipulation actions available for the planning and execution of the same manipulation actions on a dual armed robotic platform. A strong focus was also on the representations used for storing the observed information as well as the representations to assist the planning and their comparison and potential harmonization. This has been successfully demonstrated for the lower level hand trajectories (Sect. 7) observed from human demonstration and executed on a robotic hand with similar kinematic structure. For the higher levels we followed an approach where the task constraints haven been tried to extract from the motions observed. This way the kinematics of the human have been abstracted away. On the one hand this allows passing the information to the planning and execution engines of the robotic system directly but the kinematic and geometric constraints arising from the robot and the scene have to be added. If the problem is passed directly to a motion planning system situations where the task cannot be executed with the current robot position in front of the scene or due to blocking obstacles might not be regarded and end up with unwished long planning times. Therefore representations have been developed that allow reflecting the constraints posed by the robot and the influence of obstacles in the scene on those constraints. Combined with the task constraint information derived from human observation the robot is capable of generating a manipulation plan for complex tasks. To establish a real intelligent reasoning component a prediction module relying on a very fast and robust physical simulation engine would have been necessary. This way the robotic system could choose among the various solutions fulfilling all constraints the one with the highest probability to succeed with the task. In the current implementation of the DEXMART execution system such an engine is not available but could be added and therefore improve the system further more.

Acknowledgements. The research leading to these results has been supported by the DEXMART Large-scale integrating project, which has received funding from the European Communitys Seventh Framework Programme (FP7/2007-2013) under grant agreement ICT-216239. The authors are solely responsible for its content. It does not represent the opinion of the European Community and the Community is not responsible for any use that might be made of the information contained therein.

References

1. Abdel-Malek, K., Yu, W.: Placement of robot manipulators to maximize dexterity. Journal of Robotics and Automation 19, 6–15 (2004)
2. Berenson, D., Srinivasa, S., Kuffner, J.: Addressing pose uncertainty in manipulation planning using task space regions. In: IEEE/RSJ International Conference on Intelligent Robots and Systems, Saint Louis, MO (2009)
3. Bronstein, I.N., Semendjajew, K.A., Musiol, G., Mühlig, H.: Taschenbuch der Mathematik, 6th edn. Harri Deutsch (2005)
4. Butterfass, J., Grebenstein, M., Liu, H., Hirzinger, G.: DLR hand II: Next generation of a dextrous robot hand. In: IEEE International Conference on Robotics and Automation, Seoul (2001)
5. Cavallo, A., De Maria, G., Natale, C., Pirozzi, S.: Optoelectronic joint angular sensor for robotic fingers. Sensors and Actuators A: Physical 152, 203–210 (2009)
6. Cerveri, P., Lopomo, N., Pedotti, A., Ferrigno, G.: Derivation of centers and axes of rotation for wrist and finger in a hand kinematic model: Methods and reliability results. Annals of Biomechanics Engineering 33, 402–412 (2005)
7. Cerveri, P., Pedotti, A., Ferrigno, G.: Robust recovery of human motion from video using kalman filters and virtual humans. Human Movement Science 22, 377–404 (2003)
8. Chang, L.Y., Pollard, N.: Method for determining kinematic parameters of the in vivo thumb carpometacarpal joint. IEEE Transactions on Biomedical Engineering 55, 1897–1906 (2008)

9. Corato, F., Falco, P., Lösch, M., Maggio, E., Jäkel, R., Villani, L.: Original approaches to interpretation, learning and modeling, from the observation of human manipulation. In: Robotics: Science and Systems, Workshop on Understanding the Human Hand for Advancing Robotic Manipulation, Seattle, WA (2009)
10. Craig, J.: Introduction to Robotics: Mechanics and Control. Addison-Wesley, Reading (1989)
11. De Maria, G., Falco, P., Natale, C., Pirozzi, S.: Data fusion based on optical technology for observation of human manipulation. International Journal on Optomechatronics 6(1) (2012)
12. Denavit, J., Hartenberg, R.: A kinematic notation for lower-pair mechanisms based on matrices. Journal of Applied Mechanics 7, 215–221 (1955)
13. Diankov, R.: Automated construction of robotic manipulation programs, Ph.D. thesis. Carnegie Mellon University (2010)
14. Diankov, R., Ratliff, N., Ferguson, D., Srinivasa, S., Kuffner, J.: Bispace planning: Concurrent multi-space exploration. In: Robotics: Science and Systems, Zurich (2008)
15. Diankov, R., Srinivasa, S., Ferguson, D., Kuffner, J.: Manipulation planning with caging grasps. In: 8th IEEE-RAS International Conference on Humanoid Robots, Daejeon (2008)
16. Dorfmuller-Ulhaas, K., Schmalstieg, D.: Finger tracking for interaction in augmented environments. In: IEEE/ACM International Symposium on Augmented Reality, New York (2001)
17. Falco, P., Natale, C.: On the stability of closed-loop inverse kinematics algorithms for redundant robots. IEEE Transactions on Robotics 27, 780–784 (2011)
18. Fuchs, M., Borst, C., Giordano, P.R., Baumann, A., Krämer, E., Langwald, J., Gruber, R., Seitz, N., Plank, G., Kunze, K., Burger, R., Schmidt, F., Wimboeck, T., Hirzinger, G.: Rollin' Justin — Design considerations and realization of a mobile platform for a humanoid upper body. In: IEEE International Conference on Robotics and Automation, Kobe (2009)
19. Gienger, M., Toussaint, M., Goerick, C.: Task maps in humanoid robot manipulation. In: IEEE/RSJ International Conference on Intelligent Robots and Systems, Nice (2008)
20. Gienger, M., Toussaint, M., Jetchev, N., Bendig, A., Goerick, C.: Optimization of fluent approach and grasp motions. In: 8th IEEE-RAS International Conference on Humanoid Robots, Daejeon (2008)
21. Goldfeder, C., Ciocarlie, M., Dang, H., Allen, P.K.: The Columbia grasp database. In: IEEE International Conference on Robotics and Automation, Kobe (2009)
22. Guan, Y., Yokoi, K.: Reachable space generation of a humanoid robot using the monte carlo method. In: IEEE/RSJ International Conference on Intelligent Robots and Systems, Beijing (2006)
23. Han, L., Trinkle, J., Li, Z.: Grasp analysis as linear matrix inequality problems. IEEE Transactions on Robotics and Automation 16, 663–674 (2000)
24. Jung, S., Wohn, K.: Tracking and motion estimation of the articulated object: A hierarchical Kalman filter approach. Real-Time Imaging 3, 415–432 (1997)
25. Klein, C.A., Blaho, B.E.: Dexterity measures for the design and control of kinematically redundant manipulators. International Journal of Robotics Research 6(2), 72–83 (1987)
26. Konietschke, R., Frumento, S., Ortmaier, T., Hagn, U., Hirzinger, G.: Kinematic design optimization of an actuated carrier for the DLR multi-arm surgical system. In: IEEE/RSJ International Conference on Intelligent Robots and Systems, Beijing (2006)
27. Larsen, E., Gottschalk, S., Lin, M.C., Manocha, D.: Fast proximity queries with swept sphere volumes. In: IEEE International Conference on Robotics and Automation, San Francisco, CA (2000)

28. Liu, Y.: Qualitative test and force optimization of 3-D frictional form-closure grasps using linear programming. IEEE Transactions on Robotics and Automation 15, 163–173 (1999)
29. Miyata, N., Kouhci, M., Kurihara, T., Mochimaru, M.: Modeling of human hand link structure from optical motion capture data. In: IEEE/RSJ International Conference on Intelligent Robots and Systems, Sendai (2004)
30. Murray, R., Li, Z., Sastry, S.: A Mathematical Introduction to Robotic Manipulation. CRC Press, Boca Raton (1994)
31. Natale, C.: Kinematic control of robots with noisy guidance systems. In: 18th IFAC World Congress, Milan (2011)
32. Ott, C., Eiberger, O., Friedl, W., Bäuml, B., Hillenbrand, U., Borst, C., Albu-Schäffer, A., Brunner, B., Hirschmüller, H., Kielhöfer, S., Konietschke, R., Suppa, M., Wimböck, T., Zacharias, F., Hirzinger, G.: A humanoid two-arm system for dexterous manipulation. In: 6th IEEE-RAS International Conference on Humanoid Robots, Genova (2006)
33. Palli, G., Pirozzi, S.: Force sensor based on discrete optoelectronic components and compliant frames. Sensors and Actuators A: Physical 165, 239–249 (2011)
34. Palli, G., Pirozzi, S.: Miniaturized optical-based force sensors for tendon-driven robots. In: IEEE International Conference on Robotics and Automation, Shanghai (2011)
35. Park, F., Brockett, R.: Kinematic dexterity of robotic mechanisms. International Journal of Robotics Research 13, 1–15 (1994)
36. Park, J.Y., Chang, P.H., Yang, J.Y.: Task-oriented design of robot kinematics using the grid method. Advanced Robotics 17, 879–907 (2003)
37. Pettré, J., Laumond, J.P., Siméon, T.: A 2-stage locomotion planner for digital actors. In: Eurographics Symposium on Computer Animation, San Diego, CA (2003)
38. Pin, F.G., Culioli, J.C.: Optimal positioning of combined mobile platform-manipulator systems for material handling tasks. Journal of Intelligent and Robotic Systems 6, 165–182 (1992)
39. Pitarch, E., Yang, J., Abdel-Malek, K.: Santos hand: A 25 degree-of-freedom model. In: Digital Human Modeling for Design and Engineering Symposium, Iowa City, IA (2005)
40. Pressure-Profile: Fingertps (2010), http://www.pressureprofile.com/products-fingertps
41. Ristic, R., Arulampalam, S., Gordon, N.: Beyond the Kalman Filter. Artech House Publishers, Boston (2004)
42. Roa, M.A., Hertkorn, K., Borst, C., Hirzinger, G.: Reachable independent contact regions for precision grasps. In: IEEE International Conference on Robotics and Automation, Shanghai (2011)
43. Saff, E., Kuijlaars, A.: Distributing many points on the sphere. Mathematical Intelligence 19(1), 5–11 (1997)
44. Sagardia, M., Hulin, T., Preusche, C., Hirzinger, G.: Improvements of the voxmap-pointshell algorithm – Fast generation of haptic data structures. In: 53rd Internationales Wissenschaftliches Kolloquium, Ilmenau (2007)
45. Shannon, C.E.: Communication in the presence of noise. Proceedings of the Institute of Radio Engineers 37(1), 10–21 (1949)
46. Siciliano, B., Sciavicco, L., Villani, L., Oriolo, G.: Robotics: Modelling, Planning and Control. Springer, London (2009)
47. Stilman, M.: Task constrained motion planning in robot joint space. In: IEEE/RSJ International Conference on Intelligent Robots and Systems, San Diego, CA (2007)
48. Stocco, L., Salcudean, S.E., Sassani, F.: Fast constrained global minimax optimization of robot parameters. Robotica 16, 595–605 (1998)
49. Sturges, R.: A quantification of machine dexterity applied to an assembly task. International Journal of Robotics Research 9(3), 49–62 (1990)

50. Townsend, W.: The Barrett hand grasper — Programmably flexible part handling and assembly. Industrial Robot 27(3), 181–188 (2000)
51. Vaughan, J., Rosenbaum, D., Meulenbroek, R.: Planning reaching and grasping movements: The problem of obstacle avoidance. Motor Control 5(2), 116–135 (2001)
52. Veber, M., Bajd, T., Munih, M.: Assessing joint angles in human hand via optical tracking device and calibrating instrumented glove. Meccanica 42, 451–463 (2007)
53. Yoshikawa, T.: Foundations of Robotics: Analysis and Control. MIT Press, Cambridge (1990)
54. Zacharias, F., Borst, C., Beetz, M., Hirzinger, G.: Positioning mobile manipulators to perform constrained linear trajectories. In: IEEE/RSJ International Conference on Intelligent Robots and Systems, Nice (2008)
55. Zacharias, F., Borst, C., Hirzinger, G.: Capturing robot workspace structure: Representing robot capabilities. In: IEEE/RSJ International Conference on Intelligent Robots and Systems, San Diego, CA (2007)
56. Zacharias, F., Borst, C., Hirzinger, G.: Object-specific grasp maps for use in planning manipulation actions. In: German Workshop on Robotics, Braunschweig (2009)
57. Zhang, X., Lee, S., Braido, P.: Determining finger segmental centers of rotation in flexion-extension based on surface marker measurement. Journal of Biomechanics 36, 1097–1102 (2003)
58. Zhu, X., Ding, H., Tso, S.: A pseudodistance function and its applications. IEEE Transactions on Robotics and Automation 20, 344–352 (2004)

Human–Robot Interaction

Daniel Sidobre, Xavier Broquère, Jim Mainprice, Ernesto Burattini, Alberto Finzi, Silvia Rossi, and Mariacarla Staffa

Abstract. To interact with humans, robots will possess a software architecture much more complete than current robots and be equipped with new functionalities. The purpose of this chapter is to introduce some necessary elements to build companion robots that interact physically with humans and particularly for the exchange of object tasks. To obtain soft motion acceptable by humans, we use trajectories represented by cubic functions of time that allow mastering and limiting velocity, acceleration and jerk of the robot in the vicinity of the humans. During a hand-over task and to adapt its trajectory to the human behavior, the robot must adjust the time motion law and the path of the trajectory in real time. The necessity of real time planning is illustrated by the task of exchanging an object and in particular by the planning of double grasps. The robot has to choose dynamically a consistent grasp that enables both robot and human to hold simultaneously the exchanged object. Then, we present a robotic control system endowed with attentional models and mechanisms suitable for balancing the trade-off between safe human–robot interaction (HRI) and effective task execution. In particular, these mechanisms allow the robot to increase or decrease the degree of attention toward relevant activities modulating the frequency of the monitoring rate and the speed associated to the robot movements. In this attentional framework, we consider pick-and-place and give-and-receive attentional behaviors. To assess the system performances we

Daniel Sidobre · Xavier Broquère · Jim Mainprice
LAAS–CNRS, 7 av. Col. Roche, 31077 Toulouse, France
e-mail: {daniel.sidobre,jim.mainprice}@laas.fr, xavier@broquere.fr

Ernesto Burattini · Alberto Finzi · Silvia Rossi
Dipartimento di Scienze Fisiche, Università degli Studi di Napoli Federico II, via Cintia, 80126 Napoli, Italy
e-mail: {ernb,finzi,srossi}@na.infn.it

Mariacarla Staffa
PRISMA Lab, Dipartimento di Informatica e Sistemistica, Università degli Studi di Napoli Federico II, via Claudio 21, 80125 Napoli, Italy
e-mail: mariacarla.staffa@unina.it

B. Siciliano (Ed.): Advanced Bimanual Manipulation, STAR 80, pp. 123–172.
springerlink.com

introduce suitable evaluation criteria taking into account safety, reliability, efficiency, and effectiveness.

1 Introduction

Until very recently, it was impossible to consider humans and robots living together. But now, robots start to become companions or co-workers of humans, opening an important research domain to build safe and intuitive cooperation. In this chapter we intend to introduce some elements to build such robots that are able to intuitively interact with humans.

In the context of HRI, intuitive and natural exchanges of objects between robots and humans represent a canonical task. But, as we will see, such a robot system is much more complex than the current industrial robots that repeat the same tasks separated from humans by cages. We present some bricks that are necessary to give to the robots the necessary autonomy to react to the human motions and behavior. We focus the presentation on three key points. Firstly the necessity for a software architecture to coordinate and synchronize the different pieces of software. Then, we details the importance of the geometric reasoning in the case of the dynamic choice of a double grasp. In fact, to exchange an object, both the robot and the human must grasp simultaneously the object. So the robot must adapt its grasp and its motion to the human behavior in real time.

This introduces the importance of motion, which is then addressed from the geometric aspect of paths to the kinematic aspect of trajectories. Usually, motion planners compute a path, which is then executed by the robot controller, generally at a constant speed or across a dynamic simulator. But in both cases the time evolution is not taken into account at the planning level. For real time interaction with humans, the robot must master its time evolution and control where and when it hands over an object. To do this, we propose to integrate a simple model of trajectories based on series of cubic functions in a more standard random motion planner.

Finally, the human motions and the external environment should be continuously monitored by the robotic system looking for interaction opportunities while avoiding dangerous and unsafe situations. In this context, attentional mechanisms can play a crucial role: they can direct sensors towards the most salient sources of information, filter the available sensory input, and provide implicit sensory-motor coordination mechanisms to orchestrate and prioritize concurrent activities. In this work, we propose to deploy an attentional system to modulate the robotic arm motion and perception. The attentional system is expected to monitor and regulate multiple concurrent activities in order to achieve an effective coordination and interaction with the human movements in the operative space. We assume a frequency-based model of the executive attention, where each behavior is endowed with an adaptive internal clock that regulates the sensing rate and action activations. The frequency of sensor readings is here interpreted as a degree of attention towards a behavior: the higher the clock frequency, the higher the resolution at which the behavior is monitored and controlled. In this context, we consider attentional models for pick and place, give and receive, search and track (humans and salient objects).

2 Software Architecture

Clearly, to interact with humans, robots must be able to adapt in real time their movements to the behavior of the humans. Moreover, the robots must ensure safety and comfort for the humans all the while realizing socially acceptable movements. As tasks are not entirely defined in advance, but computed and adapted in real time, the robot must have all the software components to compute and adapt all the elements of an interactive manipulation task from supervision and task planning level to the hardware control one. The software architecture of such a robot is a key point for the efficiency of the communication between software modules. According to the evolution of the task and to the behavior of the human, the system should react at the right level to provide the correct response in an acceptable time. In such a context the data exchanged between the software modules must be relevant and concise to make their processing fast enough. Also we propose to build the architecture around the concept of trajectory to take into account the time and synchronize the movements.

At the lower level, the robot must respond with reflex actions like reducing velocity or recoiling when the human approaches the robot. But for more important changes, the robot must replan its action and then switch from the previous trajectory to the new one satisfying HRI constraints. For more reactive tasks like the exchange of an object with a human, the robot must be able to compute and choose a good grasp and to compute a trajectory to reach and grasp the object in real time. These different robot behaviors must be integrated in the global robot architecture.

In this section, we present a quick review of the state of the art and then introduce our architecture to control a robot interacting with human.

The introduction of robots that work among humans gives rise to new concepts and designs that were studied in recent years. The physical hardware as well as software components of the robot need to be designed by considering human's safety [1, 51]. Besides ensuring safety in robot hardware with compliant designs [39, 6, 68], the motions of the robot need to be "planned" and "executed" in a "human-aware" way by limiting the velocity at potential collision impact [33].

In [58] we have proposed a planning and control framework for synthesizing safe and socially acceptable robot motions. This framework was shown to generate human-aware motion for a static model of the human. In [44], we have extended the approach using a sampling-based "human-aware" path planner, which was based on a set of geometrical HRI constraints [60]. These constraints, taking as input the human kinematics and state, lead to safety, visibility and "arm comfort" costmaps defined over the robot configuration space. Sampling-based costmap

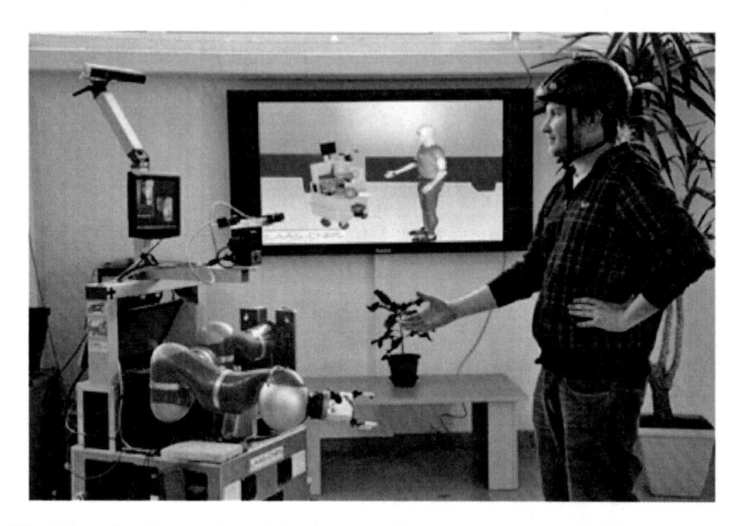

Fig. 1 The Jido robot interacting with a human. The robot model of the scene is displayed on the wall-screen. The Kinect monitors the human kinematics and the human gaze is perceived using a motion capture system. Robot motions take the human into account at planning and execution level.

planning techniques [40] were used to find good quality paths regarding the computed HRI cost criterion. Using users studies, Cakmak *et al.* [18, 17] have shown the importance of spatial and temporal pose of the robot for the exchange.

Executing motion bounded in jerk, acceleration and speed is also a way to produce human friendly robot motions. In [30], Flash and Hogan showed that the motion of the humans is by default limited in jerk and acceleration. Moreover human–robot object exchange studies [37] suggest that robot motion with minimum-jerk profile of the end-effector are preferred. In [11, 12], we have introduced a soft motion framework bounding the robot motion in jerk, acceleration and speed to ensure the human safety (speed) and comfort (jerk and acceleration).

The motion planning and execution frameworks of [58, 44] do not account for possible human motions during the trajectory execution. In motion planning literature, algorithms for dynamic environments have been introduced to take into account such changes in the robot workspace [9, 28, 69]. However, the human behaviors, which are considered in this chapter, do not lead to the same constraints as the moving obstacles taken into account by dynamic environments motion planning methods. In [9], a continuous set of homotopic paths is determined in which the initial path is deformed. Virtual forces are applied to the initial path by a control algorithm, the process can be viewed as an elastic band being stretched to gain optimality regarding clearance and length criteria. More recently in [28, 69], RRT-like algorithms have been introduced for motion planning with a limited time horizon well suited for dynamic environments. When executing the robot trajectory the human may come closer to the robot as shown in Fig. 1, changing the safety and legibility constraints that have been taken into account by the path planner. Also, in handover situations the human may want to change the object transfer position (OTP), making the target configuration irrelevant to the task.

The HRI constraints [60] modeled as cost functions represent the amount of danger and how the human feels about a given robot configuration. Hence, as the danger of injuries increases and humans are frightened with high velocities, we propose to slow down the robot motion for high cost configurations by modifying on-line the timing-law without stopping the robot motion. This reactive scheme enables a safe and legible behavior according to human movements but it is not always efficient to account for the changes in the HRI costs. Hence we also propose to use the path perturbation variant of [44] to optimize the executed solution regarding the current safety and comfort costs. In order to guarantee the jerk, acceleration and velocity bounds, we introduce an efficient way to replace a soft motion trajectory by a new one.

2.1 Architecture

From the high level decisional system that plans tasks and supervises execution to the actuators and sensors levels, the robot needs to compute many elements and disseminate data. As the robot must react at different levels, the architecture should irrigate software modules with the right flow of data, which is composed of sensors data, module results and decisions. Figure 2 shows the architecture that we propose for tasks like pick and give or receive and place. This architecture and the associated modules can be improved and extended, but the properties described are sufficient to demonstrate the proposed functionalities. At the top level, task planner and supervision are intended to plan a task like "clear the table" or "pick this object and give it to this person" and then supervise the execution of the plan.

An important part of the data exchanged represents movements, which can be described by trajectories. As the human environment is changing, the robot must adapt its trajectories in real time. For example, if the human is approaching the robot, the velocity of the trajectory should be slowed. We present further in Sect. 4.2 an interesting solution consisting in the use of series of cubic functions of time to represent trajectories.

The "path planner" uses RRT and T-RRT (see Sect. 4) to plan path as broken line for the whole robot in Cartesian or joint spaces. It is used to plan a first path and then to compute new paths in real time. For example if the robot is grasping an object from a human, and the "grasp planner" proposes a better grasp, this module computes a new path.

The "trajectory planner" transforms a broken line in soft motions satisfying the bounds in jerk, acceleration and velocity. It runs the "collision checker" and adapts jerk, acceleration and velocity limits from the costs associated to the position and behavior of the humans.

The human aware manipulation planner module (MHP) brings together the "path planner", the "grasp planner" and the "trajectory planner" to build a valid trajectory from the definition of the task and from the state of the robot provided by the SPARK module.

The SPARK module maintains a 3D model of the robot and of its environment (pose of objects, behavior of humans, position and posture of humans, visibility, etc.). This model is composed of known models and updated from data provided by the robot sensors.

The trajectory controller that monitors the execution of the trajectories is build on top of the controller provided by the robot manufacturer.

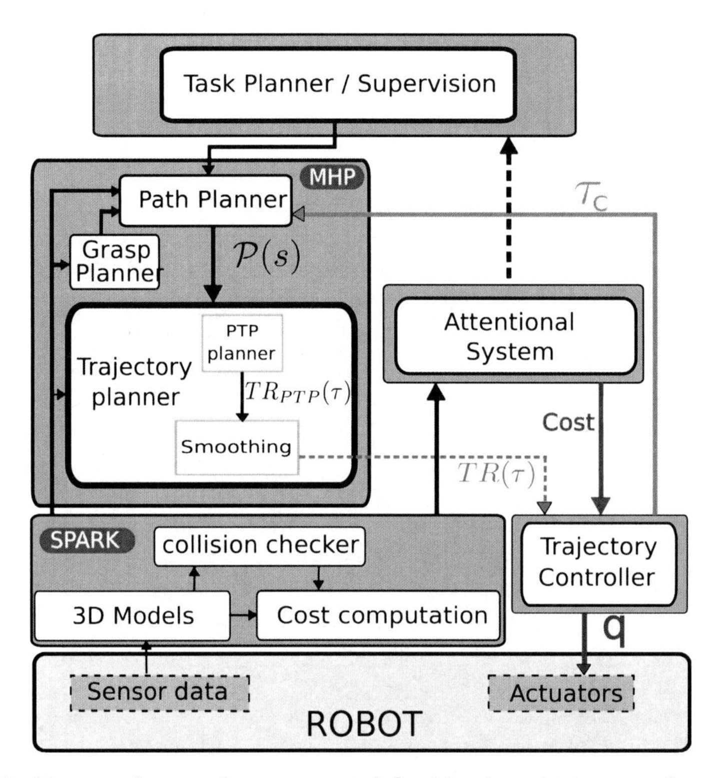

Fig. 2 Architecture: from a robot movement defined by the task planner and supervision module and the state of the robot maintained by the SPARK element, the MHP module computes in real time a trajectory for the trajectory controller and the Attentional System monitors the execution.

The "attentional system" uses sensors data that are preprocessed by the SPARK module to monitor and interpret the humans positions and behaviors. Given a model of the human behavior and the configuration of the robotic system, the attentional system can change the task that the robot is doing or the way the robot executes the task. For example, it can adapt the frequency of the clock that regulates the execution/control of the task. Finally, from the trajectory and state of the scene it can modify the trajectory, move the end position or adapt the cost to accelerate or decelerate the motion law. This possibilities are described in Sect. 5.

In this architecture, the "grasp planner" has an important place as the choice of the initial grasp impacts all the task. For example, the position of the platform of the robot must be chosen so that the arm can achieve the grasp while avoiding obstacles. In the next section we detail a grasp planner developed in this context.

3 Grasp Planning

As the choice of a grasp to grab an object greatly determines the success of the task, we present here some aspects of the grasp planner module. For a complex object and simple tasks like pick and place or pick and give to a human, a lot of constraints have to be taken into account. But one essential point for human robot interaction (HRI) is the necessity of double grasp in many situations. Of course, both hands are required to lift a heavy object, but during the exchange of an object with a human the object is also grasped by two hands. Sometimes, the robots needs to change the hand that holds an object and transitorily uses double grasp.

Grasp planning basically consists in finding a configuration for the hand(s) or end effector(s) that will allow picking up an object. If we consider a complete robotic platform, not only the grasp configuration is needed but also the configuration of the robot base and arm. To replace our work in the existing one, we give a brief overview of the state of the art concerning grasp planning in the next section.

3.1 Related Work

Most of the early grasp planning methods did not take into account finger nor arm kinematics and are often referred as contact-level techniques [50, 29, 26]. The contacts are regarded as freely-moving points with no link to any mechanical chain. Many grasp stability criteria have been introduced for this model of point/surface contact, the most common being certainly the force closure criterion [29, 5]. Force closure criterion is verified if a grasp can resist arbitrary force/torque perturbation exerted on the grasped object and is tested for a specific set of contacts (positions and normals). To integrate the notion of robustness of the grasp stability with respect to the contact positions, the concept of independent regions of contact has been introduced [50]. These regions are such that a grasp always verifies force closure as long as the contacts stay within the region. The computation of these regions has been solved for different object surfaces modelization (2D discrete surface [23], 2D polygonal surface [22], 3D polyhedral surface [54, 55]).

All these contact-level techniques were not very well-suited for real applications. Therefore, many new methods have appeared that integrate considerations on finger and/or arm kinematics.

Miller *et al.* [47] proposed to decompose the object into a set of primitives (spheres, cylinders, cones and boxes). With each primitive is associated a pregrasp configuration of the hand. A set of parameters is sampled in order to test the different possible approaches of the hand, then, for each approach, the fingers are closed on the object until collision. The quality of the obtained grasp is then computed according to the measure described in [29].

The idea of object decomposition was widely used and is still the base of many grasp planners. It offers a heuristic to reduce the possible relative palm/object poses to test. In [32], the authors decompose the object model into a superquadric *decomposition tree* employing a nonlinear fitting technique. Grasps are then planned for each superquadric with a heuristic approach close to the one in [47]. The grasps are then simulated on the original object model using the GraspIt! dynamics simulator [46], to sort them by quality.

Huebner *et al.* [38] proposed a technique to build a hierarchy of minimum volume bounding boxes from 3D data points of the object envelop. This method offers a interesting robustness with respect to the quality of the object 3D model, acquired from sensors (here laser scan).

In [36], the object is decomposed into a set of boxes called *OCP* (Object Convex Polygon). Each box of the OCP is compared to a *GRC* (Grasping Rectangular Convex), which gives an estimation of the maximum size of the object that the hand can grasp. Different GRCs are defined corresponding to different grasping styles. Xue *et al.* [66] presented a method to optimize the quality of the grasp while taking into account the kinematics of the fingers during the optimization phase. They use a swept volume precomputation associated with a continuous collision detection technique to compute, for a given hand/object relative pose, all the possible contacts of each finger on the object surface. After obtaining an initial grasp provided by the GraspIt! software [46], they locally optimize the quality of the grasp in the finger configuration space.

Some works gave more focus on arm and/or robot base inverse kinematics issues. Berenson *et al.* [4] are interested in finding grasp configurations in cluttered environments, for a given robot base position in the object range. From different object approaches, the authors precompute a set of grasps, all verifying the force closure property. Instead of trying to solve the arm inverse kinematics and checking for collisions for each grasp of the set in an arbitrary order, the authors propose to compute a *grasp scoring* function for each grasp. The function is used to evaluate the grasps that are more likely to succeed the inverse kinematics and collision tests and is based upon a force closure score, a relative object-robot position score and an environment clearance score.

The authors of [25] focused on path planning for the robot base (or body) and arm and presented a planning algorithm called *BiSpace*. Like in [4], they first compute a set of grasp configurations for the hand alone. Once one or more collision free configurations for the hand are found, they become the start nodes of several RRTs (Rapidly Random-exploring Tree [43]) that will explore the hand workspace while another RRT is grown from the robot base start configuration, that explores the robot configuration space.

Some recent works were inspired by results in neuroscience [56, 65] which have shown that humans mainly realize grasping movements that are restricted in a configuration space of highly reduced dimensionality. From a large data set of human pregrasp configurations, Santello *et al.* [56] performed a principal component analysis revealing that the first two principal components account for more than 80% of the variance. Ciocarlie *et al.* [20] called the components *eigengrasps* and use them as a base to represent the reduced configuration space of the hand. They also add the six DOF's of the wrist pose. Then, they use a simulated annealing based optimization method, in eigengrasp space, to find the best grasp according to an energy function. The energy function takes into account two parameters. First, the distance between specified points on the hand and the object surface. Secondly, a quality metric based on the one in [29].

A frequent difficulty associated with grasp planning concerns the 3D model reconstruction of the object to be grasped. This reconstruction is not an easy task and the resulting model can be very noisy. In order to avoid the need for 3D model reconstruction, Saxena *et al.* [57] proposed a method to find good grasps of objects being seen for the first time, that does not require such a model. This method is based on a learning approach that uses image features to predict good points where to grasp the object. These features are based on edges, textures and colors. A set of generated synthetic images of various objects is used to learn the feature values of region labeled as grasping points. For a novel object, a probabilistic model of the grasping point features is used to find grasping points in the image. A triangulation is then performed that uses images from different points of view to find the region where to grasp the object in 3D space.

3.2 The Grasp Planner

As explained above, for HRI, grasp planning has several uses and is not only devoted to basic pick-and-place tasks. In particular, in a planning point of view, the context is very important in order to choose a valid grasp. Therefore, the proposed approach does not rely exclusively on a heuristic that can introduce a bias on how the object is grasped. Our objective is to build a grasp list to capture the variety of the possible grasps. It will then allow finding a grasp, even in a cluttered environment, for an object with a complex shape. In the following, we illustrate the method with the Schunk Anthropomorphic Hand (SAH) depicted in Fig. 3 as it is the one used in our laboratory. It has four fingers. Each finger, except for the thumb, has four joints. Only the three first joints are actuated, the last one being coupled with the third one. The thumb has an additional actuated joint to place it in opposition to the other fingers. The method however applies to other hand kinematic structures, after some small numeric adaptations.

A single grasp is defined for a specific hand type and for a specific object. The object model is supposed to be a triangle mesh: A set (array) of vertices (three

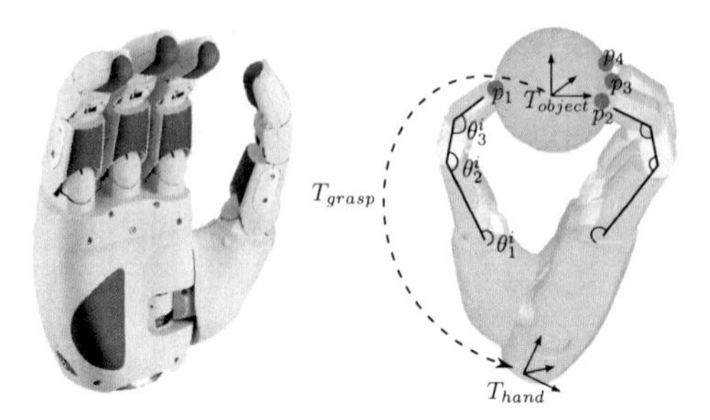

Fig. 3 Left: The Schunk Anthropomorphic Hand used to illustrate our grasp planning method. Right: A grasp is defined by a transform matrix T_{grasp}, the finger joint parameters of each finger i $(\theta_1^i, \theta_2^i, \ldots)$ and a set of contact points $(p_1, p_2, \ldots)$.

coordinates) and a set of triangles (three indices in the vertex array). It is assumed to be a minimum consistent, i.e. it has no duplicate or isolated vertices nor degenerate triangles.

3.2.1 Grasp Definition

In the following, we define a grasp by (see Fig. 3):

- A transform T_{grasp} between the object and the hand frame.
- A set of finger joint parameters $(\theta_1^i, \theta_2^i, \ldots)$ where i is the ID of the finger.
- A set of contact points $(p_1, p_2, \ldots)$ that can be deduced from the two previous items.

A contact contains the following information:

- Position: both a 3D vector and a set (triangle index + barycentric coordinates) to store the position.
- Normal: the plane normal of the triangle the contact belongs to.
- Coulomb friction: used further to compute the grasp stability.
- Finger ID: to store which finger is responsible of the contact.
- Curvature: it is interpolated from the curvature of the vertices of the triangles.

As the main concern of the grasp planner is motion planning, it is not possible to rely on the computation of an only grasp or to compute grasps according to a heuristic that could introduce a bias on the choice of the grasp. It is preferable to compute a grasp list that aims to reflect the best the variety of all possible grasps of the object. Our algorithm applies the following steps that will be detailed further:

- Build a set of grasp frame samples.
- Compute a list of grasps from the set of grasp frames.
- Perform a stability filter step.
- Compute a quality score for each grasp.

3.2.2 Grasp Frame Sampling

For manipulation planning, it is important to avoid biasing the possible approach of the hand when we compute the grasp. Therefore, we choose to sample the possible grasp frames uniformly. This is done by the mean of a grid. We have chosen, for our hand, a grasp frame that is centered on the intersection of the finger workspaces so that it is roughly centered where the contacts may occur. We set as an input the number of positions and the number of orientations, each couple position-orientation defining a frame. The positions are uniformly sampled in the object axis-aligned bounding box with a step computed to fit the desired number of position samples. The orientations are computed with an incremental grid like the one in [67]. For each grasp frame, a set of grasps will be computed.

3.2.3 Grasp List Computation

As the proposed grasp planning method does not restrict the possible hand poses or surfaces of contact on the object, it requires a lot of computation. Therefore, we have to introduce some data structures to reduce the computation times. Except for collision test, the most expensive computation is the finger inverse kinematics. One has to be able to know the fastest possible if, for a specified hand pose (relative to the object), a finger can establish a contact on the object surface and, if it is the case, where. The contacts can only occur in the intersection of the finger workspace and the object surface. For each finger, it is consequently crucial to find this intersection or at least an approximation. We use two data structures to model the object surface and finger workspaces.

3.2.4 Object Surface Model

We propose to approximate the object surface with a contact point set, keeping trace of where it is on the object mesh to be able to get some local information (surface normal and curvature) later. The set is obtained by a uniform sampling of the object surface. The sampling step magnitude is chosen from the fingertip radius. A space-partitioning tree is built upon the point set in order to have a hierarchical space partition of the points (Fig. 4). It is similar to a kd-tree. Starting from the original set of points, we compute the minimal axis-aligned box containing all the points. Such a box is usually referred as Axis-Aligned Bounding Box or AABB. This first AABB is the tree root. The root AABB is then splitted in two along its larger dimension. This leads to two new nodes, children of the root, containing each a subset of the original point set. The splitting process is then recursively applied to each new node of the tree. The process ends when a node AABB contains only one point.

We then need to find the intersection of each finger workspace with the object surface tree. So we introduce another data structure to approximate the finger workspace and compute this intersection quickly.

Fig. 4 The object mesh is uniformly sampled with a point set (top images). The point set is then partitioned using a kind of kd-tree (bottom images).

3.2.5 Finger Workspace Model

As spheres are invariant in rotation, they are interesting to build an approximation of the finger workspace. Starting from a grid sampling of the finger workspace (Fig. 5), we incrementally build a set of spheres fitting strictly inside the workspace. First, points of the grid are marked as being boundary points (on the workspace envelope) or inner points (strictly inside the workspace volume). For each inner point, the smallest distance to the boundary points is computed, referred as $dmin$. The inner point having the biggest $dmin$ is the center of the first sphere S_1, of radius $dmin$. For all the inner points that are not inside S_1, a new $dmin$ is computed, that is the minimum of the old $dmin$ and the minimal distance to S_1. The point that has the biggest $dmin$ is the center of the second sphere S_2, of radius $dmin$. This process is repeated until we have reached the maximal desired sphere number or the last computed sphere has a radius less than a specified threshold. We keep the ordering of the construction so that the sphere hierarchy starts from the biggest ones, corresponding to workspace parts that are the farthest to the finger joint bounds. These bounds were first slightly reduced (Fig. 5) in order to eliminate configurations where the fingers are almost completely stretched.

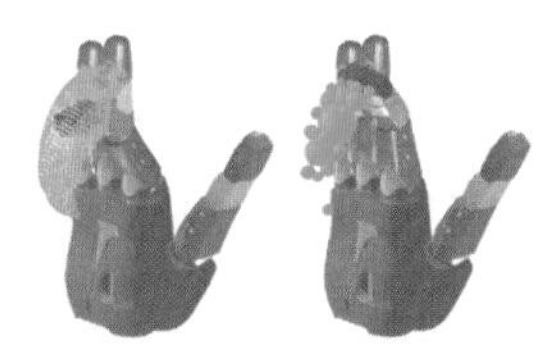

Fig. 5 The finger workspace, discretized with a grid (forefinger workspace, left image). The grid is converted to a volumetric approximation as a set of spheres (right image).

Once we have both the contacts tree and the workspace sphere hierarchy, it is very fast and easy to determine the intersection of the two sets and so the contact points.

3.2.6 Intersection between Object Surface and Finger Workspace

All the operations that have to be performed are sphere-box intersection tests. The intersection is tested from the biggest to the smallest sphere, guaranteeing that the *best* parts of the workspace will be tested first, i.e. the ones farthest to singularities due to the joint bounds. Starting from the tree root, we test if there is a non-null intersection between a AABB-node and the sphere. If not, we stop exploring this branch, otherwise we test the sphere against the two node children, until we arrive to a leaf node, i.e. a single point. We then just have to test if the point is included in the sphere volume. Figure 6 shows the contact point candidates for two different grasp frames with the same object. At this stage, we just know that the points will pass the finger inverse kinematics test. No collision tests have been performed yet. For a given grasp frame, the grasp is computed finger by finger, that means that, if

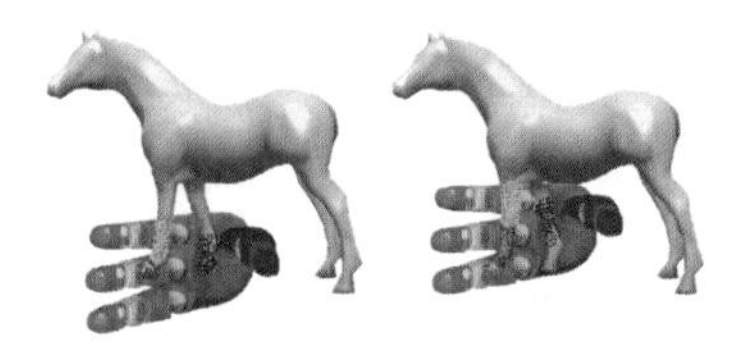

Fig. 6 The potential finger contacts, drawn in red, green, blue and magenta for the thumb, forefinger, middle finger and ring finger respectively. On the left image, no contact can be found for the ring finger because of its limited workspace.

we have the contact and configurations of the fingers 1 to $i-1$, we search a contact point for finger i and test collision only with the fingers 1 to i as the other finger configurations are not yet known. We start from the thumb as no stable grasp can be obtained without it. If a finger can not establish a contact, it is left in a *rest* (stretched) configuration. If we have three contacts or more, we can proceed to the stability test. Note that, at this stage, we have a collision-free grasp, i.e. no collision between the hand and the object and do not yet consider collision with the environment or the robot arms or body.

3.2.7 Stability Filter and Quality Score

The stability test is based on a *point contact with friction* model, that explains why at least three contacts are required. From the contact positions and normals, we compute a stability score. It is based on a force closure test and stability criterion [7]. All the grasps that do not verify force-closure are rejected. We also compute and add a second score that is the distance to the mass center of the object. The stability score is not sufficient to discriminate good grasps so we build a more general quality score.

Several aspects can be taken into account to compute a grasp quality measure [63]. A trade-off is often chosen with a score that is a weighted sum of several measures. We chose to combine the previous stability criterion with two other criteria: A finger force ellipsoid major axis score and a contact curvature score. The idea behind the first one is that it is preferable to favor contact such that the contact normal is in a direction close to the direction of the major axis of the force ellipsoid, corresponding to the better force transmission ratio. Figure 7 shows the force ellipsoids computed for a configuration of the SA Hand.

Fig. 7 Left: the finger force ellipsoids must fit the contact normal to ensure a good grasp. Right: the mean curvature of the object surface is used as a quality criterion on the contact position. Surface color varies from red (low curvature) to blue (high curvature), through green.

The curvature score is used to favor contacts where the mean curvature of the object surface is low. In real situation, it will reduce the impact of a misplaced contact as the contact normal will be susceptible to smaller change in a low curvature area than in a high curvature one. Figure 7 shows, on some objects, how low curvature areas are preferable to establish contacts. Curvature is computed for each vertex and then interpolated for each point on the surface from its barycentric coordinates. The curvature is then normalized to be always included in $[0;1]$.

3.2.8 Double Grasp Planning

A double grasp is a grasp involving both hands. It is computed from two single grasp lists L_1 and L_2, obtained for each hand. The model for double grasp simply derives from the single grasp model: it is defined by a valid grasp for each hand and the two associated quality.

Each single grasp pair sg_1 and sg_2, belonging to L_1 and L_2 respectively, is tested. All colliding pairs are rejected. The list can be filtered once we have more information about the environment or task to realize. For instance, if the task is to pick up an object with one hand, give it to the other hand before placing it on a support, we can remove all the grasps that lead to a collision with the environment for the given initial and final object poses. For instance, all the grasps that take the object from *below* will be removed as they lead to a collision between the object support (*e.g.* a table) and the hand. For each double grasp, a score is then computed based on two scores: The quality of each single grasp and an inverse kinematics (IK) score.

- The minimum of sg_1 and sg_2 quality is used as a stability score for the double grasp.
- An IK score is computed for sg_1 and sg_2. It is based on how *natural* is the way to grasp the object in its start and goal configuration using sg_1 and sg_2. The score is a distance of the arm configuration to a predefined rest configuration. For the double grasp, we take the minimum of the IK scores of sg_1 and sg_2.

After normalizing these two scores separately for all the computed double grasps, we sum them for each double grasp to obtain its score.

Figure 8 shows a double grasp computed with our algorithm.

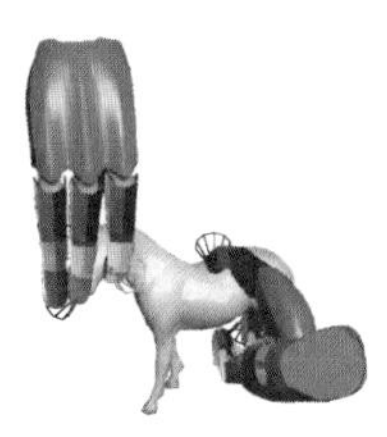

Fig. 8 An example of double grasp computed for right and left SAHs with friction cones displayed in colors.

3.2.9 Double Grasp for Object Transfer

First, a double grasp list must be computed for the object of interest. This list is computed for a hand pair composed of the robot hand and a human hand (his/her right hand *a priori*). The human hand model is of course a simplification as our modelization only deals with rigid bodies. We use the SAH as it is already available, but with a scaled kinematic structure to approximate the human hand.

Let note a double grasp of the list $dg = (sg_r, sg_h)$ where sg_r is a grasp of the robot hand while sg_h is a grasp of the human hand. For a given object, placed on a support in a particular environment, we remove, from the previously computed double grasp list, all the double grasps such that sg_r does not allow the robot to grasp the object. From each remaining double grasp, knowing the positions of both human and robot, we can deduce how to hand over the object to the human as the grasp gives the direction of the human wrist to grasp the object. The robot still has to choose a double grasp from the list and an exchange pose for the object. The choice of the double grasp is based on the notion of *intention legibility*. It must be easily interpreted as an object transmission. The best double grasps appear then to be the ones where the wrist directions of human and robot are opposed. The choice of the exchange pose is based on the notion of comfort. It must allow the human to grasp the object with a comfortable wrist/arm direction, i.e. directed from the human position to the object position (Fig. 9).

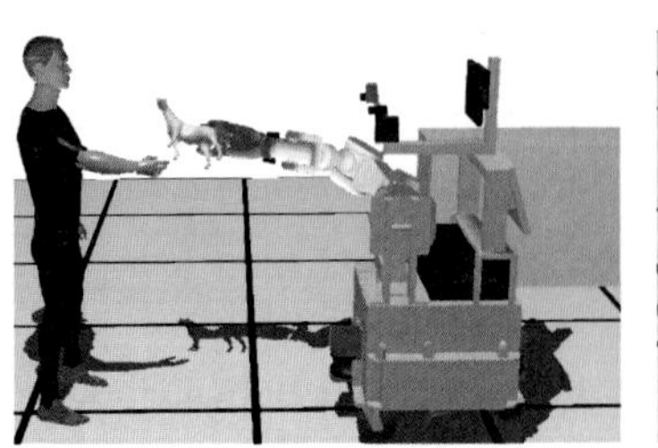 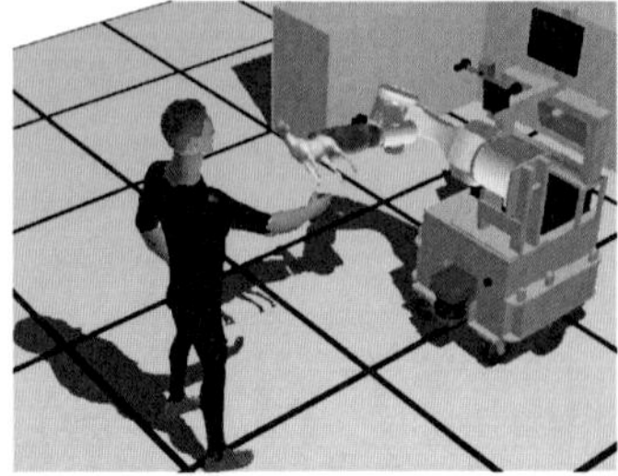

Fig. 9 A good object transmission must be easily interpretable from the human point of view and must not require an uncomfortable arm movement for the human.

3.2.10 Dual-Arm Manipulation

Dual hand/arm grasps are at least required in two situations, when the object to carry is too heavy to be carried with only one hand and when the robot has to transfer the object from one hand to the other to take advantage of the workspaces of the two arms. In this re-grasping case, a first solution consists in placing the object on a support and then picking it up again with the other hand. But a better solution is to use the second hand to realize a temporary dual-handed grasp before removing the first hand.

For a given manipulation task, the robot will start with a one-handed grasp g_i and end with a one-handed grasp g_f. The regrasping task will be achieved with the help of a double grasp obtained by combining g_i and g_f. As the hands must not collide during the regrasping task, g_i and g_f must be chosen appropriately. Grasps that were ideal in the case of single-handed manipulation are generally no more usable for dual-handed manipulation. Indeed, for stability reason it is preferable to use contacts that are close to the object center of mass. This leads to configurations where the hand is centered on the object, that do not let enough room to take the object with the other hand. When the robot uses dual-handed grasps, it will modify the initial and final single-handed grasps in order to take the object on its extremities. Such an example is depicted in Fig. 10 where the DLR's robot Justin [53] manipulates a horse statuette. The best grasps, in term of stability, are on the body of the horse. However, it is not possible to place two hands on this part of the object. Consequently, the robot has to choose to grasp the extremities of the object (leg and head on the example).

It is also possible to perform regrasping to just modify the grasp of one hand. Let us suppose the robot holds the object with the right hand and with a grasp g_{right}. The robot also needs to change the grasp. It can take the object with its left hand and a grasp g_{left}, release the first grasp, possibly re-orient the object, and grasp the object with right hand again but with a different grasp g'_{right}. Two dual-handed grasps will thus be required: $[g_{right}, g_{left}]$ and $[g'_{right}, g_{left}]$. The grasp selection uses the same principle as above but is more combinatorially complex.

Fig. 10 If the robot (DLR, [53]) has to realize a regrasping task, it must select initial and final grasps that let enough room to perform a dual-handed grasp.

This technique has been implemented for the robot Justin[1] equipped with two SAHs, within our simulator, (Fig. 10) to plan pick-and-place tasks that require regrasping.

We have presented a grasp planner for single and double grasp that allows choosing a grasp in real time according to the behavior of the human. As we have seen, grasp planning is complex and greatly determines the succes of a manipulation task. In particular, the choice of a grasp compatible with the whole task is crucial. In the next section, we introduce how to plan and adapt a displacement after the move is defined in an interactive context.

4 Motion Planning

HRI introduces real challenges for the motion planning problem. While motion planning is not yet largely used in industry where most robots are still programmed by learning, for HRI we need to plan and adapt in real time motions that take into account human movements and behaviors. Traditional motion planners plan only a path that a controller executes at constant velocity. To take into account human motion, we propose here to plan trajectories that satisfy the HRI constraints: *safety and comfort.*

[1] Justin is a robot of DLR that gracefully made the model available for LAAS-CNRS.

From an elementary task like "pick an object", "place an object" or "give an object", the motion planner must precise initial and final conditions for each move, plan a trajectory and then adapt the trajectory in real time. For example, to plan the first movement of a pick and give task, the planner must firstly choose an initial grasp that takes into account the double-grasp needed to give the object and then adapt the movement to the human behavior.

In this section, we present a first skeleton of a planner for human–robot interactive motions.

4.1 Planning the Path

The first step of the motion planning is the computation of the path $\mathscr{P}$. In our case, the motion planner is based on the planner initially proposed by Sisbot [60]. The motion planner takes explicitly into account the constraints of HRI to synthesize navigation movements (movement of the platform) and handling (fixed robot platform). HRI constraints are for example the human–robot distance or field of view of the human. This planner is based on case studies in HRI [42] and on existing theories on the proxemic behavior of humans [34]. The HRI constraints are represented by cost functions based respectively on the posture of the human, his/her field of view and accessibility. These costs are represented by costs maps defined in the working space of the robot. In [60], to solve a manipulation task like passing an object to a human, the path of the object is first planned using grids methods defined in the workspace. Then the path of the robot is planned from the inverse kinematic of the robot. However, as the path of the object is defined by the first step of the method, the original planner is not efficient in cluttered environment. We use the extension proposed by Mainprice [44]. This extension consists in extending the capabilities of the planner through the use of planning algorithms based on random sampling to compute the moves taking into account human in cluttered environments.

4.1.1 Random Path Planner

When the robot shares the workspace with humans, the path planner must take into account the costs of HRI constraints. We perform this planning with the T-RRT method [40] which takes advantage of the performance of two methods. First, it benefits from the exploratory strength of RRT-like planners [43] resulting from their expansion bias toward large Voronoi regions of the space. Additionally, it integrates features of stochastic optimization methods, which apply transition tests to accept or reject potential states. It makes the search follow valleys and saddle points of the cost-space in order to compute low-cost solution paths (Fig. 11). This planning process leads to solution paths with low value of integral cost regarding the costmap landscape induced by the cost function.

In a smoothing stage, we employ a combination of the shortcut method [3] and of the path perturbation variant described in [44]. In the latter method, a path $\mathscr{P}(s)$ (with $s \in \mathbb{R}^+$) is iteratively deformed by moving a configuration $q_{perturb}$ randomly

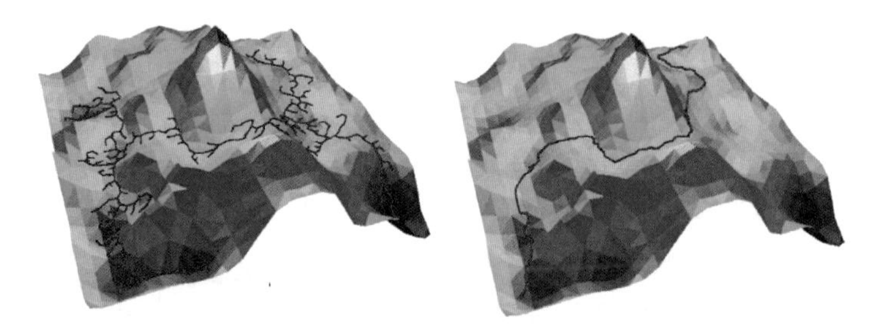

Fig. 11 T-RRT constructed on a 2D costmap (left). The transition test favors the exploration of low-cost regions, resulting in good-quality paths (right).

selected on the path in a direction determined by a random sample q_{rand}. This process creates a deviation from the current path, The new segment replaces the current segment if it has a lower cost. Collision checking and kinematic constraints verification are performed after cost comparison because of the longer computing time.

The path $\mathscr{P}(s)$ computed with the human-aware path planner consists of a set of via points that correspond to robot configurations. Via points are connected by localpaths (straight line segments). Additional via points can be inserted along long path segments to enable the path to be better deformed by the path perturbation method. Thus each localpath is cut into a set of smaller localpaths of maximal length l_{max}.

4.1.2 Taking into Account Geometric Constraints

We use costmaps to model HRI. These costs are important when the configuration of the robot is not safe or comfortable for the human. We retain three constraints:

- *Safety constraint* (Fig. 12(a)): This constraint ensures the safety of interaction by monitoring the distance between the robot and the human. The human is modeled by approximating the bounding volume of his/her body (regardless of the geometry of the arm). To reduce the risk of collision between the human and the robot, this safety constraint keeps the robot away from the head and body. The distance to be considered is the minimum distance between the robot (all parts of the robot are taken into account) and the simplified model of human (cylinder + sphere). When this distance is small, the cost is high and conversely when the distance increases the cost decreases to a minimum threshold after which it becomes null.

- *Visibility constraint* (Fig. 12(b)): This constraint aims at limiting the effect of surprise that may experience a human while the robot moves in the workspace. A human feels less surprised if the robot remains visible and the interaction is safer and more comfortable. Thus, each point of the workspace has a cost proportional to the angle between the gaze of the human and his/her position in the Cartesian space;
- *Constraint of "comfort of the human's arm"*: The third constraint is taken into account for object exchange tasks between robot and human. It allows determining an object transfer point (OTP) in the workspace. This constraint is also taken into account during the path planning of the exchange task to facilitate the exchange of the object at any time during the move. For this, the robot must reason on the kinematic and the accessibility capabilities of the human. The assumed reachable volume of the human can be pre-computed using generalized inverse kinematics. For each point inside the reachable volume of the human, the determined configuration of the torso remains as close as possible to a given resting position. Collision detection with the environment is then used to validate these postures. At each valid position, a comfort cost is assigned through a predictive model for human posture introduced in [45].

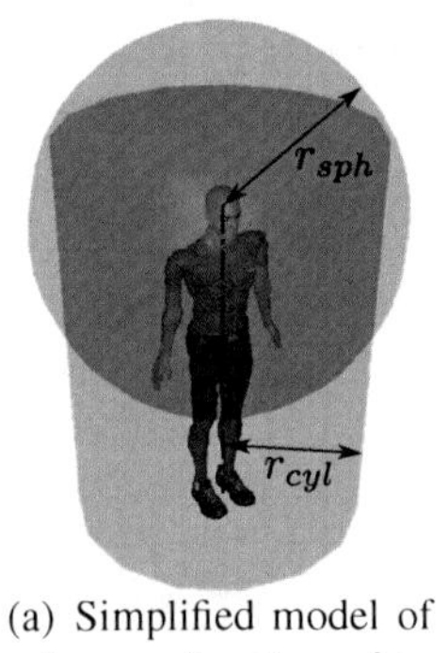

(a) Simplified model of a human for the safety cost.

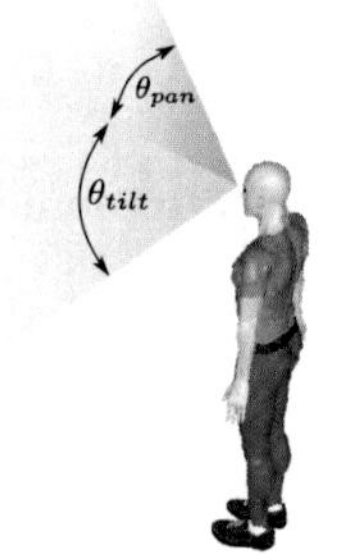

(b) Visibility model of a human.

Fig. 12 Cost models of safety and visibility.

Each constraint is represented by a three-dimensional cost map, these basic costmaps are then combined with a weighted sum:

$$cost(h,q) = \sum_{i=1}^{3} w_i c_i(h,q) \quad (1)$$

where w_i are the weights, h the human posture, q the robot configuration and c_i, the costs.

In the current implementation, the weights are empirically defined and cost functions are evaluated "on the fly" during planning.

4.1.3 Path Planning

According to the presence or the absence of the human in the scene, the path planning is performed using T-RRT if human is present or RRT if not.

In both cases, the path resulting $\mathscr{P}(s)$ ($s \in \mathbb{R}^+$) is composed of a set of robot configurations (nodes) connected by straight lines (edges). Consider the example of a two-dimensional system solved by the method RRT. Figures 13(a) and 13(b) respectively represent the initial and final positions of the yellow puck. The path obtained is shown in Fig. 13(c) (green lines connecting the spheres). The spheres represent the intermediate configurations of the path.

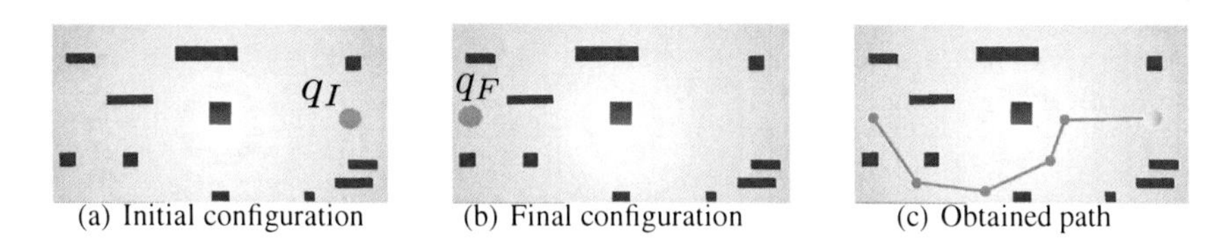

(a) Initial configuration (b) Final configuration (c) Obtained path

Fig. 13 Example of a the planning of a path in a 2D space.

4.2 *From the Path to the Trajectory*

We propose to generate a trajectory from a path using the soft motion trajectory planner designed by Broquère [12], [11], [10].

4.2.1 Trajectory Model

A trajectory $\mathrm{TR}(t)$ is represented by a combination of n series of cubic polynomial curves. The use of polynomial cubic defined by the Soft Motion Trajectory Planner provides a solution in the context of HRI where the task introduces numerous constraints. From the trajectory generation point of view the safety constraint is ensured by bounding the velocity and the comfort constraint by bounding the jerk and the acceleration.

The trajectory ${}_j\mathrm{TR}(t)$ corresponds to the evolution of the j axis and is composed of N cubic polynomial segments (curves) (Fig. 14). We consider that each axis has the same number of segments since they can be divided.

Functions ${}_jJ_k(t)$, ${}_jA_k(t)$, ${}_jV_k(t)$, ${}_jX_k(t)$ respectively represent the jerk, acceleration and velocity evolution over the k segment for the j axis. T_i is the initial time of the trajectory and T_F the final one.

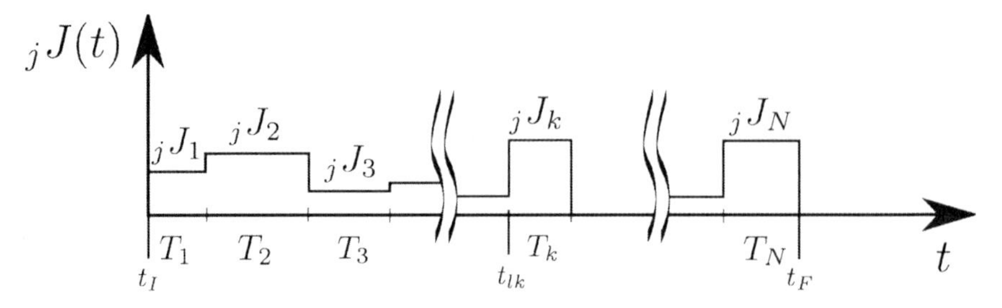

Fig. 14 The jerk evolution for the j axis of the TR(t) trajectory.

A segment is defined by Eq. (2) and depends on its duration T_k and on five parameters:

- the initial time t_{Ik} with $t_{Ik} = t_I + \sum_{i=1}^{k-1} T_i$,
- the initial conditions (3 parameters: $_jA_k(t_{Ik})$, $_jV_k(t_{Ik})$, $_jX_k(t_{Ik})$),
- the jerk value $_jJ_k$.

$\forall t \in [t_{Ik}, t_{Ik} + T_k]$:

$$_jX_k(t) = \frac{_jJ_k}{6}(t - t_{Ik})^3 + \frac{_jA_k(t_{Ik})}{2}(t - t_{Ik})^2 + _jV_k(t_{Ik})(t - t_{Ik}) + _jX_k(t_{Ik}) \tag{2}$$

where $_jJ_k$, $_jA_k(t_{Ik})$, $_jV_k(t_{Ik})$, $_jX_k(t_{Ik})$ and t_{Ik} are constant $\in \mathbb{R}$.

The initial conditions of the trajectory $_j$TR(t) are:

$$\begin{aligned} _jA_1(t_I) &= _jA_I \\ _jV_1(t_I) &= _jV_I \\ _jX_1(t_I) &= _jX_I \end{aligned} \tag{3}$$

and the final conditions:

$$\begin{aligned} _jA_N(t_F) &= _jA_F \\ _jV_N(t_F) &= _jV_F \\ _jX_N(t_F) &= _jX_F \end{aligned} \tag{4}$$

where $t_F - t_I = \sum_{i=1}^{N} T_i$.

The multidimensional trajectory is then a composition of trajectories as:

$$\mathrm{TR}(t) = [_1TR(t)\ _2TR(t)\ \ldots\ _nTR(t)]^T \tag{5}$$

where n is the number of axes.

From the N couples $(_jJ_k, T_k)$ and the initial conditions (3) of the trajectory $_j$TR(t) we can compute the kinematic state along the j axis at a given time with (6), (7) and (8). In order to simplify the notation, the j index representing the axis will be omitted.

$\forall t \in [t_{Ik}, t_{Ik} + T_k]$, with $t_I = 0$:

$$A_k(t) = J_k(t - \sum_{i=1}^{k-1} T_i) + \sum_{i=1}^{k-1} J_i T_i + A_I \tag{6}$$

$$V_k(t) = \frac{J_k}{2}\left(t - \sum_{i=1}^{k-1} T_i\right)^2 + \sum_{i=1}^{k-1} J_i T_i \left(t - \sum_{j=1}^{i} T_j\right) + \sum_{i=1}^{k-1} \frac{J_i T_i^2}{2} + A_I t + V_I \tag{7}$$

$$X_k(t) = \frac{J_k}{6}\left(t - \sum_{i=1}^{k-1} T_i\right)^3 + \sum_{i=1}^{k-1} \frac{J_i T_i}{2}\left(t - \sum_{j=1}^{i}\right)^2 + \sum_{i=1}^{k-1} \frac{J_i T_i^2}{2}\left(\sum_{j=1}^{i} T_j\right) + \sum_{i=1}^{k-1} \frac{J_i T_i^3}{6} + \frac{A_I}{2} t^2 + V_I t + X_I \tag{8}$$

4.2.2 The Kinematic Constraints

The trajectory generation method is based on constraints satisfaction (velocity, acceleration and jerk). Each constraint is supposed constant along the planned motion. In the multidimensional case, each axis can have different constraints. We also suppose that the constraints are symmetrical:

$$\begin{aligned} {}_jJ_{min} &= -{}_jJ_{max} \\ {}_jA_{min} &= -{}_jA_{max} \\ {}_jV_{min} &= -{}_jV_{max}. \end{aligned} \tag{9}$$

Hence, the jerk, acceleration and velocity must respect:

$$\begin{aligned} |{}_jJ(t)| &\leq {}_jJ_{max} \\ |{}_jA(t)| &\leq {}_jA_{max} \\ |{}_jV(t)| &\leq {}_jV_{max}. \end{aligned} \tag{10}$$

4.3 Basic Concepts of the Trajectory Generation

This section describes breifly the core of the trajectory generator bounding the jerk, the acceleration and the velocity. Details can be found in [12], [11], [10]. The three introduced methods do not use optimization steps, they are designed to be used online in a control loop to modify the trajectory and, for example, track and catch an object handled by the human.

4.3.1 The Canonical Case: The Kinematically Constrained Point–to–Point Motion

In the basic case a motion between two points where initial and final kinematic conditions are null, Figure 15 represents the optimal point-to-point motion (according to the imposed kinematic constraints). This point–to–point motion is composed of seven segments of cubic polynomial functions at most [12].

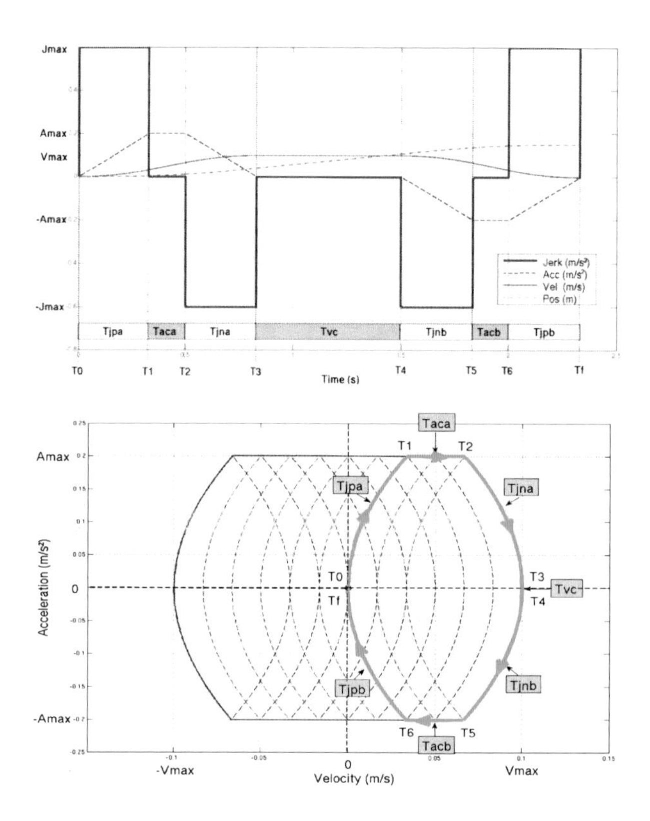

Fig. 15 Jerk, acceleration, speed and position curves and motion in the acceleration-velocity frame for a single axis.

In the multidimensional case each axis has also seven cubic polynomial segments at most. Computation details can be found in [10].

4.3.2 The Minimal Time Motion between Two Non-null Kinematic Conditions

From the canonical point–to–point case we extend the monodimensional algorithm to compute minimal time motion between two non-null kinematic states (non-null acceleration and velocity). An overview of this algorithm is presented in [12] and the details in [10]. This kind of motion is composed of a set of elementary motions saturated in jerk, acceleration or velocity. The number of elementary motions is also seven at most. For the multidimensional case, we propose in [10] a solution to synchronise the axis motions.

4.3.3 The Time Imposed Motion between Two Non-null Kinematic Conditions: The *3-Segment Method*

The method for computing a motion with an imposed duration was previouly presented in [11]. This method does not bounds the jerk, acceleration nor velocity. It uses three cubic polynomial curves to define such a motion. This simple definition provides a solution to compute analytically the motion.

4.3.4 Smoothing an Input Function

We use the method proposed in [12] to compute online a smooth movement from an input defined by acceleration and velocity. At each update of the set function, a move is computed from the current state of the system. This move is bounded by the kinematic constraints (J_{max}, A_{max} and V_{max}). Under this kinematic constraints, the minimal time motion is defined by the critical movement associated to the critical length dc [12].

Thus, in order to allow a mono-dimensional system to reach its set value in minimal time, the critical movement is computed at each iteration.

An example of a smoothed signal is plotted in Fig. 16. The blue dotted curve is the input and the green curve is the smoothed velocity. The method acts like a filter for the acceleration.

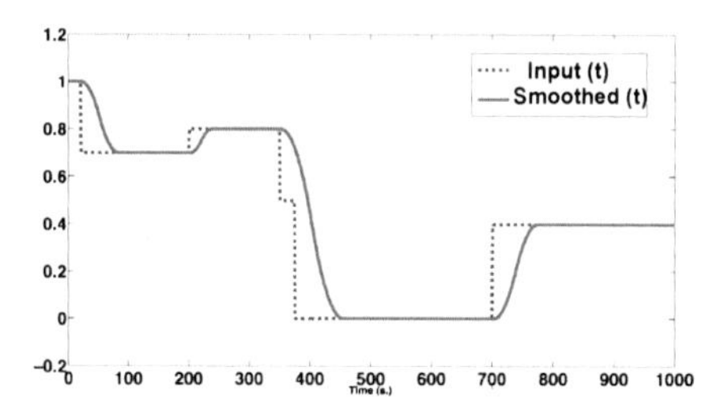

Fig. 16 Example of the smoothing of a set function.

4.4 Trajectory Generation

The trajectory generation is based on the three main methods introduced in the previous section. The input is the path $\mathscr{P}$ computed by the path planner (Sect. 4.1.1).

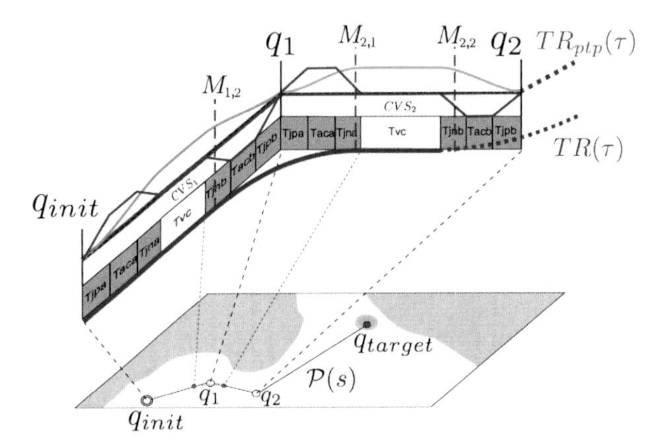

Fig. 17 From the path $\mathscr{P}$ to the smoothed trajectory TR.

The first step is to calculate a trajectory passing through all the nodes of the path $\mathscr{P}$. This trajectory, which we call TR_{ptp} consists of point–to–point movement (Sect. 4.3.1) and therefore includes stop motions at each configuration defining a node.

The second step consists in smoothing these stop motions to obtain a shorter trajectory in time TR. Smoothing uses the same 3D model than the research phase of the path. Thus, collisions are tested during the computation of the transition moves at each node. If a collision appears during the smoothing of the stop move at node q_i, then the movement will not be smoothed for this node and the stopping move will be kept.

In the following, we detail a method for smoothing stop motions based on the computation of a fixed time movement using the *3-segment* method presented in previous work.

4.4.1 Smoothing of the Stop Motions

We propose a method based on the minimum time algorithm for trajectory generation (Sect. 4.3.2) [12] and on the *3-segment* method (Setion 4.3.3) to smooth the stopping motions [11].

The trajectory TR_{ptp} (Fig. 17) between the first two nodes q_{init} and q_1 is a point–to–point motion in a straight line of duration $T_{(q_{init}q_1)}$. Similarly the motion between q_1 and q_2 is a point–to–point motion of duration $T_{(q_1q_2)}$. The stop motion is smoothed between the points $M_{1,2}$ et $M_{2,1}$.

Notation: We note the points that limit the smoothing $M_{i,j}$, the index i is the index of the point–to–point motion (the first of the trajectory has an index of 1). The index $j \in \{1,2\}$ is 1 if this point is the final extremity of the transition motion with the previous point–to–point motion and conversely for $j = 2$.

Choice of the Points $M_{i,j}$

Let us consider the transition motion in the neighborhood of q_1 located at time t_{q1}:

$$t_{q1} = t_I + T_{(q_{init}q_1)} \tag{11}$$

To simplicity, we choose $t_I = 0$ as the time origin of the trajectory.

The time positions $t_{M_{1,2}}$ and $t_{M_{2,1}}$ of the points $M_{1,2}$ and $M_{2,1}$ are determined from a given parameter τ such that:

$$M_{1,2} = TR_{ptp}(t_{q1} - max(\tau, \frac{T_{(q_{init}q_1)}}{2})) \tag{12}$$

$$M_{2,1} = TR_{ptp}(t_{q1} + max(\tau, \frac{T_{(q_1q_2)}}{2})). \tag{13}$$

So when τ is null, the movement stops at the point q_1. When τ satisfies (14), the transition motion connects the midpoints of the line segments (q_{init}, q_1) and (q_1, q_2) because of the symmetry of the velocity profile about this point.

$$\tau \geq max\left(\frac{T_{(q_{init}q_1)}}{2}, \frac{T_{(q_1q_2)}}{2}\right) \tag{14}$$

In practice, unless otherwise specified, by default we choose the points $M_{i,j}$ such that the transition movement begins at the end of the constant velocity segment of the first point–to–point movement (P_1, P_2); the transition movement ends at the beginning of the constant velocity segment of the second point–to–point movement (P_2, P_3).

Notice that, for a given value of the parameter τ, the Euclidean distance between the points $M_{i,j}$ and the corresponding point q_i varies according to kinematic parameters of the point–to–point movement.

4.4.2 Computation of the Transition Movement

Let us consider a trajectory of dimension n. The instants $t_{M_{i-1,2}}$ and $t_{M_{i,1}}$, start and end of the transition movement at the configuration q_i, are identical for all n dimensions. The computation method is described by Algorithm 6. The first step consists in computing, for each axis, the optimal time motion to determine the duration T_{imp} of the transition movement. The *3-segment* method to compute the movement in fixed time is then applied to each axis.

Algorithm 6. Computation of a transition movement near of a node q_i

begin
  Determining the switching points $M_{i-1,2}$ et $M_{i,1}$ (eq. 13 and 12)
  for *each dimension* n_i **do**
    Computation of the one-dimensional movement in minimum time (Section 4.3.2)
    Computation of the duration of the one-dimensional movement in minimum time $T_{opt}[i]$
  end
  Determination of the duration of the transition movement
  $T_{imp} = max(\forall\, i \in [1,n] \mid T_{opt}[i])$
  for *each dimension* n_i **do**
    Computation of triplets of cubic curve segments from the method *3-segments* (Section 4.3.3)
  end
end

Figure 18 illustrates an application of the method for the case of a movement defined by three points P_1, P_2, P_3 and by the kinematic constraints $V_{max} = 0.1m/s$, $A_{max} = 0.3m/s$ et $J_{max} = 0.9m/s$. The transition movements are computed for different values of the parameter τ.

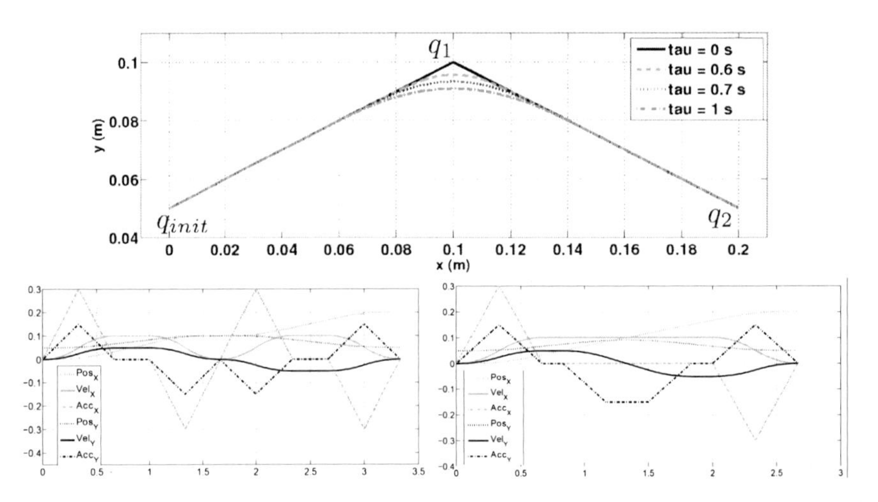

Fig. 18 Transition movement for two lines that form an angle of about 127 (top), graph of position, velocity and acceleration as function of time for a point–to–point motion for $\tau = 0$ (bottom-left) and $\tau = 1s$ (bottom-right).

The proposed method ensures the continuity in velocity and acceleration for each dimension. The initial and final velocities of the transition movements can be different and acceleration not zero. The duration of the transition movement is computed by taking into account the kinematic constraints of each dimension using the minimum time algorithm (Sect. 4.3.2). Therefore this method guarantees that changes in velocity, acceleration and jerk are limited. However, in some cases, constraints can be exceeded by the *3-segment* method. In practice, we introduce a percentage (10%) of exceeding for each constraint. If, for a transition movement, the exceeding of kinematic constraints is too large, this movement is not smoothed to comply with the constraints of human comfort.

4.5 Application to Robot Manipulators

To better explain the method, we apply it to an example of task of grasping an object, the grey tape cassette of Fig. 19. The path of the center of the end effector of the robot (hand) is described by the green line segments in Fig. 20. On this path, the spheres represent the initial, final and intermediate configurations (nodes). The path of the point–to–point trajectory TR_{ptp} is identical to the path planned. This trajectory stops at each intermediate node. The smoothed path TR is represented by the black curve. We note that the trajectory stops at the first node as a smoothing in its neighborhood would have introduced to a collision[2] between the hand of the robot and the environment. The planning of the path of the trajectory was performed in the Cartesian space of the robot by considering the platform was fixed. The following section presents the methodology to take into account the redundancy of the robot.

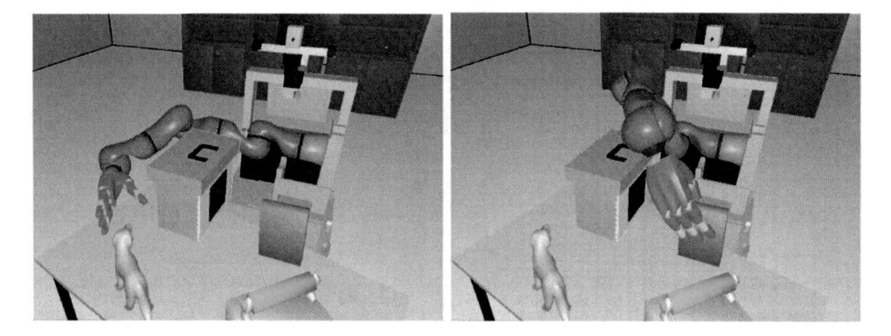

Fig. 19 Initial configuration and grasp configuration of the robot Jido.

[2] Another solution would be to compute a path that goes farer from the obstacle but it is not the purpose here.

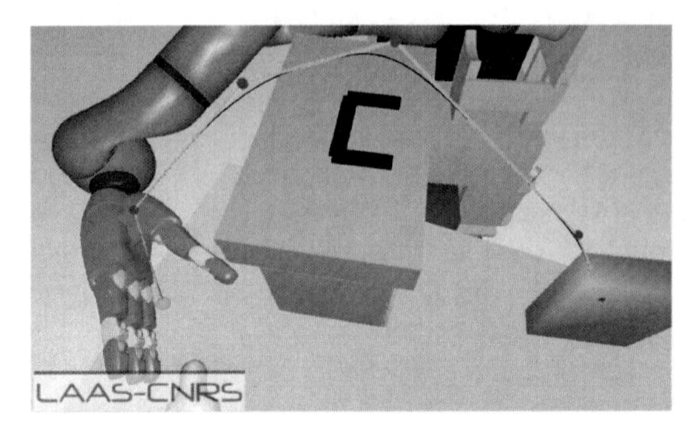

Fig. 20 Trajectories TR_{ptp} et TR in the Cartesian space to grasp the cassette.

4.6 Planning in the Cartesian Space

4.6.1 Generation of the Smoothed Trajectory TR in Cartesian Space

To represent the complete configuration of the robot in Cartesian space, we propose to use a vector X_i with:

- the position of the robot base,
- the pose of the end effector(s),
- the configuration of the redundancy axis of the arms if they have more than six degrees of freedom (DOFs),
- the configuration of the hand(s),
- the configuration of the head.

In the following, we consider that the platform is fixed. For a system operating in 3D space, six independent parameters are used to define the position of the end effector. For the planning, the system is decomposed into *passives* and *actives* parts corresponding respectively to dependent and independent variables [24], [35]. Thus a robot manipulator with six DOFs is decomposed as follows: the independent variables (active) are the six DOFs (position and orientation) of the end effector and the joint variables are the dependent variables (passive) .

In the case of our Jido[3] robot, as the robot arm is composed of seven DOFs, it is therefore redundant. In addition to the pose of the end effector, an articulation of the arm is chosen and becomes an *active* variable. Notice that, f the motion of a holonomic platform was considered, then these DOFs would be *active* variables.

During the planning of the path in the Cartesian space, only the *active* variables are sampled using, according to the circumstances, the RRT or the T-RRT algorithm. The *passive* variables are computed in a second step by solving the inverse kinematics of the arm prior to test the validity of the sampled configuration of the

[3] Jido is an MP-L655 platform from Neobotix, equipped with a KUKA LWR arm.

robot (bounds and collision) (see Sect. 4.1.1). During the test of the validity of a local path between two configurations, the inverse kinematic function is also called.

To perform the interpolation between two configurations, we represent the position of the end effector by a displacement: three parameters for the position and three parameters for the orientation (vector and angle representation with the norm of the vector equal to the angle [12]). We have implemented a local method of interpolation between two configurations. This method takes as parameters two local configurations (with their kinematic conditions) and the imposed kinematic constraints (J_{max}, A_{max} et V_{max}) for each active axis. After applying the local method between each intermediate configuration, the obtained trajectory TR_{ptp} is composed of point–to–point movement of dimension n (n is the number of active axes), that is for Jido $n = 22$ parameters (6 for the end effector, 1 for the axis of the redundant manipulator, 13 for hands and 2 for the head).

The smoothed trajectory TR in Cartesian space is then obtained by the method described in the previous section applied to the active axes (Sect. 4.4).

4.6.2 Conversion of the Trajectory in the Joint Space of the Robot

As most of the robot controllers operate in the joint space, it is important to provide a solution to convert Cartesian trajectories into joint ones. To perform this transformation, the trajectories of passive axes are obtained by discretizing the trajectory TR defined in Cartesian space and performing inverse kinematics for each sample. The trajectory TR is discretized at the period of operation of the robot controller. This allows obtaining the position, and by derivation, the velocity and the acceleration of all the DOFs of the robot.

However, this discretization removes the notion of time and requires a large amount of data to represent the trajectory.

We can use the approximation method of trajectory presented in [11] and [10] to approximate this discretized trajectory and thus obtain a compact description of the trajectory. Unlike the approximation in the Cartesian space, the trajectory error taken into account by the approximation algorithm is the maximum error of the trajectory of each DOF.

The obtained approximated trajectory TR_{app} is a function of time, it is composed of series of segments of cubic curves for each joint variable of the robot.

However, movements of the passive axes are not planned, they can exceed the kinematic limits of the robot. In this case, the trajectory cannot be directly performed. To adapt the trajectory when the task allows it, we replace the time parameter t of the trajectory by applying a function α, $\mathbb{R} \longrightarrow \mathbb{R}$. The function α will make it possible to change the time increment during the execution of the trajectory and therefore allow slowing down the execution.

The period of the trajectory controller is denoted ΔT. In the case of a classical execution, the application α is defined by:

$$\alpha(t) = t \tag{15}$$

or, in discrete notation:

$$\alpha(k\Delta T) = \alpha((k-1)\Delta T) + \Delta T \tag{16}$$

The trajectory carried out is $TR_{app}(\alpha(t))$.

The introduction of the function α makes it possible to modify the motion law of the trajectory TR_{app} and thus to adapt the evolution of each joint of the robot in a synchronized way.

To determine the function α in the case where one wishes to adapt the motion law, we first determine for each instant of the trajectory TR_{app} exceeding β the velocity of each axis relatively to the corresponding maximum velocity (maximum values used here are the default limits accepted by the system). We obtain:

$\forall k\Delta T \in [t_I, t_F]$,

$$\beta(k\Delta T) = \begin{cases} 1 & \text{if } \forall j \in [1,n],\ {}_jV(k\Delta T) \leq_j V_{max}^{mot} \\ min\left(\forall j \in [1,n] \mid \frac{{}_jV_{max}^{mot}}{{}_jV(k\Delta T)}\right) & \text{otherwise} \end{cases} \tag{17}$$

where n is the number of controlled DOFs, ${}_jV(t)$ the evolution of the velocity of the articulation j and ${}_jV_{max}^{mot}$, the maximum velocity of the articulation$j$.

Thus we obtain:

$$\alpha(k\Delta T) = \alpha((k-1)\Delta T) + \beta(k\Delta T)\Delta T \tag{18}$$

with $\alpha(0) = 0$.

However, the trajectory $TR_{app}(\alpha(t))$ cannot be executed directly because it would introduce discontinuities in velocity due to the discontinuity of β. To smooth the evolution of β, we apply a variant of the method described in Sect. 4.3.4 that anticipates the change in β. The smoothed function β is denoted by β_{smooth}. The smoothing is performed in three steps by Algorithm 7.

The method presented above allows modifying the velocity of each joint of the robot to satisfy the velocity bounds. We have supposed that the resulting path respects the constraints of acceleration. Otherwise, it is possible to identify a function β^{acc} equivalent to β to take into account overtaking accelerations. In practice, for HRI, the kinematic constraints of the trajectory are small in comparison to the capabilities of the system and it is not necessary to check for overtaking of acceleration.

In this section we have presented a method to compute a trajectory from a path that can be defined either in the joint space or in Cartesian space of the robot.

Algorithm 7. Smoothing the variation of the motion law

begin

Applying the method in Sect. 4.3.4 on the evolution of β by varying the time from t_I to t_F with a step of ΔT, the resulting curve is named $\beta_{forward}$:

$$\beta_{forward}(k\Delta T) = \begin{cases} \beta(k\Delta T) & \text{if } \beta(k\Delta T) < \beta((k-1)\Delta T) \\ f_{smooth}(\beta) & \text{otherwise} \end{cases} \quad (19)$$

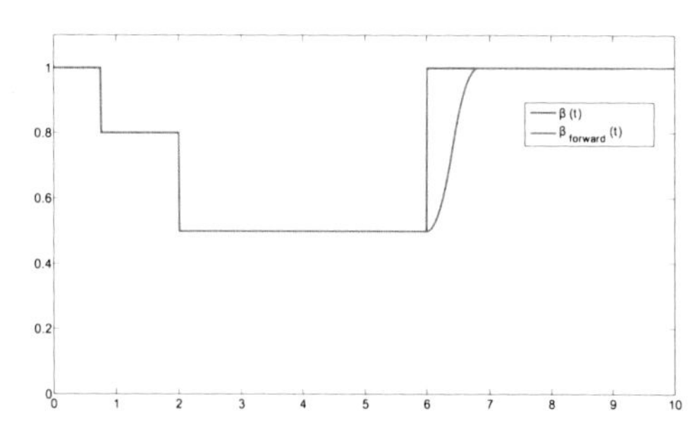

Applying the method in Sect. 4.3.4 on the evolution of β by varying the time from t_F to t_I with a step of $-\Delta T$, the resulting curve is named $\beta_{backward}$:

$$\beta_{backward}(k\Delta T) = \begin{cases} \beta(k\Delta T) & \text{if } \beta(k\Delta T) > \beta((k+1)\Delta T) \\ f_{smooth}(\beta) & \text{otherwise} \end{cases} \quad (20)$$

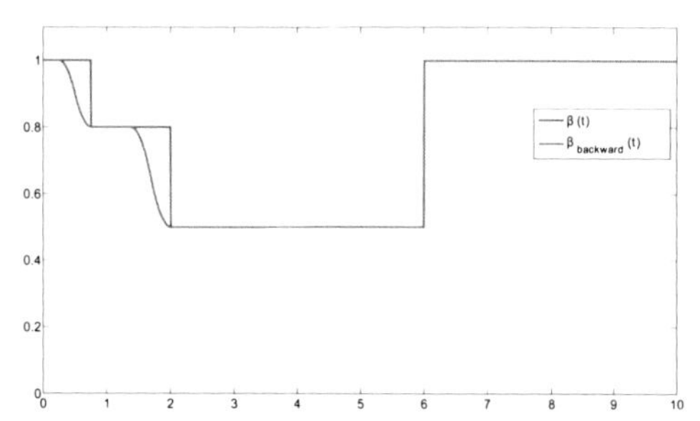

β_{smooth} is finally obtained by taking the minimum between $\beta_{forward}$ and $\beta_{backward}$:

$$\beta_{smooth}(k\Delta T) = min(\beta_{forward}(k\Delta T), \beta_{backward}(k\Delta T)) \quad (21)$$

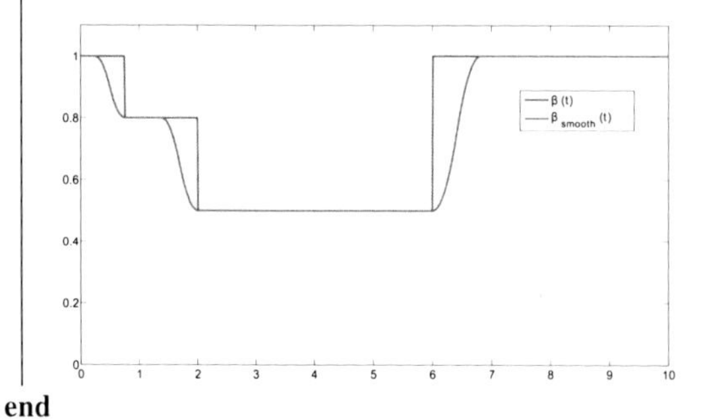

end

4.7 Adaptation of the Motion Law during the Execution

During the trajectory execution, humans located in the workspace of the robot can move, so the robot can put them in danger. We propose to use the geometric models of the robot and of the human, updated at each iteration during the execution to ensure the safety and comfort of humans. We choose to take into account the weighted average *cost* of the security and visibility constraints introduced in Sect. 4.1.2. The method to adapt the motion law is the same as the one presented in the previous section. The costs are high when the human–robot distance is short or when the robot is outside the field of view of the human, the cost taken into account is $cost_{inv} \in [0,1]$ such that:

$$cost_{inv}(k\Delta T) = 1 - cost(k\Delta T) \tag{22}$$

The cost $cost_{inv}$ is then smoothed on-line by the function f_{smooth} presented in Sect. 4.3.4.

When the trajectory $TR(\alpha(t))$ is planned in Cartesian space, the law $\alpha(t)$ is evaluated at each iteration:

$$\alpha(k\Delta T) = \alpha((k-1)\Delta T) + f_{smooth}(min(cost_{inv}(k\Delta T), \beta_{smooth}(k\Delta T)).\Delta T \tag{23}$$

In this section, we have presented a human aware motion planner. In a first part we have introduced some elements to take into account the relative position of the robot and the human, and the human behavior. Using a ramdom motion planner and cost map to represent the human constraints, this motion planner begins by computing a broken line path that is then transformed in feasible trajectories. The trajectory generator allows limiting velocity, acceleration and jerk. This generator is integrated in the motion planner and firstly presented in the case of planning in the configuration space. It is then extended to planning in Cartesian space.

The approach is general and can be applied to complex systems with two hands/arms. We have proposed an original method to convert a Cartesian trajectory in a joint trajectory. Finally, we have presented an approach to modify online the evolution of the time law of the trajectory and shown its usefulness for taking into account the presence of humans during the execution of the movement.

In the next section, we introduce an attentional system to monitor the robot activity from the perspective of the software components.

5 Attentional System

A robotic system designed to physically interact with humans should adapt its behavior to the human actions and the environmental changes in order to provide a safe, natural, and effective cooperation. The human motions and the external environment should be continuously monitored by the robotic system searching for interaction opportunities while avoiding dangerous and unsafe situations. In this context, attentional mechanisms [52, 21] can play a crucial role: they can direct sensors towards the most salient sources of information, filter the available sensory input, and provide implicit sensory-motor coordination mechanisms [41] to orchestrate and prioritize concurrent activities.

In this project, we have deployed an attentional system suitable for balancing the trade off between safe human-robot interaction and effective task execution. The attentional system is to supervise and orchestrate the human-robot interaction activities monitoring their safety and effectiveness. Our attentional execution monitoring system is obtained as a reactive, behavior-based system, endowed with simple, bottom-up, attentional mechanisms. We assume a frequency-based model of the executive attention [15, 16, 13] where each behavior is endowed with an adaptive internal clock that regulates the sensing rate and action activations. The frequency of sensor readings is here interpreted as a degree of attention towards a behavior: the higher the clock frequency, the higher the resolution at which the behavior is monitored and controlled. In particular, we consider robot manipulation tasks providing the attentional monitoring strategies for behaviors like pick and place, give and receive, search and track (humans and salient objects).

5.1 Related Work

Human aware manipulation [59, 61] and human–robot cooperation in manipulation tasks [27] are very relevant topics in HRI literature, however cognitive control and attentional mechanisms suitable for safe and effective interactive manipulation are less explored. A number of recent contributions about close HRI are based on motivational and cognitive models [8]. However, attentional mechanisms in HRI have been mainly investigated focusing on visual and joint attention [49, 8] for social interaction. In contrast, our main concern is on (supervisory) executive attention for monitoring and action orchestration [52, 21]. Attentional mechanisms applied to autonomous robotic systems have been proposed in the literature for vision-based mobile robotics (e.g. [48, 19, 31]), but here we are interested in artificial attentional processes suitable for monitoring the execution of multiple concurrent behaviors in human interaction tasks [14].

5.2 Attentional Model

Our aim is to develop an autonomous robotic system suitable for human-robot interaction in cooperative manipulation tasks. Achieving autonomy and safety in such an environment requires adaptation. For this purpose, we propose to deploy an attentional system, a kind of supervisory attentional system a la [52], to modulate the robotic arm motion and perception. The attentional system is expected to monitor and regulate multiple concurrent activities [41] in order to achieve an effective coordination and interaction with the human movements in the operative space. More specifically, our attentional model combines the following design principles:

- *Behavior-based system.* The executive control is obtained from the interaction of a set of multiple parallel behaviors working at different levels of abstraction.
- *Attentional monitoring.* Attentional mechanisms are to focus monitoring and control activities on relevant internal behaviors and external stimuli.
- *Internal and external sources of salience.* The sources of salience are behavior and task dependent; these can be dependent by either internal states (e.g. resources, processes, goals) or external stimuli (e.g. obstacles, unexpected variations of the environment).
- *Adaptive sensory readings.* For each behavior, the process of changing the rate of sensory readings is interpreted as an increase or decrease of attention towards a particular aspect of the environment the robotic system is interacting with.
- *Emergent attentional behavior.* The overall executive attention should emerge from the interrelations of the attentional mechanisms associated with the behaviors.

5.2.1 A Frequency-Based Model of Attention

The frequency-based model of the executive attention [15, 16] adopted in this chapter can be represented in a schema theory framework in terms of Adaptive Innate Releasing Mechanisms (AIRMs) [16]. In the following we briefly recall this model.

In Fig. 5.2.1, the AIRM is represented through a Schema Theory representation [2], where each behavior is composed of a Perceptual Schema (PS), which reads and processes incoming data from sensors, a Motor Schema (MS), producing commands to be given to motors, and a control mechanism, based on a combination of a *releasing mechanism* [64] and an internal *adaptive clock*. In particular, the releaser acts as a trigger signal that enables or disables the activation of the MS, according to the sensory data $\sigma(t)$. For example, a detected obstacle releases the obstacle avoidance MS. Instead, sensor readings are sampled by the adaptive clock. That is, the robot reads data just when necessary (reducing sensory readings and elaborations) with a period that can change according to the salience of the perceived stimuli. In this way, each behavior can independently monitor the environment and modulate its outputs following the clock frequency changes.

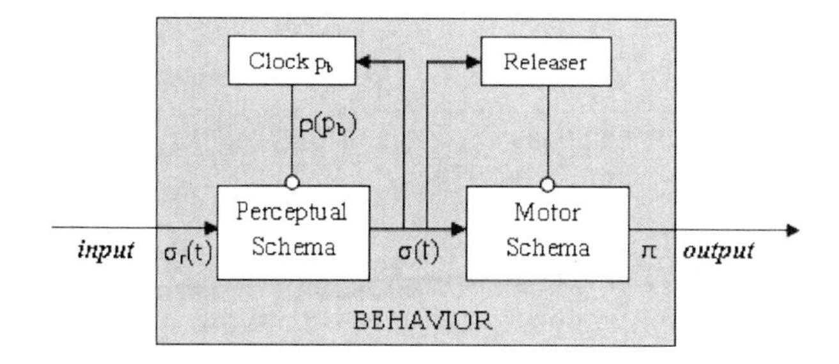

Fig. 21 AIRM Model: each behavior is composed of an adaptive clock, a releasing function, a perceptual schema and a motor schema.

Assuming a discrete time model for the adaptive clock, the way the clock adapts its period is called *monitoring strategy* and is characterized by:

- A period p^t for each behavior, ranging in an interval $[p_{min}, p_{max}]$,
- An *updating function* $f(t) : \mathbb{R}^n \rightarrow \mathbb{R}$ that changes the clock period p^t, according to the parameters the behavior depends on (sensors, internal state, environmental features, and the behavioral goal).
- A trigger function $\rho(t, p^{t-1})$, which enables/disables the data flow $\sigma_r(t)$ from sensors to PS, at every p^{t-1} time unit:

$$\rho(t, p^{t-1}) = \begin{cases} 1, & \text{if } t \bmod p^{t-1} = 0 \\ 0, & \text{otherwise} \end{cases} \tag{24}$$

- Finally, a support function $\phi(f(t)) : \mathbb{R} \rightarrow \mathbb{N}$ that maps the values generated by the updating function $f(t)$ in the allowed range for the period $[p_{min}, p_{max}]$:

$$\phi(x) = \begin{cases} p_{max}, & \text{if } x \geq p_{max} \\ \lfloor x \rfloor, & \text{if } p_{min} < x < p_{max} \\ p_{min}, & \text{if } x \leq p_{min} \end{cases} \tag{25}$$

Now, starting from the clock period at time 0, $p^0 = p_{max}$, the clock period at time t is regulated as follows:

$$p^t = \rho(t, p^{t-1}) * \phi(f(t)) + (1 - \rho(t, p^{t-1})) * p^{t-1}. \tag{26}$$

That is, if the behavior is disabled, the value of the clock period at time t remains unchanged at the previous value p^{t-1}. Instead, when the trigger function is equal to 1, the behavior is activated and, subsequently, its activation period changes according to the $\phi(f(t))$ function.

5.2.2 Attentional HRI

Based on the model introduced above, we have designed a behavior-based control system endowed with attentional monitoring strategies for human-robot interaction.

In this model, the attentional mechanisms regulates the executive system trading off between two conflicting requirements:

- safe interaction with the humans;
- effective cooperation in interactive tasks.

Each requirement is associated with a motivational drive that affects the attentional and executive state of the robotic behavior. The first one corresponds to the fear of hurting people, hence it determines caution, slow movements and intensive monitoring (in case of danger it blocks the robot motion), instead, the second one is associated with a desire to interact with people and manipulate objects, thus this attitude provides an attraction towards moving and close persons or objects.

Depending on the disposition, movements, and the attitude of a person in the robot workspace, each behavior changes its activation frequency, affecting the overall attentional state of the system. In this way, a person walking across the interaction area or a fast movement of a human head (or hand) can modify the behaviors' attentional state causing an accelerated elaboration of the associated perceptual input (human movements) and more frequent behaviors' activations.

Test-bed domain. We have considered a robotic manipulator that is to cooperate with humans in pick-and-place and give-and-receive (hand-over) tasks. Depending on the context, the robotic system should: look for an operator to interact with; give or receive an object to/from the operator; pick or place an object from/into a location. Each of these tasks are to be monitored in order to avoid dangerous/unsafe situations.

In this context, the attentional mechanisms allow us to combine the robot attraction towards human operators (to be effective and cooperative) and the robot repulsion from unexpected events and abrupt environmental changes. For each behavior, the simple perception-action response to an external stimulus may produce different patterns of interactions depending on different internal states of the robot given by the combination of the fear of hurting the user and the desire of helping him.

Environment. In our setting, the robot base is kept fixed (the mobile base is not exploited) and close to a small table where the robot can pick and place objects. Depending on the proximity, we have defined three areas in the workspace: a proximity area which is too close to the robot body and unsafe for HRI; an interaction area, where physical human-robot interaction is possible (here we refer to both visual and physical interaction in the robotic arm workspace); a far workspace area where humans and object are in the robot field of view, but too far for handover tasks.

5.3 Control Architecture

We have designed a control architecture suitable for the primitive interactive manipulation tasks introduced above. The control system integrates modules for forward, inverse kinematics, and visual servoing along with modules for face recognition, hand detection/tracking, object recognition/tracking. Given these functionalities, the attentional state of the robot is affected by the following sources of saliency: face, hands, object detection, proximity.

5.3.1 Attentional Behaviors

The behavior-based architecture is depicted in Fig. 22. This model integrates attentional behaviors for pick and place, give and receive, but also behaviors for search and track (humans and objects) as well as behaviors regulating the avoid attitude of the robotic system.

The robot attentional behavior is obtained as the combination of the following primitive behaviors (see Fig. 22): AVOID, PICK and PLACE, GIVE and RECEIVE, SEARCH and TRACK. For each behavior, we have to define the activation function and the updating policy that represents the associated attentional model.

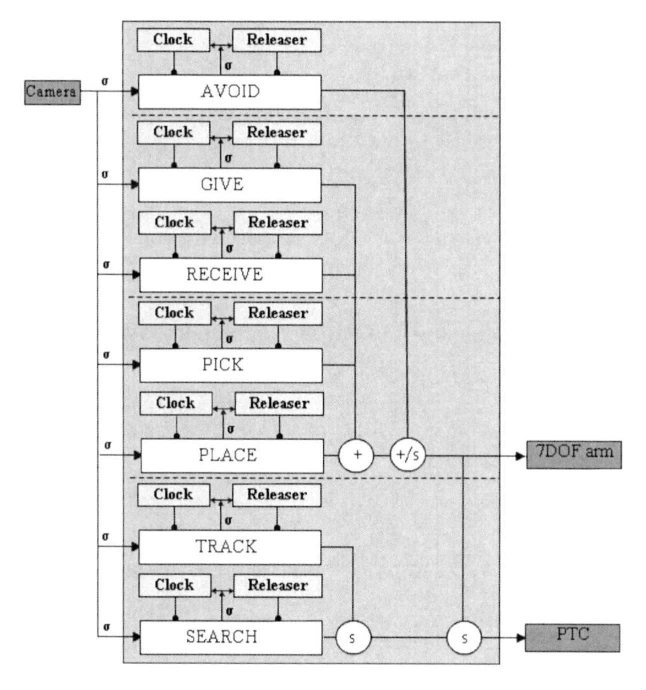

Fig. 22 Behavior-based architecture for HRI.

Behaviors settings. SEARCH controls the pan-tilt (PTU) providing an attentive scan of the environment looking for humans and objects. It is active whenever the robotic system is idling and no interesting things (objects or humans) are in the robot field of view. Its activation is periodic, but not adaptive, hence it is associated with a constant clock:

$$p_{sr}^{t} = k_{sr}. \tag{27}$$

Once a human is detected in the robot workspace (through face detection and/or hand detection), the TRACK behavior is enabled. This behavior allows the robot to monitor human motions before they enter in the interaction space. TRACK focuses the system attention on the operator movements, hence the adaptive clock should be regulated in accordance with the human motion and position. Here, the input signal $\sigma_{hm}(t)$ represents the human distance from the robot camera, in our test-bed it is the minimal distance of human faces and hands. The TRACK clock period changes according to $\sigma_{hm}(t)$ and the increment of $\sigma_{hm}(t)$, that is, the period p_{tr} is updated as follows:

$$p_{tr}^{t} = \Theta_{tr}(\sigma_{hm}(t), \frac{\sigma_{hm}(t) - \sigma_{hm}(t - p_{tr}^{t-1})}{p_{tr}^{t-1}}), \tag{28}$$

where p_{tr}^{t-1} is the period at the previous clock cycle, $\Theta_{tr}(x,y)$ is a function $\Theta_{tr}(x,y) = \phi_{tr}(\alpha x + (1-\alpha)1/y + \beta)$, where α and β are behavior-specific parameters used to weight the importance of position and velocity in the attentional model, while $\phi_{tr}(z)$ is the scaling function that introduces suitable thresholds to keep the clock period within the allowed interval $[p_{tr\_min}, p_{tr\_max}]$. Intuitively, a human that moves fast and close needs to be carefully monitored (high frequency, foreground), while a human that moves far and slow can be monitored in a more relaxed manner (low frequency, background).

The AVOID behavior checks for safety in human-robot interaction, it controls the arm motion speed and can stop the motion whenever a situation is assessed as dangerous. AVOID is enabled when a human is detected in the robot interaction area. It is endowed with an internal clock whose frequency depends on the operator proximity and motion. The associated clock frequency changes proportionally to the situation saliency. That is, if the operator is close and/or its position σ_{op} (i.e. minimal distance of face and hands) becomes closer between successive readings of sensory data, then the clock is accelerated, while it is decelerated if the operator moves away from the robot. The period of this clock changes as follows:

$$p_{av}^{t} = \Theta_{av}(\sigma_{op}, \frac{\sigma_{op}(t) - \sigma_{op}(t - p_{av}^{t-1})}{p_{av}^{t-1}}), \tag{29}$$

where Θ_{av} is defined as for TRACK. The output of this behavior results in a speed deceleration associated with high frequencies:

$$speed = \begin{cases} \frac{max\_speed \times p_{av}^{t}}{p_{av\_max}} & \text{if } prox.sp. < \sigma_{op} \leq inter.sp. \\ 0 & \sigma_{op} \leq prox.space \end{cases} \tag{30}$$

where *speed* is the current speed, *max_speed* is the maximum allowed value for the arm speed, *prox.sp.* and *inter.sp.* are the proximity and the interaction space respectively. Moreover, the arm will stop if the operator is inside the robot proximity space.

The PICK behavior is activated when the robot is not holding an object, but there exists a reachable object in the robot interactive space. PICK moves the robot's end-effector towards the object, activates a grasping procedure and, once the robot holds the object, moves this in a predefined safe position close to the robot body. For PICK, the input signal $\sigma_{obj}(t)$ represents the distance of the object from the robot end effector which can be detected by the stereo camera. In this case the clock period is associated with the distance of the object. That is, the period p^t_{pk} is updated as follows:

$$p^t_{pk} = \phi_{pk}(\alpha\,\sigma_{obj}(t)), \tag{31}$$

with $\phi_{pk}(x)$ is the scaling function used to scale and map $\sigma_{obj}(t)$ in the allowed range of periods $[p_{pk\_min}, p_{pk\_max}]$. Furthermore, the clock frequency determines also speed variations. In particular, the speed is related to the period according to the following relation:

$$speed = \frac{max\_speed \times p^t_{pk}}{p_{pk\_max}}. \tag{32}$$

In this way, the arm moves with *max_speed* at the beginning, when there is free space for movements (and a low monitoring frequency), and smoothly reduces its speed to a minimum value in order to execute a precision grip with more frequent camera information (higher monitoring frequency).

As for PLACE, it is activated when the robot is holding an object in the absence of interacting humans in the interactive space. It moves the robot end effector towards a target position, it places the object and moves the robot arm back to a predefined position close to the robot body. The clock period is regulated by a function analogous to that of (31) with the distance to the target σ_{tr} as the input signal. Also in this case, the speed is decelerated at high clock frequencies according to (32).

The GIVE and RECEIVE behaviors are activated by object and gesture detection. These behaviors are responsible for monitoring and regulating the activities of giving and receiving objects taking into account both the humans' proximity and their movements. In this case, the clock period is associated with the distance of both the objects and the speed of the operator hand. In particular, GIVE is activated when the robot holds an object and perceives a reachable human hand in its operative space. When activated, this behavior moves the end effector in the direction of the operator's hand with a trajectory and velocity which depends on the human's proximity and operator's hand movements. The GIVE sampling rate is regulated by the following function:

$$p^t_{gv} = \Theta_{gv}(\,\gamma_{obj}(\|\sigma_{obj}(t) - ee_{pos}(t)\|), \gamma_{op}(\tfrac{\sigma_{op}(t)-\sigma_{op}(t-p^{t-1}_{gv})}{p^{t-1}_{gv}})\,), \tag{33}$$

where $\sigma_{obj}(t)$ and $ee_{pos}(t)$ are the positions of the object and the end effector at time t, $\sigma_{op}(t)$ is the hand operator position, θ_{gv}, γ_{obj}, γ_{op} are suitable functions defined as follows. The function γ_{obj} sets the period proportional to the object position, i.e., the closer the object, the higher the sampling frequency:

$$\gamma_{obj} = (p_{gv\_max} - p_{gv\_min})\frac{d}{max_d} + p_{gv\_min}, \tag{34}$$

with d, max_d are, respectively, the distance $(\sigma_{obj}(t) - ee_{pos}(t))$ and the maximal distance between the end effector and the object. Instead, γ_{op} depends on the hand speed v (in terms of the incremental ratio of the hand position towards the value of the period), i.e., the higher the speed, the higher the sampling frequency. The following function is used to set and normalize the values within the allowed interval $[p_{min}, p_{max}]$:

$$\gamma_{op} = \begin{cases} (p_{gv\_max} - p_{gv\_min})(1-v) + p_{gv\_min} & \text{if } v \leq 1 \\ p_{gv\_min} & \text{otherwise} \end{cases} \tag{35}$$

Finally, the $\Theta_{gv}(x)$ combines the two functions γ with a weighted sum regulated by an α parameter

$$\Theta_{gv}(x) = \phi_{gv}(\alpha\gamma_{obj} + (1-\alpha)\gamma_{op})), \tag{36}$$

also in this case the resulting period is limited within the allowed interval $[p_{gv\_min}, p_{gv\_max}]$ by the scaling function ϕ_{gv}.

The clock frequency regulates not only the sampling rate, but also the velocity of the arm movements. More specifically, the execution speed is related to the period according to an inversely proportional relation (32). This means that the higher the sampling rate, hence the attention, the slower the hand movement. Intuitively, here we assume that when attention is needed the movement should be more carefully monitored, thus slower.

As for the RECEIVE behavior, it is activated when the robot perceives a human in the operative space holding a reachable object in his/her hand. The behavior sampling rate is regulated by a function analogous to (33) (set with different parameters) with an adaptive velocity inversely proportional to the current period, as in (32).

5.4 *Execution Example*

We now illustrate how the system works in typical interactive situations. In Fig. 23 we plotted part of the execution of the RECEIVE behavior. In particular, Figure 23a represents the variation of the distance between the end effector of the robotic arm and the operator's hand. In the execution cycle 80, the robot has almost reached the human hand, however the operator moves his/her arm away. The execution of the behavior ends at the execution cycle 162 when the robot delivers the object to the operator. Figure 23b represents the hand speed variation of the same execution, as evaluated by the RECEIVE behavior. The hand is almost stationary between the cycle 30 and the cycle 70, then it starts moving with different speeds until it stands still at cycle 162 and receives the object. Finally, Figure 23c represents the activations of the behavior at each cycle. Whenever there is a bar in the plot, this means that the behavior perceptual schema is active. Let us note that both the distance and the

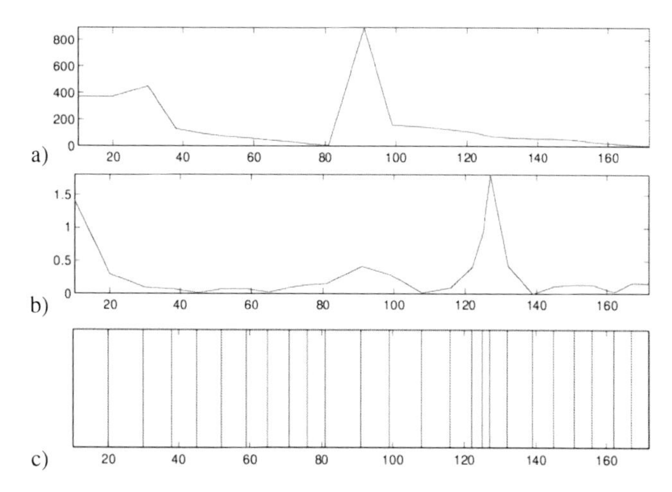

Fig. 23 a) End effector-hand distance; b) Hand speed as evaluated by the Receive Behavior; c) Activations of the Receive behavior.

hand speed are sampled and evaluated only when the behavior perceptual schema is active. The frequency of activation will increase when the distance is small (for example between cycles 40 and 80) or when the hand speed is high (for example between cycles 105 and 125) following the updating function of the behavior.

5.5 Evaluation Criteria and Experimental Results

To evaluate the performance of the attentional system and of the HRI system, we introduce some evaluation criteria considering safety, reliability, effectiveness and efficiency.

- *Safety* is measured counting dangerous human-robot interaction events (i.e. a safe robot should avoid collisions between human and a moving robot and it should minimize interactions where the two are too close).
- *Reliability* is evaluated considering unrecoverable world/robot states encountered during the tests (the robot is stuck, the object falls down, the object is not reached or located by the robot).
- *Effectiveness* is assessed considering the time needed to achieve the task (the system should minimize the time to achieve the task).
- *Efficiency* is associated with the number of behavior activations needed to achieve the task (for us, an attentional system is efficient, when it can distribute computational resources among different processes, focusing only on relevant activities).

Parameters Setting. Given the attentional model introduced in the previous section, the overall attentional behavior is obtained once we tune the parameters associated with the behaviors' monitoring strategies.

To assess the performance of the system with respect to the previous set of criteria we introduce a suitable optimization function:

$$f = M1 \times N_{Safe} + M2 \times N_{Rel} + M3 \times T_{Effe} + M4 \times N_{Effi}.$$

Here, $M_1 > \cdots > M_4$ specify the priorities in terms of weights; N_{Rel} represents the number of unrecoverable situations with respect to the number of accomplished activities (pick, place, etc.); N_{safe} the HRI unsafe situations with respect to the executed activities; T_{Effe} is for the time spent to achieve the tasks with respect to the overall mission time; N_{Effi} is for the number of behavior activations with respect to the maximal possible activations (for each behavior p_{min}).

This function can be exploited, on the one hand, to learn the system parameters and, on the other hand, to validate the overall system behavior. Different learning algorithms can be deployed for parameter learning (e.g. genetic algorithms, particle swarm optimization, simulated annealing etc.), currently, we are investigating Differential Evolution algorithms (DE) [62] which are particularly suitable for both boundedness and granularity problems, indeed DE manages unrestricted and unbounded range of values. More details about DE methods used to set attentional monitoring strategies can be found in [13].

Experimental Setup. In order to evaluate the performance of the AIRM architecture we compare it with a classical non-rhythmic architecture (P1Vmed) in which the behaviors perceptual schema are always active. For the adaptive version (AIRM) we consider adaptive concurrent clocks with $p_{min} = 1$, $p_{max} = 10$ and $speed = \frac{max-speed \times p}{p_{max}}$ for all the behaviors. For the (P1Vmed), we assume that the behaviors' perceptual schema are always active (i.e., p_{min} and p_{max} are both equal to 1) and the arm speed is set to a constant value ($speed = \frac{max-speed}{2}$). Moreover, in the case of the AIRM architecture, the updating policies of the behaviors are those specified in the previous section. The range of values for the speed is $[0;0.3]$ m/s. For the experiments, we have used the robotic platform available in the PRISCA Lab, endowed with a 7DOF robotic arm (Cyton Arm by Energid: payload 300 g, hight 60 cm, reach 48 cm, joint speed 60 rpm), a gripper (size 3.25 cm) as end effector, and a kinetic device. In this context, the proximity, interaction, and workspace distances were set, respectively, at 20 cm from the robot body, $10-50$ cm, and 50 cm to 6 m.

Empirical Results. During the empirical evaluation, we have tested each behavior 20 times with 5 different operators unaware of the robot behavior. Operators are required to observe the robot and move around in the case of Pick and Place behaviors, and interact, without any specific requirement, for the Give and Receive behaviors. In these final cases all the hand movements, made by the operators, were spontaneous. For each test we have evaluated the parameters defined above: effectiveness, efficiency, reliability, and safety.

Notice that not only are the attentional mechanisms associated with better performance in terms of effectiveness and efficiency (Fig. 24 and Tab. 1), but we have

Table 1 Evaluation of the Effectiveness, Efficiency, Reliability, and Safety criteria.

	Effectiveness		Efficiency		Reliability		Safety	
	AIRM	P1VMED	AIRM	P1VMED	AIRM	P1VMED	AIRM	P1VMED
Receive	$7.66s \pm 0.54s$	$9.69s \pm 0.31s$	14.5 ± 1.57	41.7 ± 1.42	100%	100%	100%	100%
Give	$4.87s \pm 1.4s$	$7.27s \pm 2.9s$	6.05 ± 2.65	14.65 ± 5.59	83%	80%	90%	84%
Pick	$9.14s \pm 2.07s$	$10.48s \pm 0.67s$	16.2 ± 6.58	32.65 ± 5.99	77%	54%	100%	100%
Place	$6.03s \pm 1.05s$	$8.96s \pm 0.6s$	12.95 ± 5.03	58.65 ± 10.17	100%	100%	100%	100%

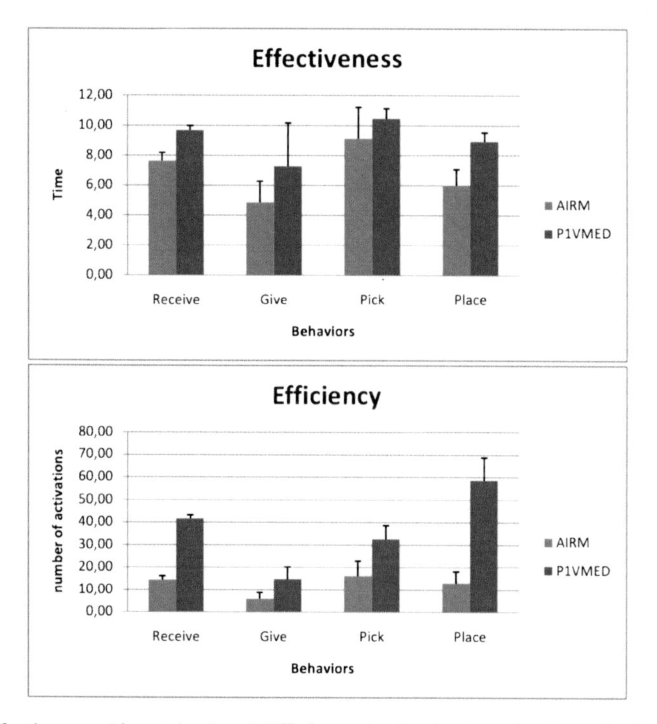

Fig. 24 Effectiveness (time taken) and Efficiency (activations) evaluation criteria.

also observed better results regarding reliability and safety (Tab. 1) compared with the non-adaptive architecture in which the perceptual schemas are always active (P1Vmed). In particular, the adaptive modulation of the robotic arm speed allows us to accomplish the task faster than keeping the speed to a constant value, furthermore the adaptive trajectory is safer and more comfortable from the operator point of view. As we expected, a small number of activations has a big impact in the efficiency for the adaptive system. This is particularly evident in the Place behavior, since interaction and precision are not requested, the task can be accomplished with minimal activations. The critical operations for the Safety and Reliability are the Give and Pick operations. As for safeness, we have observed that the Give interaction requires more care in HRI (where the robot has to pass an object to the operator) than the Receive one (where the robot has to receive an object from the operator) causing more frequent unsafe interactions. The same happens for reliability, indeed, passing an object to a human is more difficult than receiving an object. Although in these cases the success rate is not equal to 100% (as in the cases of Receive and Place behaviors), the architecture endowed with AIRMs seems more reliable than the P1Vmed standard architecture. For example, in the picking behavior the slower speed of the adaptive architectures permits a more accurate grip of the object.

In this section, we have illustrated a human–robot interactive system endowed with attention mechanisms used to coordinate simple manipulation tasks. In the proposed attentional model, each behavior is equipped with an adaptive clock and an updating policy that changes the frequency of sensory readings (focusing the attention towards relevant aspects of the external environment) and modulates the emergent behavior in terms of variations of the robot arm speed. We have defined a simple control architecture for HRI considering pick-and-place and give-and-receive attentional behaviors. To assess the system performance we have also introduced suitable evaluation criteria taking into account safety, reliability, efficiency, and effectiveness. The role of the attentional system is to find a trade-off between safety, effectiveness, and reliability in human-robot interaction and cooperation.

6 Conclusion

We have presented some concepts to build an interactive robot capable to share the workspace with humans and to become a companion or a co-worker. We have focused on the exchange of object between human and robot, which is a fundamental task for HRI. We have addressed firstly the architecture aspect and shown the importance of the communication between the different software modules. The grasp planning has then been studied to exhibit the importance of double grasp to exchange an object.

Then, we have presented how to plan and adapt robot motion to take into account the human and his/her behavior. The motion planner we have presented produces trajectories as series of cubic functions in joint or Cartesian space. This trajectory can then be adapted or modified to cope with the human changes in real time.

The last section has presented an attentional mechanism used to coordinate manipulation tasks. We have shown its interest to trade off between safety, effectiveness, and reliability in HRI and cooperation.

Of course, this approach of HRI is not complete and lots of points still need to be investigated like the exchange of information, the manipulation of more complex objects with two hands or the accomplishment of more complex tasks. For example, for a robot and a human intuitively exchange an object, they must exchange tactile information and the robot must be capable to generate and understand this information. This point is a challenge to build reliable and efficient robots that pick-and-give or receive-and-place objects in industrial environment. So, we can see that to realize simple daily tasks in interaction with human, robots need a lot of functionalities from human attention systems and supervision to tactile dialog and control. But such a robot cannot only share the space safely with humans but also do tasks for humans or help humans do tasks in an intuitive way. In this sense, the main result of this work is to have demonstrated the possibility to build intuitive and safe manipulator robots.

Acknowledgements. The research leading to these results has been supported by the DEXMART Large-scale integrating project, which has received funding from the European Communitys Seventh Framework Programme (FP7/2007-2013) under grant agreement ICT-216239. The authors are solely responsible for its content. It does not represent the opinion of the European Community and the Community is not responsible for any use that might be made of the information contained therein.

The authors would like to thank Jean-Philippe Saut for his participation to the grasp planner, Emrah Akin Sisbot for the human aware planner, Mokhtar Gharbi for the manipulation planner, Matthieu Herrb and Anthony Mallet for their help to develop software and hardware components and all others.

References

1. Alami, R., Albu-Schäffer, A., Bicchi, A., Bischoff, R., Chatila, R., De Luca, A., De Santis, A., Giralt, G., Hirzinger, G., Lippiello, V., Mattone, R., Sen, S., Siciliano, B., Tonietti, G., Villani, L.: Safe and dependable physical human-robot interaction in anthropic domains: State of the art and challenges. In: IEEE/RSJ International Conference on Intelligent Robots and Systems, Workshop on Physical Human–Robot Interaction, Beijing (2006)
2. Arbib, M.A.: Schema theory. In: Arbib, M.A. (ed.) The Handbook of Brain Theory and Neural Networks, pp. 830–834. MIT Press, Cambridge (1998)
3. Berchtold, S., Glavina, B.: A scalable optimizer for automatically generated manipulator motions. In: IEEE/RSJ International Conference on Intelligent Robots and Systems, Sendai (2004)
4. Berenson, D., Diankov, R., Nishiwaki, K., Kagami, S., Kuffner, J.: Grasp planning in complex scenes. In: 7th IEEE-RAS International Conference on Humanoid Robots, Pittsburgh, PA (2007)
5. Bicchi, A.: On the closure properties of robotic grasping. International Journal of Robotics Research 14, 319–334 (1995)
6. Bicchi, A., Tonietti, G.: Fast and soft arm tactics: Dealing with the safety-performance trade-off in robot arms design and control 11(2), 22–33 (2004)
7. Bounab, B., Sidobre, D., Zaatri, A.: Central axis approach for computing n-finger force-closure grasps. In: IEEE International Conference on Robotics and Automation, Pasadena, CA (2008)
8. Breazeal, C.: Designing Sociable Robots. MIT Press, Cambridge (2002)
9. Brock, O., Khatib, O.: Real-time re-planning in high-dimensional configuration spaces using sets of homotopic paths. In: IEEE International Conference on Robotics and Automation, San Francisco, CA (2000)
10. Broquère, X.: Planification de trajectoire pour la manipulation d'objets et l'interaction homme-robot, Ph.D. thesis, LAAS-CNRS and Université Paul Sabatier (2011)
11. Broquère, X., Sidobre, D.: From motion planning to trajectory control with bounded jerk for service manipulator robots. In: IEEE International Conference on Robotics and Automation, Anchorage, AK (2010)

12. Broquère, X., Sidobre, D., Herrera-Aguilar, I.: Soft motion trajectory planner for service manipulator robot. In: IEEE/RSJ International Conference on Intelligent Robots and Systems, Nice (2008)
13. Burattini, E., Finzi, A., Rossi, S., Staffa, M.: Attentive monitoring strategies in a behavior-based robotic system: An evolutionary approach. In: International Conference on Emerging Security Technologies, Canterbury, UK (2010)
14. Burattini, E., Finzi, A., Rossi, S., Staffa, M.: Attentional human-robot interaction in simple manipulation tasks. In: 7th ACM/IEEE International Conference on Human Robot Interaction, Boston, MA (2012)
15. Burattini, E., Rossi, S.: A robotic architecture with innate releasing mechanism. In: 2nd International Symposium on Brain, Vision and Artificial Intelligence, Napoli (2007)
16. Burattini, E., Rossi, S.: Periodic adaptive activation of behaviors in robotic system. International Journal of Pattern Recognition and Artificial Intelligence 22, 987–999 (2008)
17. Cakmak, M., Srinivasa, S.S., Lee, M.K., Forlizzi, J., Kiesler, S.: Human preferences for robot-human hand-over configurations. In: IEEE/RSJ International Conference on Intelligent Robots and Systems, San Francisco, CA (2011)
18. Cakmak, M., Srinivasa, S.S., Lee, M.K., Kiesler, S., Forlizzi, J.: Using spatial and temporal contrast for fluent robot-human hand-overs. In: 6th International Conference on Human–Robot Interaction, New York (2011)
19. Carbone, A., Finzi, A., Orlandini, A., Pirri, F.: Model-based control architecture for attentive robots in rescue scenarios. Autonomous Robots 24, 87–120 (2008)
20. Ciocarlie, M., Goldfeder, C., Allen, P.: Dimensionality reduction for hand-independent dexterous robotic grasping. In: IEEE/RSJ International Conference on Intelligent Robots and Systems, San Diego, CA (2007)
21. Cooper, R., Shallice, T.: Contention scheduling and the control of routine activities. Cognitive Neuropsychology 17, 297–338 (2000)
22. Cornellà, J., Suárez, R.: Fast and flexible determination of force-closure independent regions to grasp polygonal objects. In: IEEE International Conference on Robotics and Automation, Barcelona (2005)
23. Cornellà, J., Suárez, R.: Determining independent grasp regions on 2D discrete objects. In: IEEE/RSJ International Conference on Intelligent Robots and Systems, Edmonton (2005)
24. Cortés, J., Siméon, T.: Sampling-based motion planning under kinematic loop-closure constraints. In: 6th International Workshop on Algorithmic Foundations of Robotics, Utrecht (2004)
25. Diankov, R., Ratliff, N., Ferguson, D., Srinivasa, S., Kuffner, J.: Bispace planning: Concurrent multi-space exploration. In: Robotics: Science and Systems, Zurich (2008)
26. Ding, D., Liu, Y.H., Wang, S.: The synthesis of 3D form-closure grasps. In: IEEE International Conference on Robotics and Automation, San Francisco, CA (2000)
27. Edsinger, A., Kemp, C.C.: Human–robot interaction for cooperative manipulation: Handling objects to one another. In: 16th IEEE International Symposium on Robot and Human Interactive Communication, Jeju Island, Korea (2007)
28. Ferguson, D., Stentz, A.: Anytime RRTs. In: IEEE/RSJ International Conference on Intelligent Robots and Systems, Beijing (2006)
29. Ferrari, C., Canny, J.: Planning optimal grasps. In: IEEE International Conference on Robotics and Automation, Nice (1992)
30. Flash, T., Hogan, N.: The coordination of arm movements: an experimentally confirmed mathematical model. Journal of Neuroscience 5, 1688–1703 (1985)
31. Frintrop, S., Jensfelt, P., Christensen, H.I.: Attentional landmark selection for visual slam. In: IEEE/RSJ International Conference on Intelligent Robots and Systems, Beijing (2006)

32. Goldfeder, C., Allen, P., Lackner, C., Pelossof, R.: Grasp planning via decomposition trees. In: IEEE International Conference on Robotics and Automation, Roma (2007)
33. Haddadin, S., Albu-Schäffer, A., Hirzinger, G.: Requirements for safe robots: Measurements, analysis and new insights. International Journal of Robotics Research 28, 1507–1527 (2009)
34. Hall, E.T.: A system for the notation of proxemic behavior. American Anthropologist 65, 1003–1026 (1963)
35. Han, L., Amato, N.: A kinematics-based probabilistic roadmap method for closed chain systems. In: Donald, B.R., Lynch, K.M., Rus, D. (eds.) Algorithmic and Computational Robotics: New Directions, pp. 233–246. A.K. Peters, Wellesley (2001)
36. Harada, K., Kaneko, K., Kanehiro, F.: Fast grasp planning for hand/arm systems based on convex model. In: IEEE International Conference on Robotics and Automation, Roma (2008)
37. Huber, M., Rickert, M., Knoll, A., Brandt, T., Glasauer, S.: Human–robot interaction in handing-over tasks. In: 17th International Symposium on Robot and Human Interactive Communication, München (2008)
38. Huebner, K., Ruthotto, S., Kragic, D.: Minimum volume bounding box decomposition for shape approximation in robot grasping. In: IEEE International Conference on Robotics and Automation, Roma (2008)
39. Ikuta, K., Ishii, H., Nokata, M.: Safety evaluation method of design and control for human-care robots. International Journal of Robotics Research 22, 281–297 (2003)
40. Jaillet, L., Cortés, J., Siméon, T.: Sampling-based path planning on configuration-space costmaps. IEEE Transactions on Robotics 26, 635–646 (2010)
41. Kahneman, D.: Attention and Effort. Prentice-Hall, Englewood Cliffs (1973)
42. Koay, K.L., Sisbot, E.A., Syrdal, D.A., Walters, M.L., Dautenhahn, K., Alami, R.: Exploratory study of a robot approaching a person in the context of handling over an object. In: Association for the Advancement of Artificial Intelligence Spring Symposia, Palo Alto, CA (2007)
43. LaValle, S.M., Kuffner, J.J.: Rapidly-exploring random trees: Progress and prospects. In: 4th International Workshop on the Algorithmic Foundations of Robotics, Hanover, NH (2000)
44. Mainprice, J., Sisbot, E., Jaillet, L., Cortés, J., Siméon, T., Alami, R.: Planning human-aware motions using a sampling-based costmap planner. In: IEEE International Conference on Robotics and Automation, Shanghai (2011)
45. Marler, R., Rahmatalla, S., Shanahan, M., Abdel-Malek, K.: A new discomfort function for optimization-based posture prediction. In: Digital Human Modeling for Design and Engineering Conference, Seattle, WA (2005)
46. Miller, A., Allen, P.: Graspit! a versatile simulator for robotic grasping. IEEE Robotics & Automation Magazine 11(4), 110–122 (2004)
47. Miller, A., Knoop, S., Christensen, H., Allen, P.: Automatic grasp planning using shape primitives. In: IEEE International Conference on Robotics and Automation, Taipei (2003)
48. Mitsunaga, N., Asada, M.: Visual attention control for a legged mobile robot based on information criterion. In: IEEE/RSJ International Conference on Intelligent Robots and Systems, Lausanne (2002)
49. Nagai, Y., Hosoda, K., Morita, A., Asada, M.: A constructive model for the development of joint attention. Connection Science 15, 211–229 (2003)
50. Nguyen, V.D.: Constructing force-closure grasps. In: IEEE International Conference on Robotics and Automation, San Francisco, CA (1986)

51. Nonaka, S., Inoue, K., Arai, T., Mae, Y.: Evaluation of human sense of security for coexisting robots using virtual reality. 1st report: Evaluation of pick and place motion of humanoid robots. In: IEEE International Conference on Robotics and Automation, New Orleans, LA (2004)
52. Norman, D., Shallice, T.: Attention in action: Willed and automatic control of behaviour. In: Davidson, R., Schwartz, R., Shapiro, D. (eds.) Consciousness and Self-Regulation: Advances in Research and Theory IV, pp. 1–18 (1986)
53. Ott, C., Eiberger, O., Friedl, W., Bäuml, B., Hillenbrand, U., Borst, C., Albu-Schäffer, A., Brunner, B., Hirschmüller, H., Kielhöfer, S., Konietschke, R., Suppa, M., Wimböck, T., Zacharias, F., Hirzinger, G.: A humanoid two-arm system for dexterous manipulation. In: 6th IEEE-RAS International Conference on Humanoid Robots, Genova (2006)
54. Ponce, J., Sullivan, S., Sudsang, A., Boissonnat, J.D., Merlet, J.P.: On computing four-finger equilibrium and force-closure grasps of polyhedral objects. International Journal of Robotics Research 16, 11–35 (1997)
55. Roa, M., Suárez, R.: Independent contact regions for frictional grasps on 3D objects. In: IEEE International Conference on Robotics and Automation, Pasadena, CA (2008)
56. Santello, M., Flanders, M., Soechting, J.F.: Postural hand synergies for tool use. Journal of Neuroscience 18, 10105–10115 (1998)
57. Saxena, A., Driemeyer, J., Ng, A.Y.: Robotic grasping of novel objects using vision. International Journal of Robotics Research 27, 157–173 (2008)
58. Sisbot, E.A., Marin-Urias, L., Broquère, X., Sidobre, D., Alami, R.: Synthesizing robot motions adapted to human presence. International Journal of Social Robotics 2, 329–343 (2010)
59. Sisbot, E.A., Clodic, A., Alami, R., Ransan, M.: Supervision and motion planning for a mobile manipulator interacting with humans. In: 3rd ACM/IEEE International Conference on Human Robot Interaction, Amsterdam (2008)
60. Sisbot, E.A., Marin-Urias, L.F., Alami, R., Siméon, T.: Human aware mobile robot motion planner. IEEE Transactions on Robotics 23, 874–883 (2007)
61. Sisbot, E.A., Marin-Urias, L.F., Broquère, X., Sidobre, D., Alami, R.: Synthesizing robot motions adapted to human presence — A planning and control framework for safe and socially acceptable robot motions. International Journal of Social Robotics 2, 329–343 (2010)
62. Price, K., Storn, R.: Differential evolution — A simple and efficient heuristic for global optimization over continuous spaces. Journal of Global Optimization 11, 341–359 (1997)
63. Suarez, R., Roa, M., Cornella, J.: Grasp quality measures, Technical Report, Universitat Politecnica de Catalunya (2006)
64. Tinbergen, N.: The Study of Instinct. Oxford University Press, Oxford (1951)
65. Todorov, E., Ghahramani, Z.: Analysis of the synergies underlying complex hand manipulation. In: 26th Annual International Conference of the IEEE Engineering in Medicine and Biology Society, San Francisco, CA (2004)
66. Xue, Z., Zöllner, J.M., Dillmann, R.: Grasp planning: Find the contact points. In: IEEE International Conference on Robotics and Biomimetics, Sanya, China (2007)
67. Yershova, A., LaValle, S.: Deterministic sampling methods for spheres and so(3). In: IEEE International Conference on Robotics and Automation, New Orleans, LA (2004)
68. Zinn, M., Khatib, O., Roth, B., Salisbury, J.K.: Playing it safe [human-friendly robots]. IEEE Robotics & Automation Magazine 11(2), 12–21 (2004)
69. Zucker, M., Kuffner, J., Branicky, M.: Multipartite RRTS for rapid replanning in dynamic environments. In: IEEE International Conference on Robotics and Automation, Roma (2007)

Innovative Technologies for the Next Generation of Robotic Hands

Gianluca Palli, Claudio Melchiorri, Gabriele Vassura, Giovanni Berselli, Salvatore Pirozzi, Ciro Natale, Giuseppe De Maria, and Chris May

Abstract. With the aim of reproducing the grasping and manipulation capabilities of humans, many robotic devices have been developed all over the world in more than 50 years of research, starting from very simple grippers, normally used in industrial activities, to very complex anthropomorphic robotic hands. Unfortunately, the reduced functionality and/or reliability of the devices developed so far prevent, together with the cost, their usability in unstructured environments, and in particular in human everyday activities. The adoption of design solutions inherited from conventional mechanics and the lack of purposely developed sensors and actuators are among the main causes of the partial fail in achieving the final goal of reproducing human manipulation capabilities. Our research activity aims at developing innovative solutions concerning the mechanical design, the sensory equipment and the actuation system for the implementation of anthropomorphic robotic hands with improved reliability, functionality and reduced complexity and cost, considering also aspects related to safety during human–robot interaction, paving the way toward the next generation of robotic hands.

Gianluca Palli · Claudio Melchiorri
Dipartimento di Elettronica Informatica e Sistemistica,
Alma Mater Studiorum Università di Bologna, viale Risorgimento 2, 40136 Bologna, Italy
e-mail: {gianluca.palli,claudio.melchiorri}@unibo.it

Gabriele Vassura · Giovanni Berselli
Dipartimento di Ingegneria delle Costruzioni Meccaniche Nucleari Aeronautiche e di Metallurgia, Alma Mater Studiorum Università di Bologna, viale Risorgimento 2, 40136 Bologna, Italy
e-mail: {gabriele.vassura,giovanni.berselli}@unibo.it

Salvatore Pirozzi · Ciro Natale · Giuseppe De Maria
Dipartimento di Ingegneria dell'Informazione, Seconda Università degli Studi di Napoli, via Roma 29, 81031 Aversa (CE), Italy
e-mail: {salvatore.pirozzi,ciro.natale,giuseppe.demaria}@unina2.it

Chris May
Lehrstuhl für Prozessautomatisierung, Universität des Saarlandes, Campus A5 1, 66123 Saarbrücken, Germany
e-mail: c.may@lpa.uni-saarland.de

B. Siciliano (Ed.): Advanced Bimanual Manipulation, STAR 80, pp. 173–218.
springerlink.com

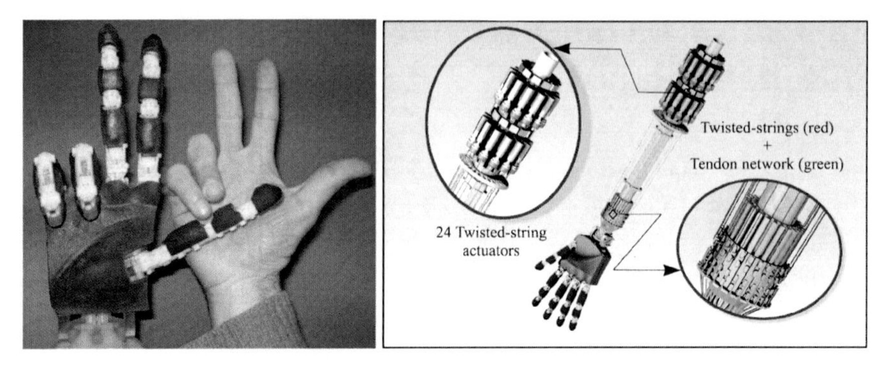

Fig. 1 The DEXMART Hand prototype and its CAD virtual design including 24 twisted string actuators.

1 Introduction

Historically, the reproduction of the human functional capabilities and appearance can be considered among the main reasons for the development of robotic hands. Nevertheless, despite the relevant efforts for the study and design of robotic hands developed all over the world [46], the reproduction of human capabilities in terms of dexterous manipulation seems still far from being achieved.

The aim of the DEXMART project is to develop a robotic bimanual manipulation system able to operate in a unstructured human-like environment and to interact safely with humans. In such a case, the robotic end-effector is expected to provide high flexibility and adaptability to the environment and to the manipulated objects, ideally replicating the overall functionality of the human hand. Therefore, issues such as dexterity, anthropomorphism, sensing capability and human-like motion become fundamental. The general perception, however, is that current hands, especially those aiming at an anthropomorphic aspect, are too complex, bulky and unreliable to really represent effective solutions. Note that, besides their mechanical or electronic/sensing features, also general tools for their use and programming are missing, making their application to real tasks even more difficult.

Many robot hands have been designed in the past, trying often to reproduce or enhance specific features of the human hand. Researches have been interested in designing hands with reduced number of actuators [12], with anthropomorphic aspect [16], with high dexterity potentialities [15], with very high speed [49], or compliance [8], or many other specific aspects. As a matter of fact, dozens of different designs have been proposed, and it is difficult to learn some lessons from this wide scenario. Some of the robot hand prototypes developed so far possess rigid and hard structures and complicated sensori-motor systems, design solutions being mainly based on non-biologically inspired mechanics, with abundance of gears, pulleys, bearings, and similar hardware, nonetheless the sensing devices are often inherited from other fields and not purposely suited for this particular application. This "classical" approach leads to efficient devices that are yet very complex, expensive and many times not sufficiently reliable.

Within the DEXMART project, the simplification of the robot hand mechanics is obtained by the introduction of non conventional joints, compliant structures and transmission systems based on tendons, together with the use of innovative actuation systems like the twisted strings principle [53]. This innovative anthropomorphic robot hand is the fourth version of the UB Hand and has been called DEXMART Hand, see Fig. 1 where a prototype of this innovative device is shown. The adopted design choices imply the use of an appropriate sensory equipment, purposely designed to fit into the DEXMART Hand structure from the point of view of both the mechanical and the electronic integration. With the aim of improving the grasping capabilities of the DEXMART Hand, purposely designed soft pads have been developed for the integration into the hand structure. Moreover, the adoption of suitable control strategies for the compensation of the side effects given by these design choice must be considered. This fact is a direct consequence of the needs in terms of increased reliability and reduced costs, that moves the complexity of the system from the time-consuming mechanical design to the easy-reprogrammable device control strategies.

In this chapter, a general overview of the design solutions and innovative devices and technologies that have been developed within the DEXMART project for the implementation of a new generation of robotic hands will be given.

2 DEXMART Hand Design

The general design philosophy that has been followed within the DEXMART project is to aim at reduction of the device complexity and the costs. In particular, the DEXMART Hand has been inspired by the following driving issues:

- To adopt an endoskeletal structure articulated by means of non conventional joints, sliding or compliant.
- To actuate the joints by means of remotely located actuators with tendon-based transmissions routed by sliding paths (*sliding tendons*) integrated within the finger structure.
- To exhibit surface compliance through a purposely designed soft cover mimicking the human dermal-epidermal layers.
- To reduce manufacturing and assembly complexity by systematic parts integration adopting proper advanced materials and technologies (e.g. polymers and additive manufacturing technologies like Fused Deposition Manufacturing or stereo-lithography).
- To reduce weight and cost of the overall hand system, increasing its "affordability".

A general view of the present DEXMART Hand prototype is shown in Fig. 1, whereas in Fig. 2 an insight view of the design of this innovative robot hand is shown. The development of this innovative anthropomorphic robot hand has been

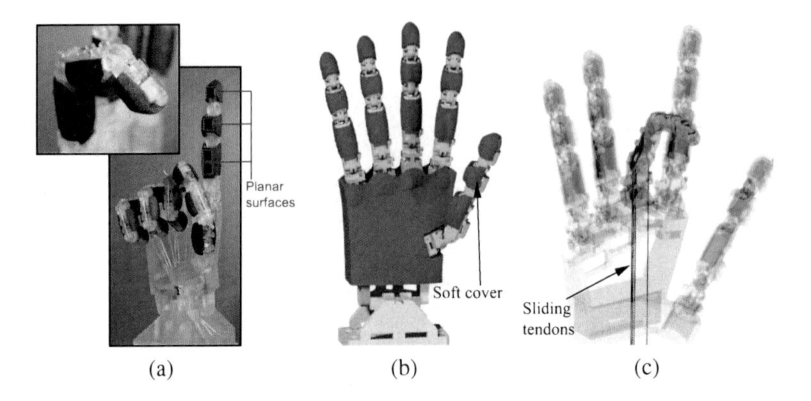

Fig. 2 Prototype of the DEXMART Hand with enhanced surface compliance (a). Hand and palm soft cover (b). Sliding tendons network (c).

started from the design of the finger searching for the maximum achievable integration between the various components, like the mechanical structure, the sensors, the electronics, the actuation and the soft pads, in the perspective of a structural simplification, allowing one-step monolithic manufacturing and consequent reduction of the assembly complexity.

Particular attention has been given to the design of the joints to simplify this fundamental component, avoiding the use mechanical parts, such as bearings and similar hardware, that may cause problems in the integration of the sensors and the tendon network in the proximity of the joint itself. In general, joint design is heavily dependent on the inseparable binomial material-technology. At present, different technologies and a wide range of materials (including lightweight metal alloys) can be used in order to produce the articulated finger structure in a single production step (*fully integral finger*). Such technologies include CNC machining, plastic molding (such as Shape Deposition Manufacturing (SDM)), Selective Laser Sintering (SLS), Fused Deposition Modeling (FDM), Stereo-Lithography (SLA) and Electron Beam Melting (EBM). Nevertheless, recent advances in the plastic materials technology suggest that the use of polymers might be well suited for the production of artificial hands once a lightweight, relatively economical solution is sought. For instance, plastic materials recently developed for SLA and FDM are beginning to offer acceptable performance and costs allowing the production of complex joint shapes (as it is remarked by the introduction of plastic grippers obtained though FDM within the robotic industry).

In detail, two concepts for the development of fully integral fingers have been explored:

- Monolithic fingers with integral Compliant Joints (CJs) made of the same material of the phalanx structure (Fig. 3(b)). It can be recalled that a CJ consists

in a flexible region that provides displacement (rotational and/or translational) between two rigid parts through material deformation.
- Fingers with pin joints integrated into the phalanx body (Fig. 3(c)) simply consisting in a plastic shaft which slides on a cylindrical surface (Integrated Pin Joints (IPJs)).

In both cases, tendons are routed through a series of sliding paths which are obtained directly within the finger structure (sliding tendons, Fig. 2(c), 3(b), 3(c)) as will be detailed in Sect. 4.1. A complete analysis of the tendon transmission modeling, control and material selection is reported in [51]. As for the IPJs, in spite of the sliding contacts, the joint shows very good reliability. On the other hand, stiction and dynamic friction may deteriorate the open-loop position control of the finger and can lead to mechanism locking as the contact pressure between the shaft and the hub increases (due to increased tendon traction). To prevents this undesired effects, suitable control strategies will be adopted.

Clearly, the use of *large-displacement* CJs [29, 39] within the hand endoskeleton is very attractive as long as it can allow the generation of very slender and light mechanisms that are more safe, robust to impact and better respect the goal of reproducing biological structures. In addition, the CJ can store energy during the actuation phase, restituting it during the return stroke. In such a way, as demonstrated in [46, 48], the joint can be actuated by means of one single-acting backdrivable actuator instead of a couple of agonistic-antagonistic ones (with obvious advantage in terms of weight and cost). Despite this consideration and after the evaluation of several solutions for the implementation of the finger joints, the IPJ solution has been selected because of its simplicity and reliability, see Fig. 3(c) where a detail of the joints implemented in the DEXMART Hand finger's is shown. The benefits of IPJ when compared to traditional kinematic pairs (like bearing couplings) include the simplification of the manufacturing and assembly process ensuring size and weight reduction in spite of a larger friction that will be compensated by control [14].

Since the tendons go from the forearm, where the actuators are placed, to the fingers traversing the wrist, particular attention has been given in the design of this crucial component of the hand. Also in this case both CJs and IPJs have been evaluated, together with a solution based on a combination of CJs and conjugated profiles, but also in this case IPJs have been preferred because of their simplicity and reliability.

3 Sensorial Apparatus

Generally speaking, a robotic hand can be equipped with position/velocity sensors for the measurement of the joints and/or actuators configuration, with force sensors for measuring the force applied on the wrist, the palm, the phalanges and/or the joints, and with tactile sensors for reconstructing the pressure map during the contact (usually restricted to the fingertip) by means of an array of sensible elements. In some limited cases, also proximity and single pressure sensors are part of the robot

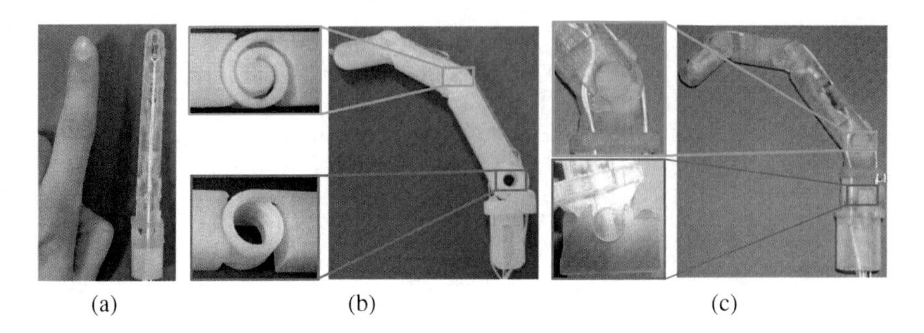

Fig. 3 Fully integral finger: dimensions when compared to human finger (a), finger with integrated compliant joints (b), finger with pin joints integrated into the phalanx.

hand sensory equipment. Since the first prototypes of robotic hands, it was clear how important were the requirement of limited invasiveness for every kind of sensor to be integrated into the device. From the first solutions provided for tendon-driven fingers [35] only limited progress has been made [33] and the sensor solutions adopted appear still cumbersome and quite invasive.

For many reasons, the use of sensible elements with intrinsic high immunity to electromagnetic disturbances and with limited requirements in terms of both conditioning electronics and amplification is preferable to improve the reliability, to ease the miniaturization and the integrated design, and to simplify the sensory subsystem. This is the main reason why, within the DEXMART project, the use of optoelectronic components has been widely studied for all the sensors needed in the DEXMART Hand. In particular a LED and PhotoDetector (PD) couple with wide angle-of-view has been adopted for the implementation of the joint position sensors [18] described in Sect. 3.1, whereas components with narrow angle-of-view have been used for the force sensors [54, 55] reported in Sect. 3.2. Finally, many different principles can be used for the implementation of tactile sensors [40], but the problem is still the integration of all the electronics and acquisition system within the limited space of the fingertip. To solve this issue, a tactile sensor based on discrete SMD optoelectronic components has been developed [58]. This sensor allows acquiring information directly on the deformation, due to the contact forces, of the soft pads mounted above the sensible grid without any amplification circuit. A detailed description of this sensor is reported in Sect. 3.3.

The adoption of optoelectronic components for the implementation of all the sensors allows achieving a fundamental simplification of the condition and acquisition electronics of the robotic hand, exploiting the same conditioning circuit for all the sensors and avoiding the use of amplifiers. The scheme of the acquisition electronics adopted for all the sensors developed for the DEXMART Hand is reported in Fig. 4: the data collected from the sensors are, in this way, directly digitalized and transmitted through a digital SPI bus to the hand control system.

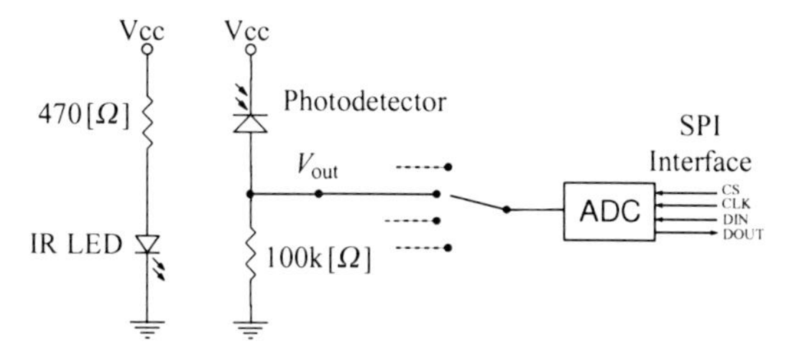

Fig. 4 Measuring circuit for the sensors based on optoelectronic components.

3.1 *Angular Sensor Based on LED-Photodetector Couple*

The proposed angular position sensor is based on a LED-PD couple, mounted to two contiguous phalanges of a DEXMART Hand's finger. When the joint between the considered phalanges flexes, the photocurrent measured by the PD changes with the angular displacement, since it depends on the distance among the two components and on their mutual orientation. In order to reduce sensitivity to ambient light in actual applications of the sensor, the components have selected with relative radiant intensity and relative spectral sensitivity located in the infrared wavelength range. The reduced working distance prevents from using the optical radiation patterns of the devices reported by the datasheets and thus an experimental model is proposed. The obtained model is then used to select the optimal placement of the components over the phalanges. A calibration technique, based on an optical motion capture system, is described and applied. Finally, the sensor performance is assessed by estimating its repeatability and the obtainable linearity by inverting the calibration curve via software. Also, the noise characteristic is evaluated showing a signal-to-noise level higher than 60dB.

3.1.1 Working Principle

Figure 5(a) shows a single joint of a finger in the rest position. The LED and the PD are placed respectively on the first and second phalanx attached to the joint, facing each other with mechanical axes overlapping.

In this state a certain amount of light emitted by the LED reaches the PD and it is proportionally converted into an electrical current, I_0. As the joint starts to flex (see Fig. 5(b)), the mechanical axes of the emitter and of the receiver experience some angular displacement. In this new condition a different amount of light will be sensed by the PD and converted into a current different from I_0. This happens because the radiation pattern of the LED varies with the observation angle, so that the receiver detects different values of radiant flux in the two cases. At the same time also the way the PD senses received light varies, as its relative position and orientation respect to the source change according to its responsivity pattern. The combination of these two effects leads to the observed variations of the photocurrent.

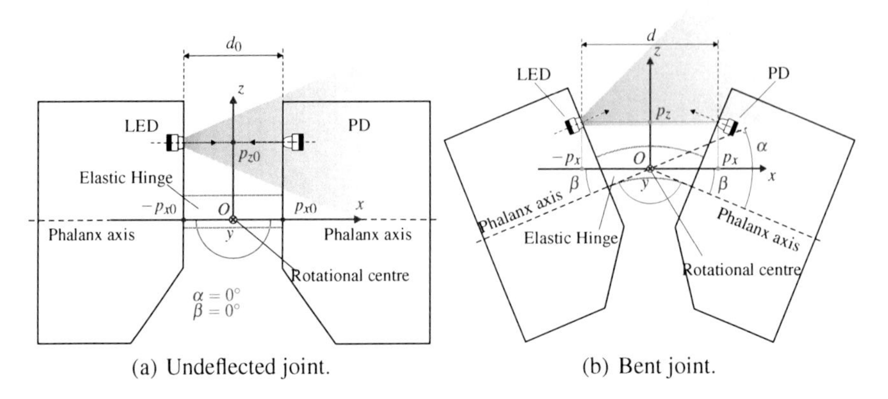

(a) Undeflected joint. (b) Bent joint.

Fig. 5 Schematics of the angular sensor using a LED-PD couple.

Being the electrical current of the PD (the photocurrent) variable with the flexure angle of the joint, it is possible to introduce the function $I(\alpha)$. If $I(\alpha)$ is a monotonic function of its argument, a measure of its value is uniquely associated to a value of the joint rotation angle α. Thus if the power of light emitted by the LED is kept constant and the PD current is measured, the flexure angle of the joint at every instant can be reconstructed. A LED-PD couple is needed for each joint whose angular displacement has to be detected.

Figure 5 shows that the LED and the PD have a fixed displacement from the axis of the phalanx to which they are attached. The position of the emitter and the detector can be specified with respect to a fixed Cartesian coordinate system $Oxyz$. The origin O of the system is chosen to coincide with the center of rotation of the joint, the x-axis overlapping the axes of the phalanges when $\alpha = 0°$. Assuming the devices to lie symmetrically with respect to the z-axis, their angle-dependent coordinates are (p_x, p_z) for the PD and $(-p_x, p_z)$ for the LED. Their variation with α has been subtended. When the joint is in its rest position ($\alpha = 0°$), it is $p_x = p_{x0}$ and $p_z = p_{z0}$. As α varies both p_x and p_z vary, together with the distance d between the tips of the LED and PD. In particular

$$d(\beta) = 2p_x(\beta) = 2\left[p_{x0}\cos(\beta) + p_{z0}\sin(\beta)\right], \tag{1}$$

which for $\beta = \alpha = 0°$ gives the initial distance between the devices $d = d_0 = 2p_{x0}$. The initial positioning of the devices (in terms of p_{x0} and p_{z0}) represents a degree of freedom that can be used to alter the d-β characteristic and in particular its monotonicity, consequently changing the sensitivity of the sensor.

Owing to the symmetry of the system, with elementary geometric considerations on the basis of Fig. 5, when the joint exhibits a certain bending angle α, the mechanical axes of both the LED and the PD form an angle β with the x axis. The symmetric positioning of the components with respect to the center of rotation of the joint is not a mandatory condition to have a well working sensor. The

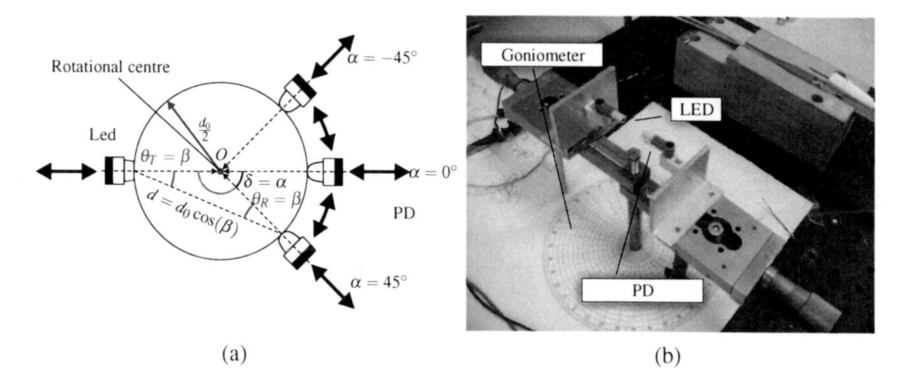

Fig. 6 (a) Sketch of the measurement apparatus used to sample the function $F(d,\beta)$. (b) Measurement apparatus used to sample the function $F(d,\beta)$.

choice made here only simplifies the geometric considerations used to optimize the positioning and the calibration of the sensor. In general the couple LED-PD can be used as an angular displacement sensor also without a symmetric positioning.

3.1.2 Modeling

Due to the limited distance between the LED and the PD, the modelling of the proposed sensor has been carried out on an experimental basis rather than on a physical basis. In detail, the photocurrent has been measured for different values of the geometrical parameters characterizing the distance and the mutual orientation of the LED and the PD. A function interpolating measured data is then found. In general, the photocurrent is a function of three variables: the distance d between the tips of the devices and the two angular displacements θ_T and θ_R. Let this function be $G(d,\theta_T,\theta_R)$. As observed before, the symmetry of the system imposes the constraints $\theta_T = \beta$ and $\theta_R = -\beta$. Hence, the problem reduces to look for another function of the sole variables d and β, let it be $F(d,\beta)$.

A preliminary testing phase has been performed to select the components among various commercially available devices. The selection has been made taking into account the nominal beam angle of the LED and the acceptance angle of the PD so as to guarantee an acceptable sensitivity in the whole angular range of interest. The selected devices used for experimental implementation are branded Avago Technologies Inc. and are spectrally matched with an infrared peak wavelength of 875 nm. Such a choice in terms of wavelength range guarantees a sufficient robustness against ambient light. The LED (manufacturer code number HSDL-4400) is an AlGaAs flat-top light emitting diode featuring a nominal beam angle of 110°, an on-axis radiant intensity of 6 mW/sr corresponding to a 100 mA maximum forward current, and a bandwidth of about 7 MHz. The PD (manufacturer code number HSDL-5420) is a domed PIN photodiode characterized by a 28° acceptance angle and a nominal optical bandwidth of 50 MHz.

A suitable apparatus has been realized allowing to alter the spatial configuration of the devices and register the correspondent photocurrent values, i.e. samples of the function $F(d,\beta)$. A schematic of the measurement system is depicted in Fig. 6(a) while in Fig. 6(b) a picture is shown. As these figures show, the tips of both devices are kept on a circumference of diameter d_0 and center O. The diameter d_0 can be varied by means of two micropositioning stages to which the LED and PD are respectively attached. At every measurement step the same amount of linear displacement Δ is applied on both stages, resulting into an increment of 2Δ over d_0. However, while the LED is simply translated, the PD is also rotated around O by an angle δ. The mechanical axes of both devices pass through the center of the circumference so that the source and the receiver are both looking at O. It can be noted that in such a system $\theta_T = \theta_R$ for every value of δ. This is exactly the constraint needed for the sampling of $F(d,\beta)$. Moreover, from Fig. 6(a) it is evident that $\delta = \alpha = 2\beta$, which can be measured using a goniometer. The distance d between the tips of the devices is instead indirectly measured by knowing d_0 as $d = d_0 \cos(\beta)$. In conclusion, by imposing the values of the parameters d_0 and α, the presented apparatus allows setting a couple of values of the independent variables d and β. Recalling that $\alpha = 2\beta$, it follows that

$$F(d,\beta) = H(d_0,\alpha)|_{d_0=d/cos(\beta),\alpha=2\beta} \tag{2}$$

In correspondence to each couple (d_0,α) a value of the photocurrent is measured and point by point the function $H(d_0,\alpha)$ is sampled over a grid of points obtained by scanning $\alpha \in [-90°,90°]$ with a step of 2°, and $d_0 \in [4.5,14.5]$ mm with a step of 1 mm. Finally the function $F(d,\beta)$ is reconstructed by means of (2). The measured samples have been interpolated through a universal approximator based on a fuzzy system with 3 rules for the input d_0 and 7 rules for the input α using Gaussian membership functions [74]. Figure 7(a) shows the measured data and the interpolating surface, where the photocurrent has been normalized with respect to its maximum. Negative values of α were included to investigate the symmetry of the radiation process. The slight asymmetry of the function with respect to $\alpha = 0°$ is attributable to a small asymmetry present in the measurement apparatus itself and to the misalignment between mechanical and optical axes of the components.

3.1.3 Optimal Positioning

The proposed model is general enough to allow for an optimization of the geometrical parameters p_{x0} and p_{z0}, in order to obtain the best sensitivity for the sensor over the range of interest for angular displacements. In view of (1), for fixed values of p_{x0} and p_{z0}, the surface $F(d,\beta)$ is reduced to a curve $F(d(\beta),\beta;p_{x0},p_{z0})$, where the design parameters p_{x0} and p_{z0} appear explicitly. In order to optimize the placement of the devices, a cost function has to be defined taking into account the sensor sensitivity, i.e.

$$\sigma(\beta) = -\frac{1}{2}\frac{d}{d\beta}F(d(\beta),\beta;p_{x0},p_{z0}). \tag{3}$$

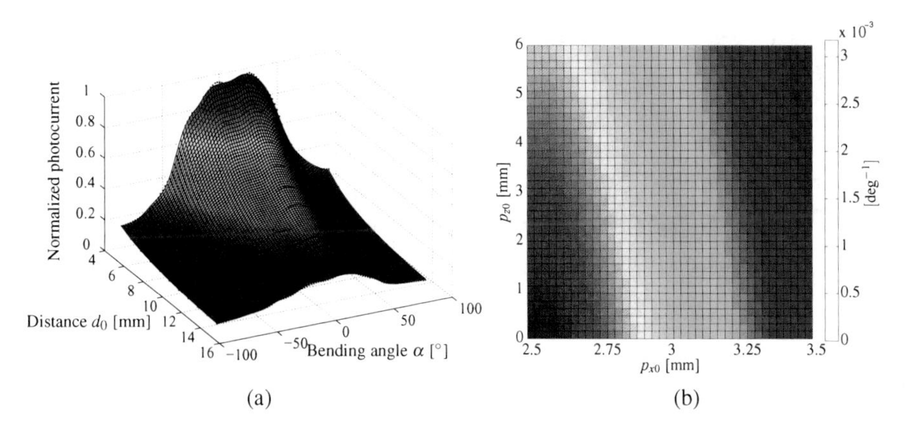

Fig. 7 (a) Measured data and $H(d_0,\alpha)$ interpolating surface. (b) Cost function J values for optimal positioning.

Note that for the optimal positioning of the devices only a symmetric half of the surface $F(d,\beta)$ is necessary and the sign of $\sigma(\beta)$ has to be chosen in order to obtain positive values. Therefore defining $m(p_{x0},p_{z0}) = \min_{\beta\in[0°,45°]}\sigma(\beta)$, the cost function to be maximized can be defined as

$$J(p_{x0},p_{z0}) = \begin{cases} 0, & \text{if } m(p_{x0},p_{z0}) \leq \overline{\sigma} \\ m(p_{x0},p_{z0}) - \overline{\sigma}, & \text{if } m(p_{x0},p_{z0}) > \overline{\sigma} \end{cases} \tag{4}$$

where $\overline{\sigma}$ is the minimal desired sensitivity. As a consequence, the optimal location is identified as $(p^*_{x0},p^*_{z0}) = \arg\max_{p_{x0},p_{z0}} J(p_{x0},p_{z0})$. Figure 7(b) shows a color scaling representation of the cost function J with the couple (p_{x0},p_{z0}) that varies for all values physically admissible to mount the devices on a joint of the DEXMART Hand. It is evident that the optimal position is $p_{x0} = 2.5$ mm and $p_{z0} = 0$ mm.

3.1.4 Sensor Characterization

The components have been mounted to the DEXMART Hand's joint according to the optimal position computed in the previous section. Fig. 8(a) reports a picture of the mounted sensor. The sensor has been calibrated using a Vicon 460 optical motion capture system. An optical marker has been attached to each phalanx contiguous to the joint with the mounted devices. A set of experiments has been made with the motion capture system in order to reconstruct the 3D space position of each marker when the joint is flexed. First of all, the center of rotation for the considered joint has been estimated using a least mean square algorithm to identify the circumference in the three-dimensional space that is the best fit for the measured marker position (see Fig. 8(b)). The same measured marker coordinates have been transformed into the joint angle α using the estimated center of rotation and elementary trigonometric

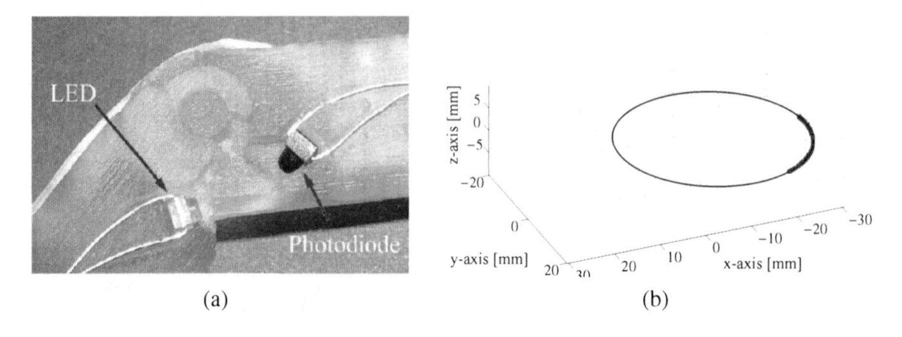

Fig. 8 (a) Detail of the sensor mounting. (b) Circumference that best fits measured marker coordinates.

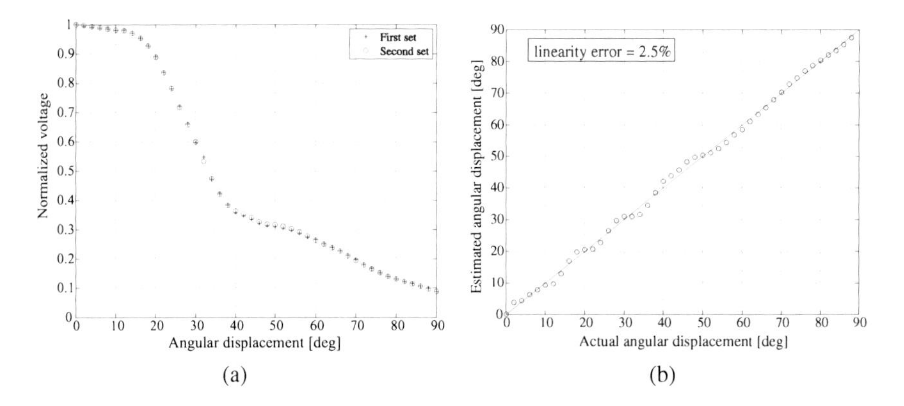

Fig. 9 (a) Calibration curve of the sensor. (b) Evaluation of sensor linearity.

relations. Two different measurements have been carried out to calibrate the sensor varying the angle from 0° to 90° and vice versa so as to evaluate also repeatability of the sensor. The obtained measurements are reported in Fig. 9(a) showing the reconstructed angle α versus the voltage measured on the output of the PD circuit. The repeatability appears quite satisfactory and also no drift is experienced after several repetitions of the measurement. The sensor linearity can be easily improved by inverting the calibration curve via software. In fact, estimating the angular displacement through the 7-th order polynomial interpolating the calibration data, a maximum linearity error of 2.5% is obtained as shown in Fig. 9(b).

3.2 Optoelectronic Force Sensor

Since the dawn of robotics, the availability of joint torque sensors improve the dynamic performance of the servomechanisms that are basic components of any robot, from the industrial manipulators to antropomorphic robotic hands [61, 31, 41, 8, 3].

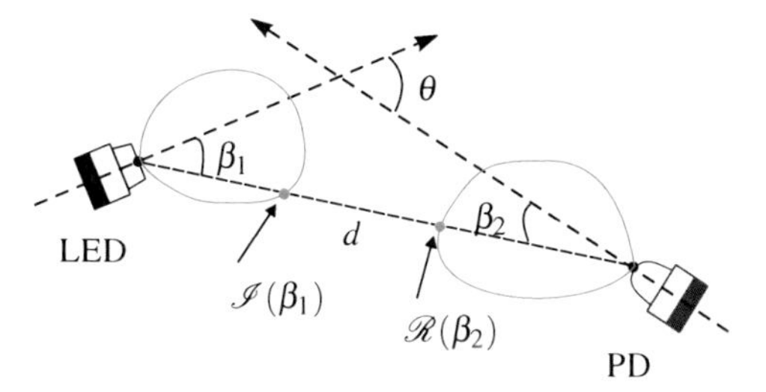

Fig. 10 Working principle of the proposed force sensor.

The main reason can be attributed to the rejection capabilities of disturbance torques acting on the transmission chain of the motion from the motor to the load via torque feedback. The most relevant disturbance affecting transmission systems, and in particular tendon-based ones, is known to be the dry friction, and when a force/torque feedback is not available, the only possibility is the friction compensation usually based on more or less accurate friction models [1]. Unfortunately, most of the friction compensation algorithms require a good model of the phenomenon whose parameters are usually very difficult to identify.

An important field of application of force/torque sensors is the control of robotic hands, in fact such complex systems are specifically designed to allow the robot to interact with the environment, usually very unstructured and so generic that a safe interaction can be ensured only if the mechanism possesses a compliant behaviour. Such a compliance has to be provided by force/torque control, hence finger joint force/torque sensors appear mandatory [43, 15, 26]. The same objective exists for tendon-driven artificial hands which are the most used solutions for prosthetic applications [17] and when the torque/force control could be useful not only for overcoming friction and other disturbances but specifically also for reproducing nonlinear characteristic of human-like tendons [28].

The proposed solution makes use of an optoelectronic components couple, an infrared LED and a PD, mounted on a compliant frame that is deformed under the action of the tendon tension. The compliant frame is a monolithic metallic structure suitably designed to obtain an angular displacement of the optical axes of the optoelectronic components linearly proportional to the force applied in the direction of the tendon. Compliant mechanisms are often used also for the implementation of linear actuators using smart materials [7], for the development of compliant transmission systems [50] or as displacement/force magnifiers [25, 42]. The designed compliant frame has been mounted between the actuation module and the tendon that drives the phalanx.

3.2.1 Working Principle

Figure 10 shows a LED and a PD positioned with an initial relative angle between their mechanical axes $\theta = \beta_1 + \beta_2$. In this figure, β_1 represents the angle between

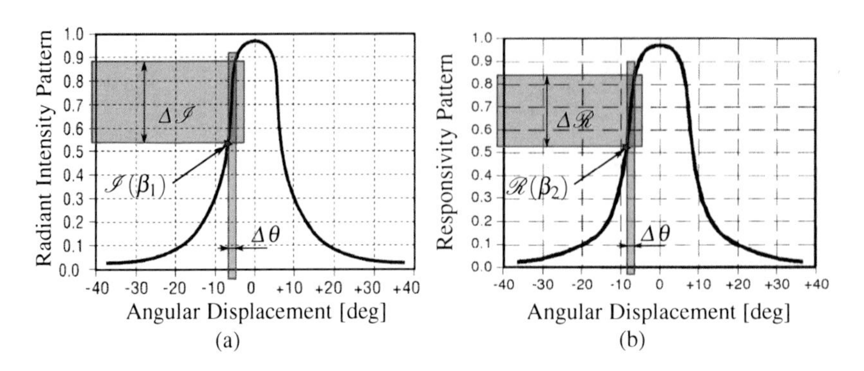

Fig. 11 Characteristics of the Honeywell optoelectronic components taken from datasheets: normalized LED radiant intensity pattern (a) and normalized PD responsivity pattern (b).

the LED mechanical axis and the segment that indicates the distance d from the tip of the PD to the tip of the LED, and β_2 representing the angle between the PD mechanical axis and the same segment. In this state a certain amount of light emitted by the LED reaches the PD and it is proportionally converted into an electrical current I_θ.

When the angle θ between the mechanical axes of the emitter and of the receiver experiences a variation with respect to its initial value, a different amount of light will be sensed by the PD and converted into a current different from I_θ. This happens because the radiation pattern of the LED varies with the angle β_1, so that the receiver detects different values of radiant flux for different values of β_1. At the same time, also the way the PD weights received light varies, according to the variations of its responsivity pattern with β_2. The combination of these two effects leads to the observed variations of the photocurrent. Recalling the theory on LED radiation patterns [37], it is possible to model the system in order to optimize the design of the sensor, selecting initial relative angle θ between the mechanical axes of the two devices. In particular, if the distance d is large enough to render the far-field approximation valid, the LED could be regarded as a point source. In this case, the photocurrent I_θ (and thus the received radiant flux by the PD) will be proportional to the product between the radiant intensity pattern of the LED, evaluated in β_1 (denoted as $\mathscr{I}(\beta_1)$) and the responsivity pattern of the PD, evaluated in β_2 (denoted as $\mathscr{R}(\beta_2)$[1]), and inversely proportional to the square of the distance d

$$I_\theta \approx \frac{\mathscr{I}(\beta_1)\mathscr{R}(\beta_2)}{d^2}. \tag{5}$$

For the specific sensor here presented, optoelectronic components with a very narrow angle of view have been chosen with the aim of obtaining a large sensitivity of the sensor with a very limited angular variations. In particular the selected LED,

[1] The φ-dependence of the radiation and responsivity patterns is here omitted since the devices only move within a plane at constant φ.

manufactured by Honeywell (code SEP8736), is an aluminum gallium arsenide infrared emitting diode molded in a side-emitting smoke gray plastic package. The selected PD, manufactured by Honeywell (code SDP8436), is an NPN silicon PD molded in a black plastic package. The LED and PD have a nominal beam angle of 10° and 18°, respectively. They are mechanically and spectrally matched with a peak wavelength of 880 nm. Figure 11 reports the characteristics of the LED and of the PD, taken from the datasheets. In these figures, the large variation of radiant intensity pattern and responsivity pattern of the selected optoelectronic components over a very limited variation of the angular displacement in a suitably selected region has been highlighted. This characteristics have been exploited to implement the force sensor based on these optoelectronic components. As shown in Fig. 13(a) that will be detailed later, the LED and the PD has been mounted with an initial relative angle $\theta = 15^\circ$, such that the no-load working point ($\mathscr{I}(\beta_1)$ and $\mathscr{R}(\beta_2)$ for the LED and the PD respectively) is located in the lower part of the response characteristics indicated by the blue stars in Fig. 11. By rotating the axes of the optoelectronic components of an angle $\Delta\theta$, the relative response of the LED and of the PD changes of $\Delta\mathscr{I}$ and $\Delta\mathscr{R}$ respectively, as highlighted in windows of Fig. 11. These changes can be detected by measuring the output voltage V_{out} of the simple circuit shown in Fig. 4.

As simplifying assumption adopted during the compliant frame design, the characteristics of the optoelectronic components have been considered linear within the region of interest highlighted in Fig. 11 and the compliant frame has been designed to achieve a variation of the angle between the optical axes of the optoelectronic components linearly proportional to the external traction force applied to the sensor within the range of interest ($0 \sim 80$ N). To achieve the symmetry of the structure, the compliant frame has been split into two identical parts, each of those can be seen as a compliant Slider-Crank Mechanism (SCM), see Fig. 12(a), and the optoelectronic components have been mounted on only one of this two parts. Each part can be schematized as reported in Fig. 12(a), where the scheme of the half compliant frame equivalent pseudo-rigid SCM is reported. In the compliant frame, the linear torsional springs K_1, K_2 and K_3 have been implemented by means of corner-filleted Flexural Hinges (FHs) [45], anyway due to the very small fillet radius, the FHs have been designed considering the simple beam model (flexural pivot [29]). From Fig. 12(a) it is clear that the tendon tension causes the slider movement in the upward direction. For a detailed description of the compliant frame design please refer to [54].

A suitable Finite Element Method (FEM) analysis has been performed with the aim of both verifying the accuracy of the design procedure and to check if the limits of the adopted materials (Aluminum 7075-T6) in terms of yield strength are satisfied. The measured elongation is reported in Fig. 12(b): it is worth noting that also the FEM analysis confirms the effectiveness of the proposed design. The final dimensions of the selected compliant frame are reported in Fig. 13, whereas a picture of the manufactured frames and a sensor prototype provided with the optoelectronic components are reported in Fig. 14(a) and 14(b) respectively. The frames has been manufactured by means of wire electro-erosion adopting aluminum 7075-T6. The

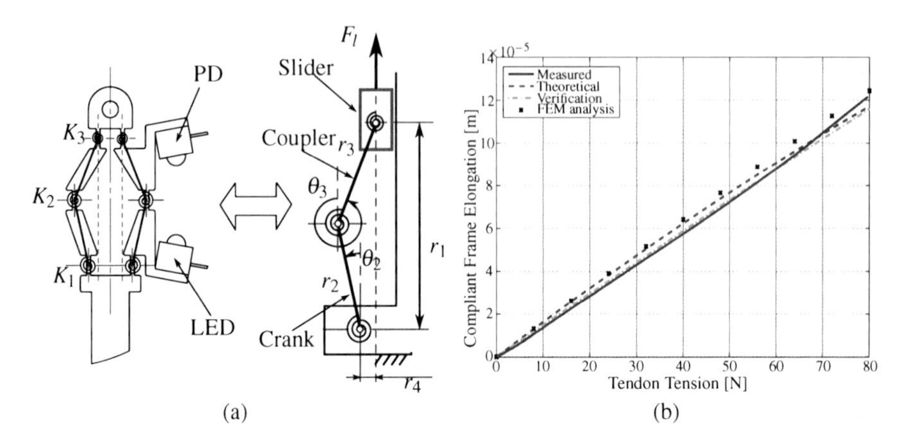

Fig. 12 (a) Scheme of the pseudo-rigid SCM and equivalence with half model compliant frame. (b) Linear displacement of the compliant frame Vs. Tension force.

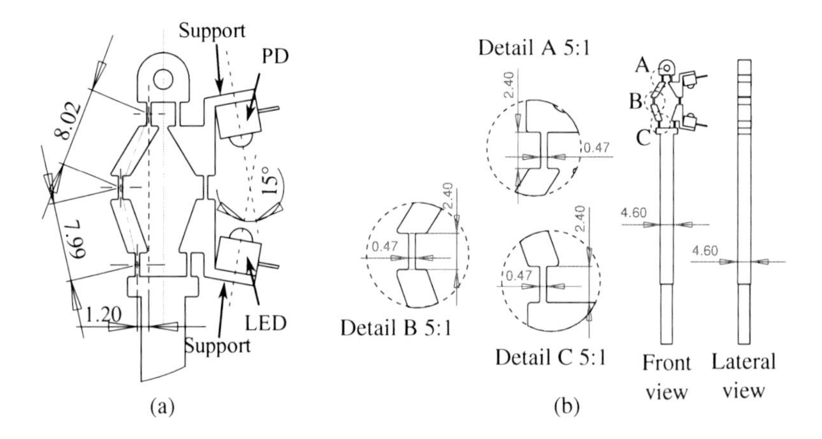

Fig. 13 (a) Detail of the compliant SCM and (b) detail of the compliant hinges implementation (all dimensions are in [mm]).

slender parts marked as "Support" in Fig. 13(a) are designed to place in the correct initial position the optoelectronic components. These component are simply bonded with a cyanoacrylate-based glue on the support and aligned using as reference the compliant frame borders.

3.2.2 Sensor Characterization

During the sensor calibration, a suitable load has been applied to the force sensor and the corresponding output voltage V_{out} of the circuit shown in Fig. 4 has been measured. These data have been then compared with the information coming from a strain gauge load cell. The calibration curve obtained during the experiments is

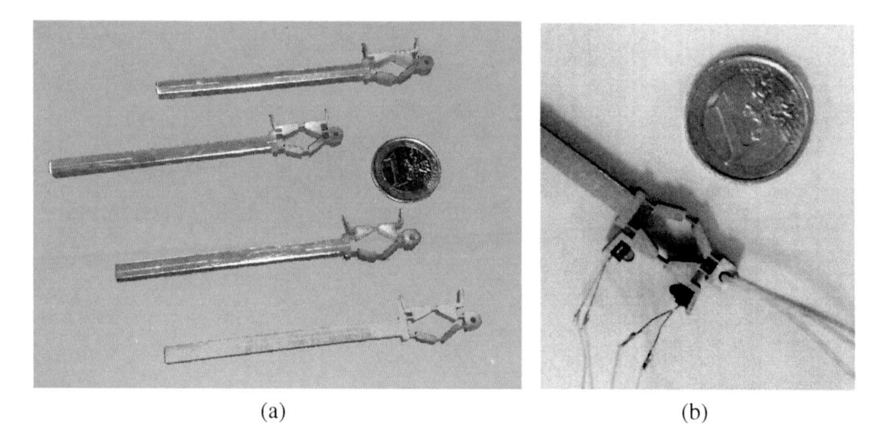

(a) (b)

Fig. 14 (a) Prototypes of the compliant frame. (b) A prototype with the optoelectronic components.

reported in Fig. 15(a), together with the characteristics obtained by means of a 3-rd order polynomial interpolation. This plots show that this interpolation is suitable for describing the response of the sensor. In Fig. 15(a) the output signal of the sensor is shown to highlight that, since it is of 1 V order, amplification is not strictly required. The shape of the signal is given by the controller and control objective used during this experiments. This particular signal has been selected because of its non-trivial harmonic content. By using a 16-bit analog-to-digital conveter over the 5 V supply, it is then possible to achieve a sensor resolution of about 0.01 N, that is suitable for our application. With the aim of showing the effect of the electromagnetic noise on the sensor, in Fig. 15(b) the tendon force measured (without any filtering) close to the maximum force range by means of both the strain-gauge and the optical load cell is reported: from this comparison it is quite evident the effects of the electromegnetic noise generated by the linear motors on the strain-gauge load cell, due also to the high amplifier gain needed to achieve a suitable signal level with respect to the optical based load cell. It follows that, for our application, other than the reduced dimensions, the reduced cost and the simplified conditioning electronics achievable with the proposed sensor, the signal-to-noise ratio of the optical load cell is more or less an order of magnitude better that the one of the strain-gauge load cell.

3.3 *Force/Tactile Sensor for Robotic Applications*

Tactile sense is used by humans to grasp and manipulate objects avoiding slippage, or to blindly operate in a dynamic environment. An artificial tactile sensor, by mimicking the human touch, should possess the capability to measure both dynamic and geometric quantities, i.e. contact forces and torques as well as spatial and geometrical information about the contacting surfaces. Each of these may be measured

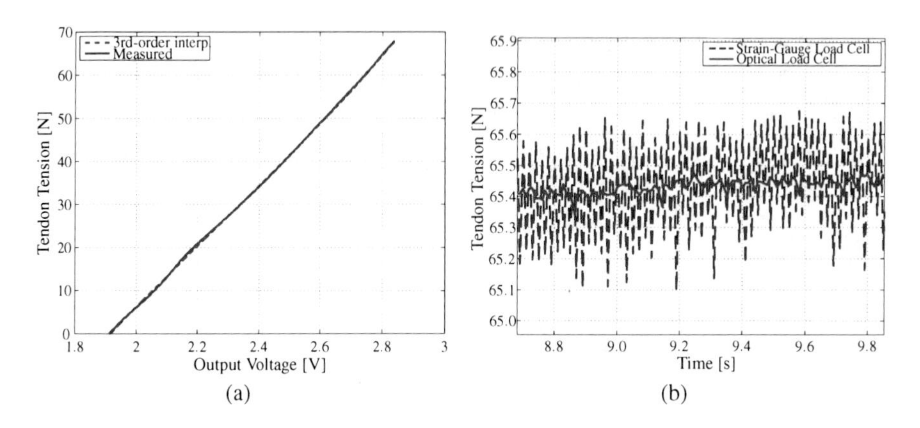

Fig. 15 (a) Calibration curve of the of the load cell based on optoelectronic components. (b) Comparison between the signals acquired from the strain-gauge load cell and the load cell based on optoelectronic components.

either as an average quantity for some part of the robot or as a spatially resolved, distributed quantity across a contact area [67]. A definition of tactile sensor is given by Lee and Nichols [40]: *a device or system that can measure a given property of an object or contact event through physical contact between the sensor and the object.* The one above is probably the best, and at the same time the broadest definition of a tactile sensor.

The force/tactile sensor here proposed exploits the thorough study based on Finite Element (FE) modelling conducted in [23] where the working principle has been presented for the first time. There, only a simplified prototype with limited sensing capabilities was tested with the aim of showing only the feasibility of the approach, whereas the main focus was on the mechanical characterization and optimization of the device. The sensor is based on the use of optoelectronic technologies and it aims to overcome most of the problems encountered in the works cited above, mainly: difficulty of the integration into small spaces, high costs, repeatability and complex conditioning electronics. The sensor has different capabilities, i.e. it can measure the six components of the force and torque vectors applied to it, and it can be used as a tactile sensor providing a spatial and geometrical information about the contact with a stiff external object. In fact, an approximated analytical model of the physical contact is derived, that is usefully exploited to extract information on the contact geometry from the sensor signals. Experimental characterization results are presented to both validate the model and to show how the sensor, with a proper algorithm, can be used to provide a complete characterization of the contact between the sensor and a stiff external object.

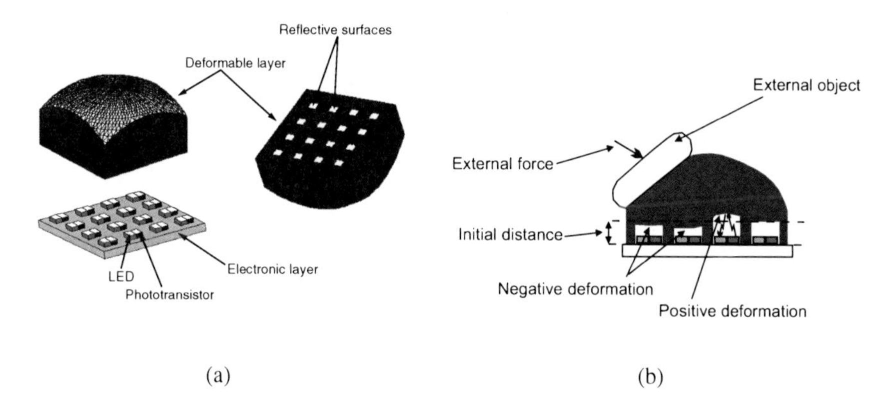

Fig. 16 (a) Structure of the force/tactile sensor. (b) Sketch of the working principle.

3.3.1 Working Principle

The proposed tactile sensor is based on the use of LED-PD couples and a deformable elastic layer positioned above the optoelectronics devices (see Fig. 16). The optoelectronic components are organized in a matrix structure. For each couple, the LED illuminates the reflecting surface which coincides with the bottom facet of the deformable layer. Practically, the deformable layer transduces an external force and/or torque into a deformation of its bottom facet through its stiffness. An external force applied to the deformable layer produces local variations of the bottom surface of the elastic material and the couples of optical devices measure the deformations in a discrete number of points. In particular, these deformations produce a variation of the reflected light intensity and, accordingly, of the photocurrent flowing into the PD. The deformations can be positive or negative, i.e. the photocurrent can locally increase or decrease (see Fig. 16), depending on amplitudes of tangential and normal force components, as well as on torque components. FE analysis, which has been carried out in [23], demonstrated that the latter relationship can be used to actually reconstruct the external force and/or torque components by measuring the elastic layer deformations in a discrete number of points. The paper cited above reports an experimentally identified model of the material used to realize the elastic layer, and then adopted in a FE analysis aimed at optimizing the shape of the deformable layer in order to obtain a satisfactory sensitivity for both normal and tangential components of the contact force vector.

3.3.2 Sensor Prototype

The realization of the sensor prototype (see Fig. 17(c)) took into account the results of the FE analysis to manufacture the deformable layer and some observations to select the optoelectronic components.

Concerning the deformable layer of the realized prototype, it is made of black silicone in order to avoid cross-talk problems between taxels and ambient light disturbances, since the black colour guarantees the maximum absorption for every wavelengths. Only the surface which faces each devices pair is white to increase the sensor sensitivity (see Fig. 17(b)), ensuring the maximum reflection for every wavelengths. According to the FE analysis results, the aspect ratio of the black walls between taxels has been selected in order to reduce the horizontal deformations with respect to the vertical ones. In particular, for the presented prototype, the thickness of the black walls is 1 mm, whereas the extension of the white reflecting surfaces is 1.6 mm, which results in a total size for the deformable layer of 11.4×11.4 mm. The height of the reflecting surfaces from the base of the deformable layer, in order to respect the necessary aspect ratio, is 1.5 mm. The top of the deformable layer is a section of a sphere with a radius of 11.4 mm. With the silicone choice modelled above, according to the numerical simulations, the expected measurement range of the sensor prototype is $[0,4]$ N. The maximum force level can be adapted by changing the hardness of the deformable layer. The maximum measurable force is limited by the maximum vertical deformation of the reflecting surface of each taxel, so the former can be changed by acting on the deformable layer geometry. A linear relation between the Shore hardness and the logarithmic of the Young's modulus has been derived in [2] for elastomeric materials. Using this relation, the maximum predictable force level, with the current geometry, goes from 2 N to 40 N by changing the hardness of the deformable layer from 4A to 60A.

Recalling the theory on LED and PD radiation pattern [37] described also in Sect. 3.2.1, the optoelectronic components suitable for this sensor should have very large viewing angles in order to minimize the effects of LED radiation pattern and PD responsivity pattern on the photocurrent and to leave only the dependence with the distance. Considering these aspects, the realized prototype uses optoelectronic components manufactured by OSRAM (see Fig. 17(a)). The LED (code SFH480) is an infrared emitter with a typical peak wavelength of 880 nm, whereas the PD is a silicon NPN phototransistor (code SFH3010) with a maximum peak sensitivity at 860 nm wavelength. Both the components have a viewing angle of $\pm 80°$. The conditioning electronics of the sensor is depicted in Fig. 4.

3.3.3 Prototype Characterization as Force/Torque Sensor

The objective of this section is to show all the potentiality of the presented prototype sensor and a calibration procedure necessary to use it as a 6-axes force/torque sensor. The characterization of the sensor has been made in the hypothesis that the contact surface can be approximated by a plane with an high stiffness with respect to the deformable layer. The hypothesis that the contact surface is a plane can be considered verified each time the external object has a curvature radius larger than that of the deformable layer. This condition is true for a large number of objects used in everyday manipulation and grasping tasks. Taking into account the hardness of the silicone, estimated in [23], used to realize the presented prototype, also the

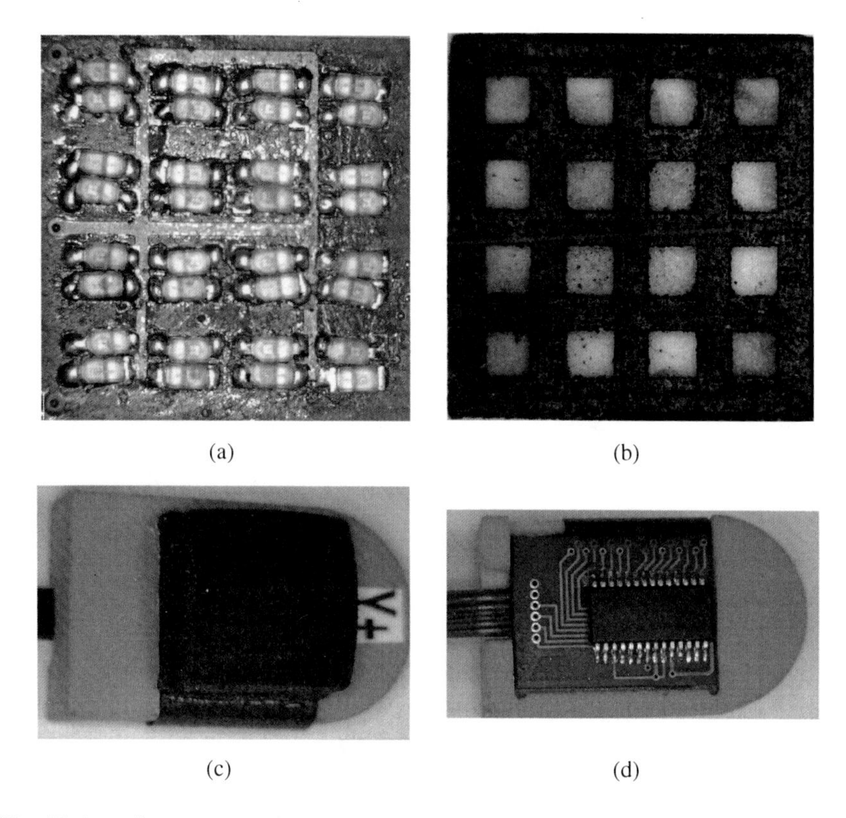

Fig. 17 A tactile sensor prototype: (a) electronic layer, (b) bottom view of deformable layer, (c) top view of assembled sensor, (d) bottom view of assembled sensor.

hypothesis of the high stiffness for the contact surface can be considered true for most of the daily use objects.

Figure 18(a) shows a sketch of the sensor, where the position of each cell with respect to the reference axes is indicated. The position of the k-th taxel can be identified with the (x_k, y_k) coordinates of the center position of the taxel. At rest position, for each taxel, a certain amount of light emitted by the LED is reflected from the white surface and reaches the PD, generating an initial voltage value on the collector. When an external force and/or torque is applied to the sensor, the distance of the reflecting surface of each cell from the corresponding LED-PD couple on the electronic layer can be subjected to a positive or a negative variation. These distance variations imply changes of the reflected light and, accordingly, of the voltages measured on the PD collector. Denoting with v_k the voltage variation of the k-th taxel, $v_k > 0$ denotes an increasing distance (and then a decreasing photocurrect), whereas $v_k < 0$ denotes a decreasing distance (and then an increasing photocurrect) between the reflecting surface and the electronic layer (obviously $v_k = 0$ denotes no variation). Figure 18(b) reports typical values of a voltage v_k, for the realized prototype.

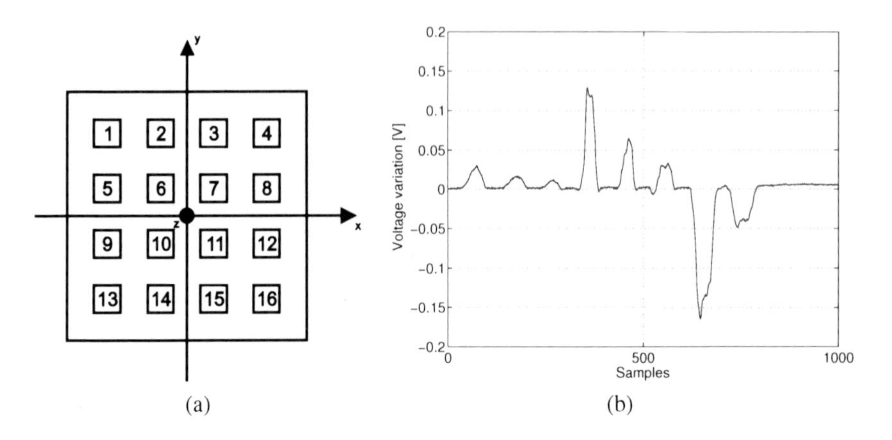

Fig. 18 (a) A scheme of the sensor taxels with respect to reference axes. (b) Voltage variations v_k for a generic taxel.

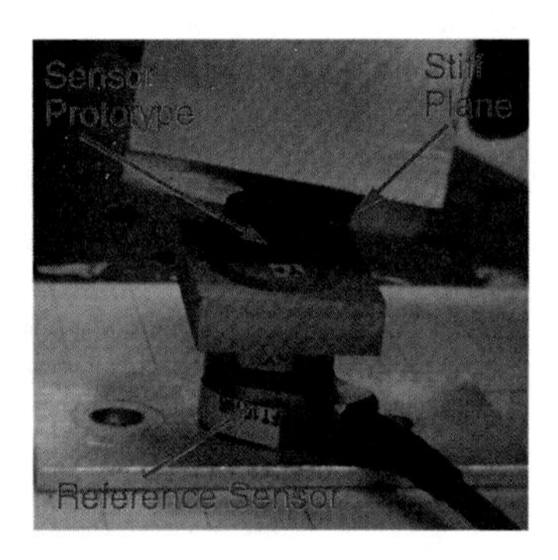

Fig. 19 Calibration system as force/torque sensor.

To calibrate the prototype for use as a six-axes force/torque sensor, the proposed approach is based on the use of a neural network to interpolate a number of data sufficient to model the relationship between the applied forces and torques and the PD measurements. The sensor has been mounted on a six-axes load cell used as reference sensor. The model used is the FTD-Nano-17 manufactured by ATI, with a measurement range equal to ± 12 N and ± 17 N for horizontal and vertical force components, respectively, whereas the measurement range for all torque components is equal to ± 120 Nmm. The reference axes, reported in Fig. 18(a), are located on the plane that separates the prototype from the reference sensor. Figure 19 shows the calibration system used to collect data for neural network training. An operator carried out various experiments, using a stiff plane, applying different external forces and torques and simultaneously acquiring all the voltage variations on the PD and all the forces and torques components measured by the reference load cell. These data, acquired at a sample rate of 100 Hz, have been organized in a training set and a

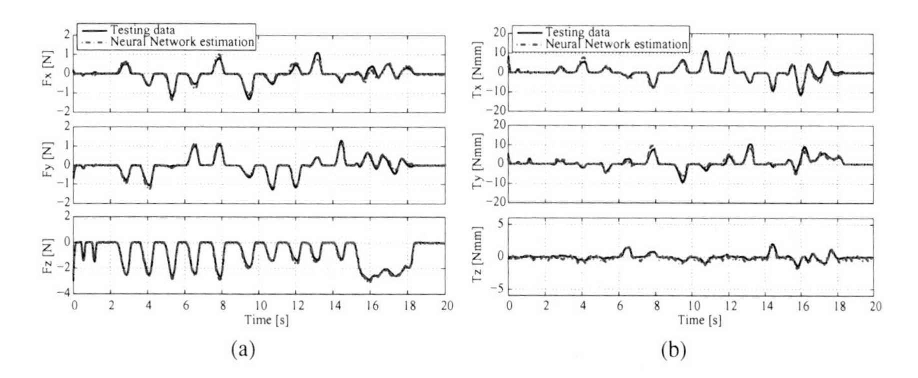

Fig. 20 Neural network testing with the stiff plane: force components estimation (a) and torque components estimation (b).

validation set to be used as input data (voltage variations) and target data (forces and torques components) of the neural network. A testing set has been prepared using data from experiments other than those used to collect the training and the validation sets, in order to assess the trained network.

A standard two-layer feed-forward neural network f_{NN}, trained with the Levenberg-Marquardt method, has been used to fit training data

$$H = f_{NN}(V), \tag{6}$$

where $H = [F_x\ F_y\ F_z\ T_x\ T_y\ T_z]^T$ is the output vector, F_i and T_i with $i = x, y, z$ are force and torque components with respect to the reference axes and $V = [v_1\ v_2\ \dots\ v_{16}]^T$ is the input vector containing the voltage variations of the taxels. The network is constituted by 24 neurons for the hidden layer and 6 neurons for the output layer. The trained network testing results are reported in Fig. 20(a) for the force components and in Fig. 20(b) for the torques.

The estimation is good for all force and torque components, especially when force and torque values are high enough. In fact, in the performance analysis it must be also taken into account that the training data are very noisy when the measured values are small compared to the full scale of the reference sensor.

In different experiments the trained neural network has been tested with objects that have a finite curvature radius. In particular, Figure 21 shows a standard bottle and a classic can used to collect additional testing data.

Figure 22 shows the real and the estimated forces and torques for the bottle and the can. The estimation is accurate also in these cases, obviously with a minimal reduction in performance when the curvature radius of the contact surface decreases. It is important to underline that these performances have to be evaluated along with the fact that compared to commercially available sensors, the proposed one is more compact, low cost, low power consumption, provided with a digital interface, and the deformable layer guarantees good adaptability and stability during grasping and manipulation application.

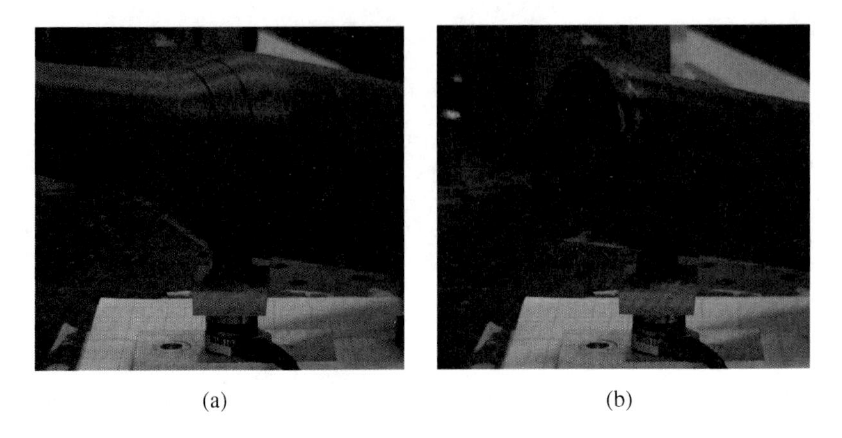

(a) (b)

Fig. 21 Pictures of testing objects: (a) a standard bottle and (b) a classic can.

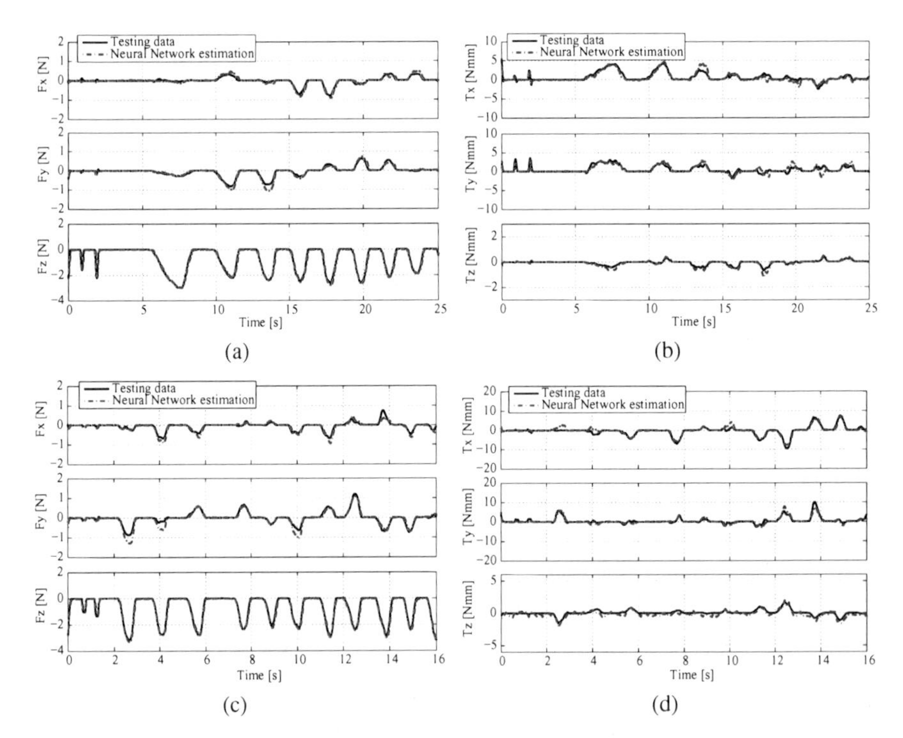

Fig. 22 Neural network testing with a standard bottle: force components estimation (a) and torque components estimation (b). Neural network testing with a classic can: force components estimation (c) and torque components estimation (d).

3.3.4 Prototype Characterization as Tactile Sensor

The proposed prototype can be also used as tactile sensor, estimating not only force and torque components as described in Sect. 3.3.3, but also the contact geometry. In some applications, e.g., complex robotic manipulation tasks, the availability to the control system of an estimate of contact plane position and orientation together with the interaction forces exchanged by the sensor and the external object are fundamental to successfully execute the task. To obtain this information from the sensor, an approximated physical model is first derived and validated to describe the contact between a stiff surface and the deformable layer.

The deformable layer is considered as an elastic homogeneous hemisphere of known radius R and the external object as a stiff plane. Figure 23(a) shows a generic contact where the stiff external object deforms the elastic hemisphere until, at equilibrium, the contact plane coincides with the plane π_2. The full characterization of the contact geometry means to estimate the position and orientation of the plane π_2 and the direction of the force F with respect to π_2. In particular, the plane π_2 can be uniquely defined by θ and ϕ angles and by its distance $\overline{TO}$ from the origin of the reference frame. Considering the axial symmetry of the deformable layer, the angle ϕ can be directly estimated from the measured F_x and F_y components, reducing the three-dimensional problem in a two-dimensional one in the zp-plane, where the p axis is defined by the direction of the vector $(F_x, F_y, 0)$ with magnitude $F_p = \sqrt{F_x^2 + F_y^2}$. Thus it is $\phi = \mathrm{atan2(F_y, F_x)}$,where $\mathrm{atan2(b,a)}$ is the argument of the complex number $a + ib$.

Note that the force F direction in Fig. 23(a), from the contact geometry point of view, is related only to the position of point H, for a fixed contact plane. As a consequence, being F related to the v_k taxel voltage variations, as described in Sect. 3.3.3, it is possible to relate point H coordinates directly to the v_k voltage variations. The set of voltage variations represents a tactile image that, for example, could be used to estimate a pressure map using an appropriate reference sensor. In addition, the information contained in the tactile image is used to estimate the coordinates of the point H in the xy-plane, i.e. the coordinate couple (x_H, y_H). From this information and by exploiting the knowledge of the F_z and F_p force values, that can be calculated by the neural network (6), the variables θ, d_n and d_t, that determine the contact geometry, can be estimated. For more details about the methodology for reconstructing the tactile image from the sensor information please refer to [24].

Forces with different directions and magnitudes have been applied to the prototype sensor using different plane contacts, whose positions and orientations have been fixed with the mechanical structure shown in Fig. 23(b). In particular, the goniometer has been used to fix the θ angle values, while the micrometric stage to fix the d_n and d_t deformations values. In a first set of experiments, the θ, d_n and d_t values, measured with the mechanical structure, have been collected and used to estimate the tactile image parameters.

A second data set has been collected from experiments to test the model performance. Figures 24(a) and 24(b) compare the reconstructed F_z and F_p with the force values measured by the reference sensor. The error is less than 0.15 N for the

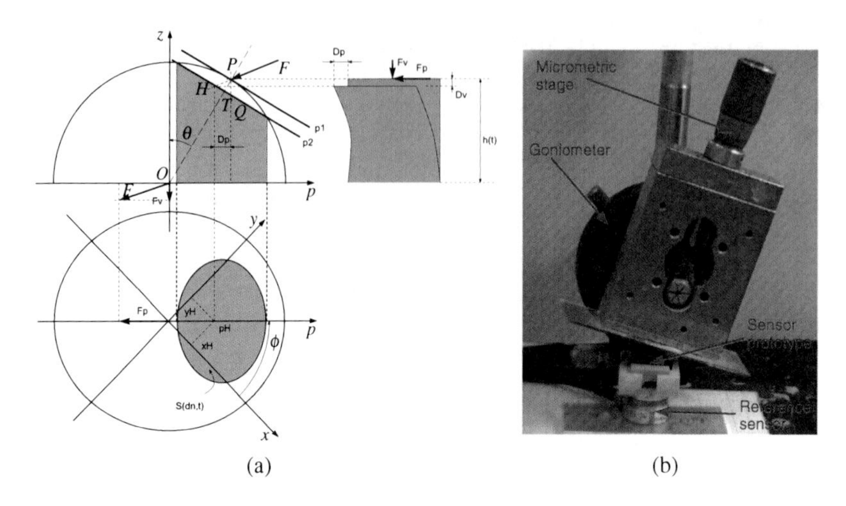

Fig. 23 (a) Geometric characteristics of the contact model. (b) Mechanical system for contact geometry characterization.

vertical force component and less than 0.08 N for the horizontal force component. The same procedure can be applied using the force information reconstructed by means of the neural network (6) on the basis of the proposed sensor measurements. The comparison between measured and modelled force components is presented in Figs. 24(c) and 24(d). For this case the reconstruction error is only slightly worse than the previous case, due to the additional error introduced by the neural network.

Then, the algorithm for the reconstruction of the tactile image, and in particular of the parameters x_H and y_H, has been tested with different measurements and the results are reported in Figs. 25(a) and 25(b). The maximum estimation error is less than 0.5 mm for x_H and less than 0.6 mm for y_H. In order to verify the effectiveness of the algorithm for the reconstruction of the tactile image, the procedure has been tested with experimental input data collected from the reference sensor (regarding F_z and F_p) and from the mechanical structure shown in Fig. 23(b), (regarding p_H). The results of the θ, d_n and d_t variables estimation are reported in Fig. 25(c), where the high accuracy obtained in the contact geometry reconstruction is evident. In particular, the maximum error is less than 0.2° for θ, less than $100\,\mu m$ for d_n and less then $20\,\mu m$ for d_t. Afterwards, the proposed procedure has been tested also with force data computed by the neural network (6). The results are shown in Fig. 25(d). In this case, the maximum error is less than 2° for θ, less than $200\,\mu m$ for d_n and less then $60\,\mu m$ for d_t, due to errors in the estimation of the force components and of the (x_H, y_H) coordinates.

4 The Robotic Hand Actuation

The actuation system of an anthropomorphic dexterous hand should enable the whole system to move with human-like agility and speed as well as empower the

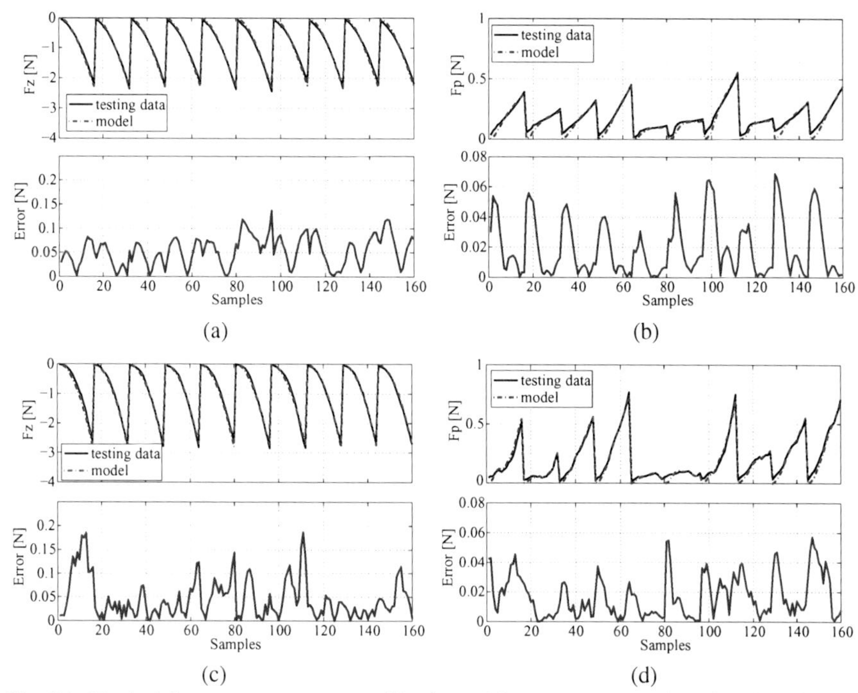

Fig. 24 Vertical force component (a) and horizontal force component (b) using the reference load cell measurements. Vertical force component (c) and horizontal force component (d) using prototype sensor measurements.

hand to grasp with human-like strength. These performance criteria should be fulfilled by a robotic construction resembling the human hand and forearm at least in terms of dimensions and weight but if possible also in terms of form and aesthetics. These constructional guidelines translate into challenging specifications for the actuation system. In particular, the actuation system should fit within the dimensions corresponding to the human hand and arm, with an overall weight of less than 1 kg.

Current technology does not allow arranging twenty or more actuators in a robot hand with dimensions similar to those of a human hand and with suitable requirements in terms of speed and forces. The tendon-based transmission system allows placement of the actuators within the forearm, which corresponds with the location of the most powerful muscles in the biological model. This placement simplifies the hand construction, frees up space in the fingers for the integration of the sensors as well as achieves a more anthropomorphic weight distribution. Consequently, the tendon-based transmission system represents the most promising solution for dexterous anthropomorphic robotic hands.

As will be detailed in the next section, the DEXMART Hand fingers are actuated by means of four tendons each. Four additional tendons are necessary for the wrist actuation, resulting in a requirement of 24 independent actuators; refer to Fig. 1 where a general overview of the DEXMART Hand design in shown. In the

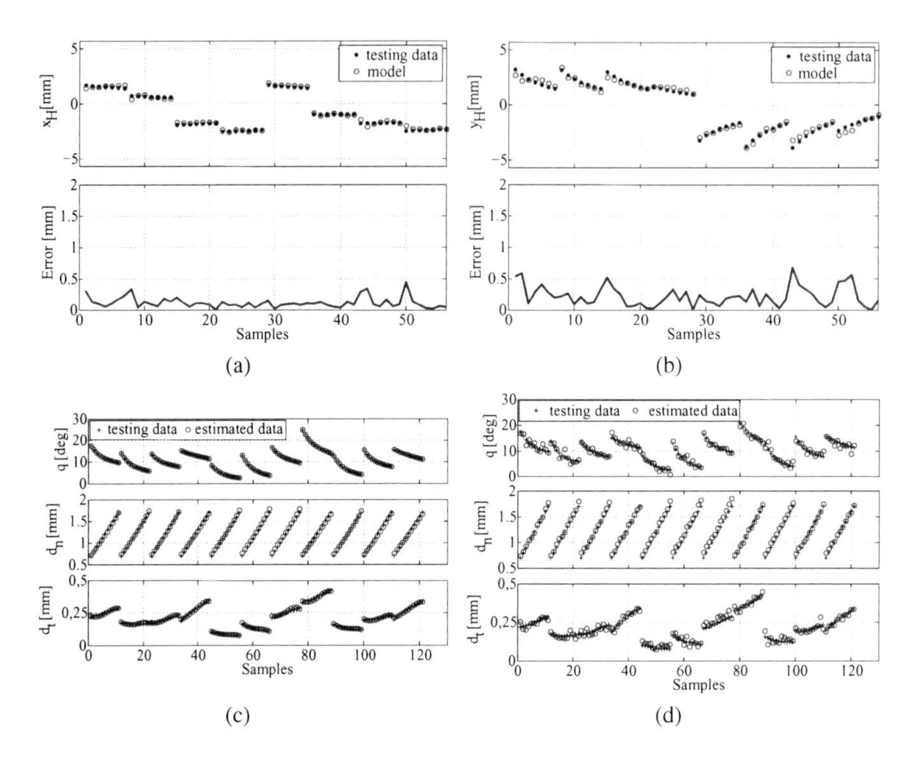

Fig. 25 (a) Estimation performance for coordinate x and(b) estimation performance for coordinate y. (c) Estimation of contact variables. (d) Estimation of contact geometry information for a generic experiment.

following, the tendon-based transmission system of the DEXMART Hand will be detailed and the twisted string actuators will be described, showing also the results of the tests performed to show the effectiveness of this actuation system.

4.1 Tendon Network

Due to the quite complicated human tendon network, instead of directly imitating the biological model, many different simplified solutions have been proposed in the literature. Moreover, whereas in the biological model the tendons slide around the bones, for the optimization of the transmission system in terms of reducing both the friction and the coupling among the hand movements, it is preferable for the tendon to traverse the endoskeleton and to be routed close to the center of rotation of the joints by means of suitable channels. In tendon actuation systems, different movements can be easily coupled by simply connecting two or more tendons to the same actuator, thus obtaining a defective actuation, avoiding additional mechanisms, and reducing both costs and complexity.

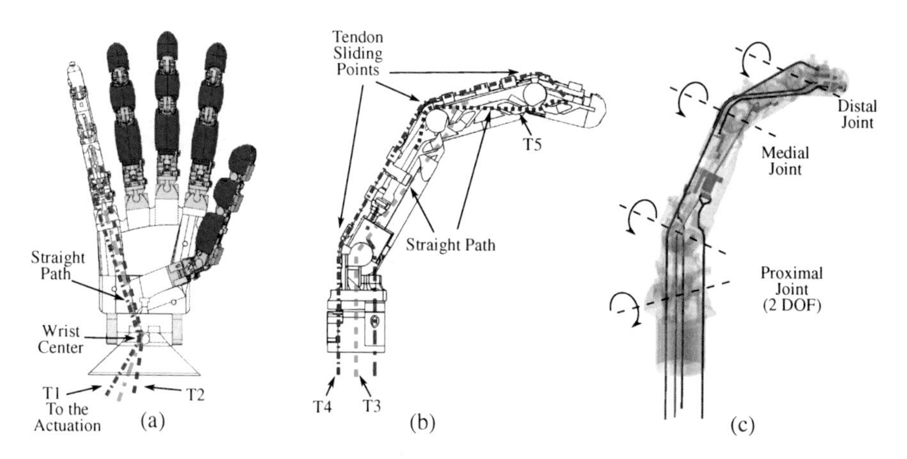

Fig. 26 DEXMART Hand: Details of the tendon network.

A fundamental problem in tendon-based actuation is the way the tendons are routed from the motors to the joints. Usually, tendons are routed by means of pulleys, sheaths or sliding surfaces: whereas pulleys reduce at minimum the friction forces acting along the tendon, this approach implies a more complicated mechanical design due to the presence of bearings and similar hardware partially reducing the advantages introduced by the use of tendons. The use of sheaths is a convenient solution due to its simplicity, but it introduces distributed friction along the tendon, that means hysteresis and dead-zones in the transmission system characteristic [34, 51].

The selection of the tendon material plays a crucial role. Usually, very thin steel ropes are used allowing to obtain a linear force-elongation behavior of the tendon but introducing some design and assembly constraints due to the limited curvature radius of steel cables. In the last years, polymeric fibers have been largely used to improve the design flexibility of tendon transmissions. Despite the comparable elastic module with respect to alloy cables, polymeric tendons present hysteresis in the force-elongation characteristic that introduces stability problems in the transmission system control [51]. Different studies confirm that the total amount of friction acting along the tendon depends only on the friction coefficient and on the total curvature of the tendon path from the motor to the joint [34, 51]. Whereas the friction coefficient can be reduced by a suitable selection of the path coating and tendon materials (other than introducing lubricants), the path curvature minimization is a non-trivial design problem.

A detailed view of the *N+1* tendon network of the DEXMART Hand is presented in Fig. 26. In an *N+1* tendon network, each finger (and its joints) is actuated by a number of tendons equal to the number of joints plus one. This configuration allows using of the minimum number of actuators and, at the same time, to avoid pretension mechanisms. Moreover, the *N+1* configuration allows both the independent regulation of the joint torques and the regulation of the internal forces. Fig. 26(a) shows the straight path of the tendons from the finger base to the wrist center so as to avoid tendon curvature inside the palm (no friction is introduced) and minimize the coupling between the wrist and the finger movements. The optimization of the tendon path inside the finger is reported in Fig. 26(b). With reference to this figure, the tendons that actuate the base joint (T1 and T2) are connected directly to the proximal phalanx, whereas the antagonistic tendon (T4) slides over the joints on the back finger side. The tendon that actuates the medial joint (T3) is routed very close to the center of the proximal joint to limit as much as possible the coupling between the movements of these two joints. The path of the tendon (T5) that connects the proximal to the medial joint is straight to avoid contacts between the tendon and the endoskeleton and to limit friction. Finally, Fig. 26 reports a 3D view of the finger and a detail of the joints location (refer to [13] for a complete description of the DEXMART Hand's finger kinematics and tendon network characteristics). Note that, in order to reduce the path curvature, part of the tendon path is formed directly within the endoskeletal structure of the finger.

4.2 The Twisted String Actuation System

Different solutions concerning the actuation system of robotic hands have been proposed in the past, based essentially on rotative electric motors [44, 38, 22] or linear pneumatic actuators [30, 63, 60], usually McKibben motors [19]. Although actuation solutions adopted in robotic hands developed so far each have their own benefits and shortcomings, the so-called twisted string actuation system has been developed within DEXMART, aiming at fitting with project requirements.

In contrast to these similar actuation concepts reported in literature [32, 59, 47, 69, 68], the twisted string actuation system adopts very thin and long strings twisted around themselves allowing a displaced location of the motors (with respect to the joints) and the use of very small high-speed motors without speed-reducer, facts that make the actuation concept proposed here particularly suited for the development of innovative tendon-driven robotic hands.

As a design parameter, the actuation force was derived from the consideration of a 10 N load applied perpendicularly to and at the tip of an outstretched finger. Taking into consideration the kinematic design of the fingers and the tendon routing, this load translates into a tendon force as high as 80 N. Moreover, the tendon displacement corresponding to full closure of the hand is on the order of 20–25 mm for each tendon. As a matter of fact, with an appropriate choice of the rotative electric motors and of some design parameters of the strings (in particular the radius and length), the actuation system presented here can satisfy all of the tight requirements for the implementation of miniaturized and highly-integrated mechatronic devices, paving the way for the next generation of multifingered robotic hands. With respect to conventional solutions, the main advantages of this actuation system consist in the direct connection between the motor and the tendon without any intermediate mechanisms such as gearboxes, pulleys or ballscrews, in the direct transformation from rotative to linear motion, in the extremely reduced friction (only an axial

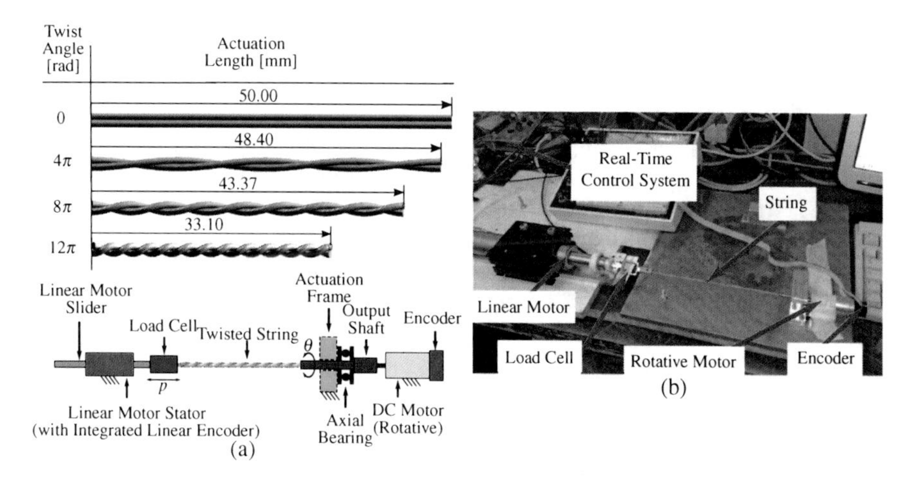

Fig. 27 (a) Basic concept (top) and schematic representation of the twisted string actuation system (bottom). (b) Experimental setup.

bearing is needed), in the very high reduction ratio, in its intrinsic compliance and in the use of very small high-speed motors.

The basic idea of the quite simple actuation system is illustrated schematically in Fig. 27(a): two or more strands are connected in parallel on one end to a rotative electrical motor and on the other end to the load to be actuated. Twisting the strands at the one end by means of the motor reduces the length of the transmission, resulting in a linear motion of the other end.

This actuation concept, because of its high (though configuration dependent) reduction ratio, permits the use of very small and lightweight electric motors and therefore is very interesting in applications where size and weight are of crucial importance. This concept was firstly implemented in an experimental setup for verifying its main properties, see Fig. 27(b). The setup consists of a small rotative DC motor (Faulhaber 2233) and a string pair aligned along the rotation axis of the motor and connected at one end to the motor output shaft (without any speed reducer), as schematically shown in Fig. 27(a), and at the other end to the load, which in the test bed is emulated by a linear motor (LinMot P01-37×120) able to apply a force up to 160 N along the motion axis of the slider and equipped with a load cell for measuring the actuation force. The rotative motor is equipped with an optical encoder to measure the rotation angle, whereas the encoder integrated in the linear motor is used to measure the actuation elongation with a resolution of 1 μm. A suitable controller is used to drive the load (the linear motor) so that it behaves as a mass-spring-damper system with adjustable parameters —refer to [52] for additional details on the linear motor driving technique. Note that the encoders are used only for monitoring purposes and serve as position reference for the linear motor load controller.

From the control point of view, an important aspect here considered is the possibility of controlling the system using only force feedback. For this reason, the design of a controller based on the measurement of the actuation force only is faced. In fact, neither measurement of the system state (motor angular displacement and velocity, load position and velocity) nor accurate knowledge of the system parameters (e.g. string or actuation length, string radius, motor or string preload angle, load parameters) are required. Moreover, due to the finite stiffness of the strings and to the particular implementation, a non-negligible configuration-dependent compliance of the proposed transmission systems was observed during the early experimentation. This phenomenon has been measured and modeled to allow future evaluation for control and safety purposes [11].

4.2.1 Modeling of the Actuation System

As a simplifying hypothesis, it is assumed here that some strands do not contribute to the total axial force: these fibers form the core of radius r_c of the helix, see Fig. 28(a). The load force F_z is balanced by the n external strands of radius r_s which form n coaxial helices of radius $r = r_s + r_c$. As a limit case, Fig. 28(b) shows a string formed by a pair of twisted strands, for which $r_c = 0$ and thus $r = r_s$, considering the helices formed by the strand axes.

In order to obtain the relationships describing the statics of the actuation system, assume that the strands constituting the string form an ideal helix of constant radius $r = r_c + r_s$ along the whole range of the motor angular position θ. The kinematic relationship between the motor angle and the load position can be easily derived from the geometry of the helix formed by the strands —see in particular Fig. 28— which implies the following straightforward relations:

$$L = \sqrt{\theta^2 r^2 + p^2}, \quad \sin\alpha = \frac{\theta\, r}{L}, \quad \cos\alpha = \frac{p}{L}, \quad \tan\alpha = \frac{\theta\, r}{p}, \tag{7}$$

where α is the helix slope, L is the strand length and p is the length of the transmission system or, in other words, the load position. Note that (7) can be easily obtained by "unwrapping" the helix of total length L and radius r and applying Pythagoras' theorem to the resulting triangle in Fig. 28(d). From (7) it follows that $\dot{L} = \dot{p}\cos\alpha + \dot{\theta}\, r \sin\alpha$.

The static model of the actuation system can be easily obtained looking at Fig. 28(c), where the external torque τ_L is balanced by the tangential force F_τ, i.e. $F_\tau = \tau_L / r$. Assuming that the load is equally distributed over the n strands that form the string, one obtains

$$F_\tau / n = F_i \sin\alpha \quad \Rightarrow \quad \tau_L = r\, n\, F_i \sin\alpha \tag{8}$$

where F_i is the longitudinal feasible force in each strand. The resulting total axial force F_z acting on the transmission system is

$$F_z = n\, F_i \cos\alpha \quad \Rightarrow \quad \tau_L = F_z\, r \tan\alpha = F_z\, \theta\, r^2 / p. \tag{9}$$

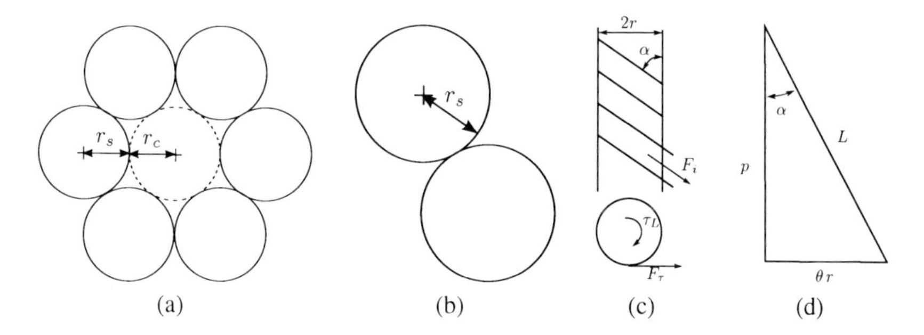

Fig. 28 Schematic representation of the helix formed by the strands that compose the twisted string. (a) String section with $n = 6$. (b) Two-string section with null core radius ($n = 2$). (c) Lateral and axial view. (d) Unwrapped helix.

In order to take into account the finite stiffness of the string, the strands are assumed to act as linear springs, with the capability of resisting tensile (positive) forces only and not compressive (negative) forces, as is usual for cable-based transmission systems. Therefore, the total length of a strand L (each strand has the same length of the untwisted string) changes with respect to the unloaded length L_0, according to the fiber tension F_i and the strand stiffness K (normalized with respect to the length unit), i.e.

$$F_i = \frac{K}{L_0}(L - L_0) = \frac{K}{L_0}\left(\sqrt{p^2 + r^2\theta^2} - L_0\right). \tag{10}$$

From (10), it is possible to note that each strand acts as a spring whose deformation is defined as $\sqrt{p^2 + r^2\theta^2} - L_0$ and which can be modulated through the motor angular position θ. It follows that the actuation elongation p (i.e. the load position) is given by

$$p = \sqrt{L_0^2\left(1 + \frac{F_i}{K}\right)^2 - \theta^2 r^2} \tag{11}$$

whereas, from the helix geometry, the pitch q of the helix is related to the string length p and to the motor rotation angle θ by the relation $2\pi p = q\theta$.

The stiffness S at the load side of the twisted string transmission system can be modeled by considering (7) and (10), and computing the derivative of the load force F_z in (9) with respect to the load position p:

$$F_z = n\frac{K}{L_0}\left(\sqrt{p^2 + r^2\theta^2} - L_0\right)\frac{p}{L} = n\frac{K}{L_0}\left(p - \frac{L_0 p}{\sqrt{p^2 + r^2\theta^2}}\right),$$

$$S = \frac{\partial F_z}{\partial p} = nK\left(\frac{1}{L_0} - \frac{1}{\sqrt{p^2 + r^2\theta^2}} + \frac{p^2}{(p^2 + r^2\theta^2)^{3/2}}\right). \tag{12}$$

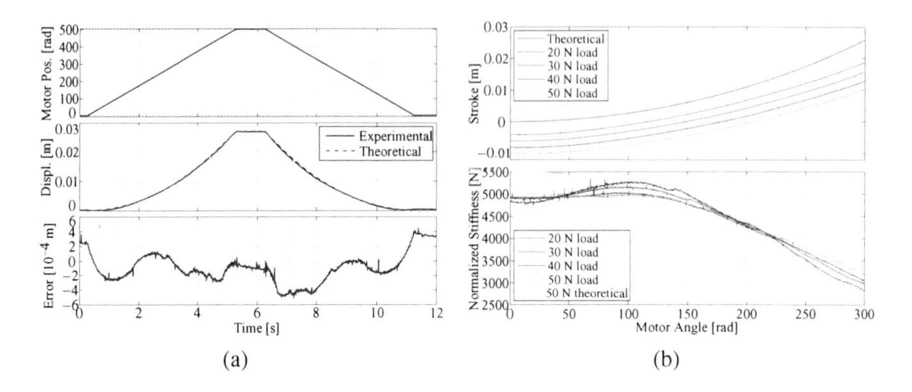

Fig. 29 (a) Evaluation of the static relation (1 N constant load). (b) Transmission contraction (top) and stiffness (bottom) vs motor angle for different load conditions.

4.2.2 Experimental Validation

The string used during the experiments is a commercial ⌀0.2 mm Dyneema fishing line with $L_0 = 0.2$ m. Figure 29(a) reports the comparison between the theoretical model (11) and the experimental data obtained in very low load (1 N) condition. Note that the error between the theoretical model and the experimental data is very small. The first plot of Fig. 29(b) shows the effect of the load force F_z on the transmission contraction observed during experiments in comparison with the unloaded theoretical model result: it is evident that the string compliance is not negligible with respect to the expected transmission load force. Anyway, the capability of moving the load by twisting the string is preserved. Figure 29(b) also shows the corresponding transmission stiffness variation with respect to the motor angular position measured during the experiments: these data have been computed numerically by forming the ratio between the transmission load force and the incremental variation in the measured load position. It can be clearly noted how the stiffness of the actuation system varies with the motor angle and the transmission load. The theoretical value of the stiffness of the twisted string transmission system, computed according to (12), for a load of 50 N is also reported in this plot for comparison with the experimental results.

The results of the experiments performed on the twisted string actuation system driven by the controller proposed in [53] are reported in Fig. 30(a). In these plots, it is possible to distinguish between three different phases: 1) the peaking avoidance action is active from $t = 0$ s when the experiment starts, since the initial reference value differs from the initial conditions and the system output reaches the desired value very smoothly at $t \simeq 4$ s. Then, the response to a setpoint step variation at $t = 6.5$ s can be appreciated and, finally, the system is evaluated with a 20 N amplitude sinusoidal setpoint at 0.5 Hz. The very good fitting between simulation results and experiments can be clearly seen, even if the nominal value of the system parameters are considered in the control system design and no identification procedure has

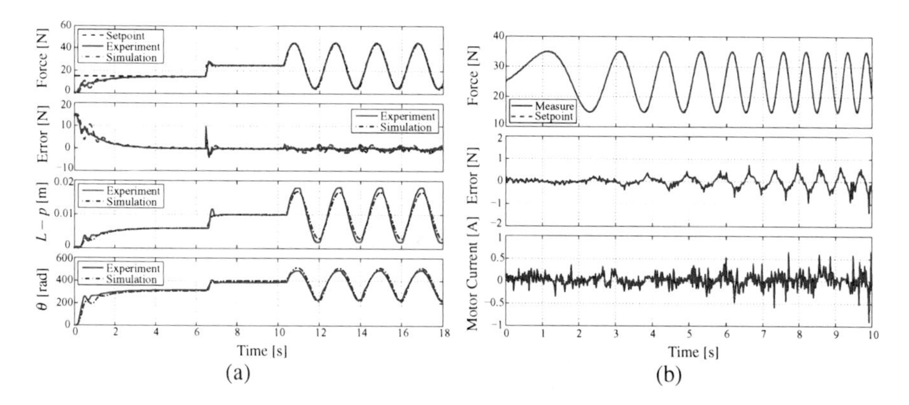

Fig. 30 (a) Comparison between simulation and experimental results. (b) Experimental evaluation of the dynamic performance of the twisted string actuation: the setpoint is a sinusoidal signal with time-varying frequency from 0.1 to 2 Hz.

been used to evaluate the real value of these parameters. Figure 30(a) also reports the comparison between the simulated and the experimentally measured actuation displacement $L - p$ and motor position θ (string twist angle): in the case of these two variables, the differences between the theoretical and the real values are more noticeable due to the uncertainties affecting the system parameters.

With the aim of showing more clearly the dynamic performance of the twisted string controller, an experiment in which the force setpoint is a sinusoidal signal with time-varying frequency (also called chirp signal) in the range 0.1–2 Hz has been performed, and the results are shown in Fig. 30(b). The force setpoint presents a mean value of 25 N and the amplitude of the setpoint oscillations is 10 N. From the plots reported in Fig. 30(b) it is possible to note that the system presents good performance in terms of setpoint tracking over the whole signal frequency range, and in particular within the controller design bandwidth that is 1 Hz. Also the commanded motor current has been reported in Fig. 30(b) to highlight the quite limited control effort (the maximum motor current is less than 1 A).

5 Soft Covers Based on Differentiated Layer Design

The adoption of soft covers (pads) for artificial hands and fingers is important primarily for three reasons: *functionality* in some specific tasks, *safety*, and *acceptance* by the users.

Concerning functionality, the presence of a surface compliance can highly influence the performance of the hand when contacting the environment during force/position controlled task, similarly to what happens in human fingers or feet which are covered by pulpy tissues [66, 71, 76, 36]. First of all, the presence of a passive compliant surface is beneficiary in terms of contact effectiveness. In fact an increased pad compliance (or, inversely, a low pad stiffness) means larger

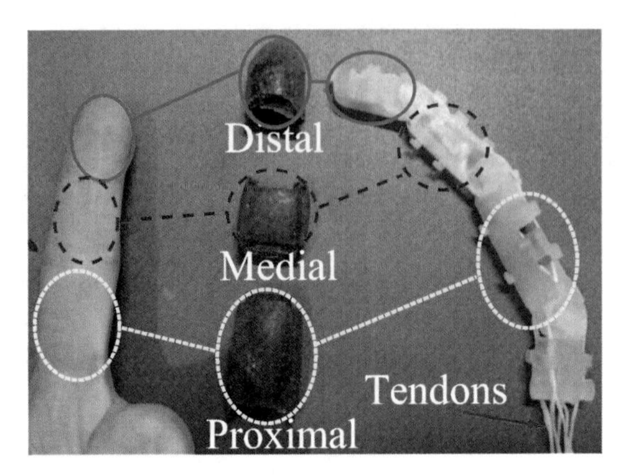

Fig. 31 DEXMART Hand's finger endoskeletal structure and soft layer (proximal, medial, distal phalanx cover).

contact areas for a given load and therefore reduced contact pressure, reduced material stress and better contact stability [20, 65, 21]. Furthermore, the soft pad allows local shape adaption in case of contact with sharp edges or objects with morphological irregularities and can contribute to vibration damping [70]. At last, a compliant covering surface helps protecting both the mechanical structure of the hand (including the transmission system) and the delicate sensory apparatus (if present).

Concerning safety, as demonstrated and quantified in [10], the soft pad can be seen as a passive device which reduces possible injuries in case of accidental impacts with humans.

Finally, concerning acceptance, a soft-touch feeling can be important in the case of human-machine interaction. This issue is particularly evident in the prosthetic industry where hand-like gloves providing enhanced functionality and increased cosmetic appeal are usually chosen at the expense of efficiency, cost and weight of the overall prosthesis.

In terms of design requirements, the properties of an ideal soft cover are hardly definable. For instance, the overall stiffness of a robotic fingertip, which is designed for manipulation purposes, can be different if compared to the stiffness of an arm soft cover, whose main functionality is limited to safety issues. In addition, a complete characterization of a robotic pad must include investigation on many properties and behavioral aspects [72]. However a primary role is played by the behavior of the pad under normal contact load, in interaction with a rigid planar object. Therefore, in the following, the investigation will regard the contact behavior of soft fingerpads pressed against a rigid flat surface.

Concerning robotic hands, the majority of soft pads studied so far were made by viscoelastic polymers homogeneously shaped over an internal rigid core mimicking the bone or the robotic finger inner rigid structure [64, 73, 5]. In such a case [9], the parameters that mainly contribute to the pad compliance, for a given external geometry, are:

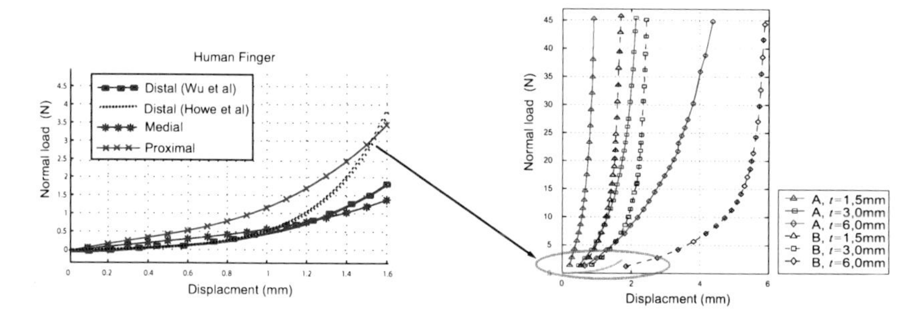

Fig. 32 Experimental results [9]. Displacement (mm) versus normal load (N) for the human fingertip and for soft pads with different hardness (materials a and b) and different thickness, t. Material A = soft silicon rubber (hardness 18 shore a). Material B = very soft silicon rubber (hardness 20 shore 00).

- The material hardness. An higher material hardness, which is beneficiary in terms of surface reliability, signifies lower compliance.
- The layer thickness. An higher thickness signifies higher compliance which is beneficiary in terms of safety and increases grasp stability/sustainability. On the other hand, high pad thickness signifies high overall limb dimension. As a matter of fact, thickness reduction is a significant goal for the robotic limb designer, that cannot easily reduce the overall size of the internal rigid core (hosting actuators, transmissions, sensors, etc.) but wants to obtain slender bio-mimetic limbs at the same time.

As an example, let us consider the behavior of the human fingertip [56, 75, 62] (distal phalanx, Fig. 31) which is shown on the left diagram in Fig. 32. In order to replicate the compression behavior of the human fingertip, it is necessary to employ a very soft silicon rubber (hardness 20 Shore 00) with very high thickness (6,0 mm).

Usually the adopted pad design is a trade-off between the need of slender robotic limbs and good material properties. Still, it is sometimes impossible to tailor the pad properties to the specific application by simply using an homogeneous viscoelastic layer.

Looking for alternative solutions to homogeneous soft covers, the authors have previously proposed the concept of Differentiated Layer Design (DLD) [6, 57] which allows both increasing the pad compliance and minimizing its thickness. The concept of DLD consists in the adoption of a single solid material dividing the overall thickness of the pad into layers with different structural design (i.e. an external continuous skin layer coupled with an internal layer with voids). Figure 33 shows a DLD soft pad.

In particular, a methodology have been proposed in [4] which allows minimizing the designer effort when trying to replicate the non-linear relationship $F = f(\delta)$ between the applied normal Load, F and the contact Deformation, δ (LD curve), which is representative of endoskeletal structures covered by pulpy tissues.

Given the allowable pad thickness and the overall contact area, the purpose is to tailor the pad properties to the specific application by:

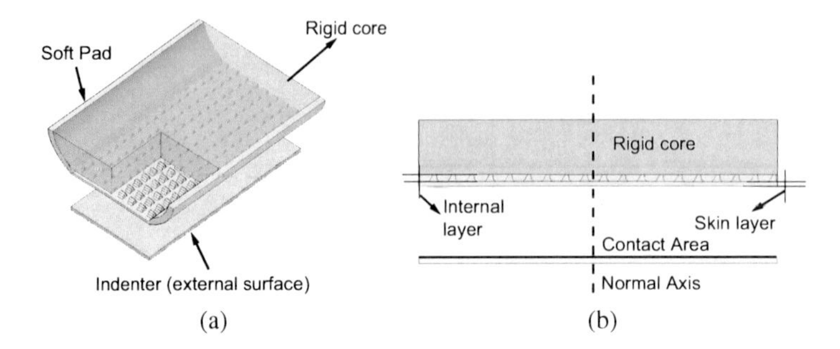

Fig. 33 Differentiated layer design concept.

- Selecting a skin material characterized by proper tribological features (hardness);
- Designing an internal layer geometry (Fig. 33) so as to obtain a specific static compliance (increased with respect to a non structured pad).

The methodology adopted for designing the internal layer is composed of two steps:

- Firstly, the cover surface (overall contact area) is conceptually split into finite elementary triangular sub-regions;
- Secondly, the internal layer of each Triangular Element (TE) is designed in order to replicate the shape of the given non-linear relationship $f(\delta)$. A series of symmetrically disposed inclined micro-beams is used for the purpose.

Once the compression law of each triangular element is known, the overall pad compliance can be modulated by correctly choosing the number and, consequently, the size of the elements composing the pad.

5.1 Selection of a Skin Material with Proper Hardness

Following the conceptual procedure outlined in the previous section, the design of the pad starts with the selection of a suitable polymer with proper hardness. Two solutions have been considered:

- Silicone rubber Wacker ELASTOSIL RT 623 A/B: two component silicone that vulcanizes at room temperature whose hardness can be varied in a very wide range by adding a third component (silicone fluid AK). Various pad geometries can be obtained through injection moulding.
- Tango Plus Fullcure 930 (hardness 27 Shore A): polymeric resin used for stereo-lithography. This stereo-lithographic technique allows to get complex shape in a short producing time.

By using Rapid Prototyping or injection moulding, the intermediate layer can be obtained with various geometries with exception of closed-cell structures. In fact, concerning rapid prototyping, a removable wax must be deposited as a sustaining additional material in case of negative slope of the lateral surfaces. Concerning

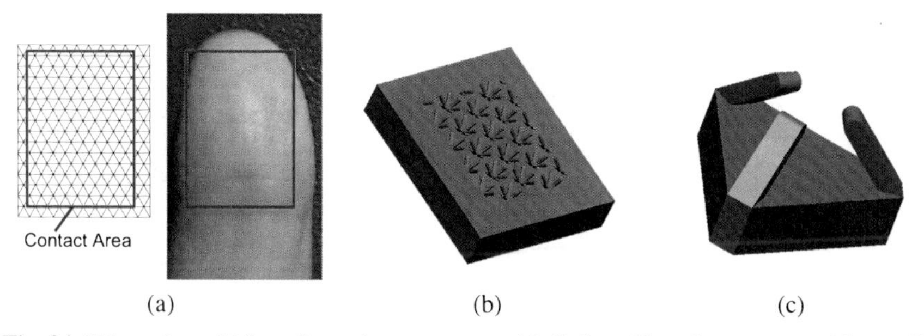

(a) (b) (c)

Fig. 34 Triangular grid for a fingertip contact area (a). Soft pad based on pattern with equally spaced micro-beams (b) and associated te (c).

injection moulding the possible geometries are limited by the extraction of the mould. In addition, silicon pad must be carefully cured in order to avoid the presence of air within the mould. Nonetheless, beside the technological limitation and production difficulties, different DLD pads depicted are realizable with both materials.

5.2 *Design of the Structured Pad Inner Layer*

The basic idea concerning the choice of the inner layer geometry is that it is simpler to design and analyze a simple shape element and then to replicate it as many times as needed. Therefore, it is suggested to conceptually divide the overall contact planar area into finite elementary regions. Once the element LD curve is known (by means of numerical analysis or experiments), the number of elements, N, can be chosen such that:

$$F = N \cdot F_t \tag{13}$$

where $F_t = f_t(\delta)$ is the non-linear LD curve of each element. Hence, it is proposed to divide the pad contact area into finite TEs by using a *triangular grid* [27].

A triangular grid is defined as an isometric grid formed by tiling the plane regularly with equilateral triangles. The grid cells that fall outside the object are removed. The result is a mesh with equal interior TEs. If needed, the grid cells that intersect the object boundary can be adjusted or trimmed so that they fit into the object. Nevertheless, deformed TEs which are located on the boundary might present an LD curve which slightly differs from the LD curve of interior TEs. An example of a triangular grid for meshing a fingertip contact area is shown in Fig. 34(a).

Obviously, smaller TEs are beneficiary for two reasons:

- The object boundaries can be better captured by a fine mesh than by a coarse mesh.
- Whatever will be the inner layer geometry of each TE, the contact pressure will tend to be a uniform function (i.e. a continuous function) as $N \to \infty$.

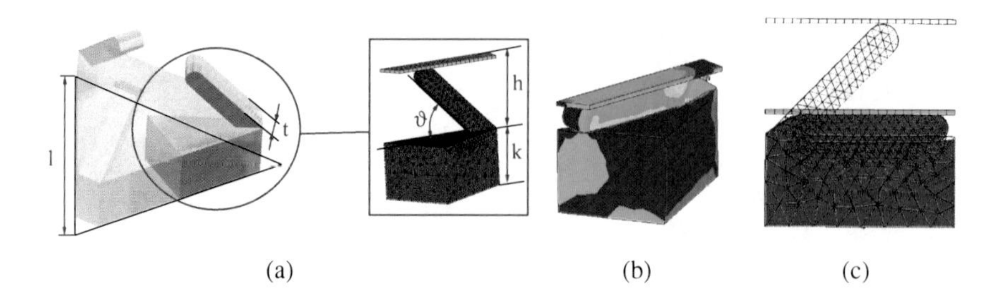

Fig. 35 Numerical analysis of te based on a series of inclined micro-beams. model mesh (a), collapsed micro-beam (b,c). material: *tango plus* rubber [57].

Note that: 1) the procedure outlined in the following regards the definition of an overall pad contact force (i.e. overall pad compliance) which is an integral (rather than a local) property of the Pad; 2) N elements are involved in the contact simultaneously and the contact area is displacement independent. The smallest size, A_{TE}^{min}, of feasible TEs is determined by the technological feasibility of the pads. On the other hand, Equation 13 constraints the number of elements which must be contained within a given contact area and therefore the element size $\overline{A}_{TE}$. If the size $\overline{A}_{TE}$ required by 13 is lower than A_{TE}^{min}, a practical solution cannot be achieved for the given TE inner-layer geometry. On the other hand, if $\overline{A}_{TE} > A_{TE}^{min}$, the TE outer-layer (skin) can be enlarged without altering the TE LD curve. As for the TE internal layer, it is designed in order to replicate the qualitative shape of the non-linear compression law which is typical of endoskeletal structures covered by pulpy tissues. This behavior is well exemplified by the LD curve of the human finger shown in Fig. 32: it is possible to note an initial, quasi-linear LD curve for small displacement followed by a rapid load increase. In order to replicate this particular compression law, it is proposed to use a series of micro-beams inclined of $\vartheta = 45°$ with respect to the normal to the external surface (normal axis, Fig. 33(b)), thus transforming normal loads acting on the contact into bending actions applied on each beam. The micro-beams are placed on the edge of the TE as depicted in Fig. 34 (artificial pad internal layer surface and associated TE). This peculiar geometry presents a quasi-linear LD curve for small displacements which is characterized by a very low stiffness. On the other hand, the load rapidly increases once the micro-beams collapse on the outer skin. In such situation, the TE behave similarly to a pad made of a uniform soft material. A finite element model of the TE is shown in Fig. 35. In particular, Figs. 35(b) and 35(c) depicts one collapsed micro-beam.

5.3 Soft Pad Realization

The smallest TE which is considered realizable by means of SLA is an equilateral triangle having surface area of 6.9 mm^2 (i.e. 4mm side). The pad thickness is chosen

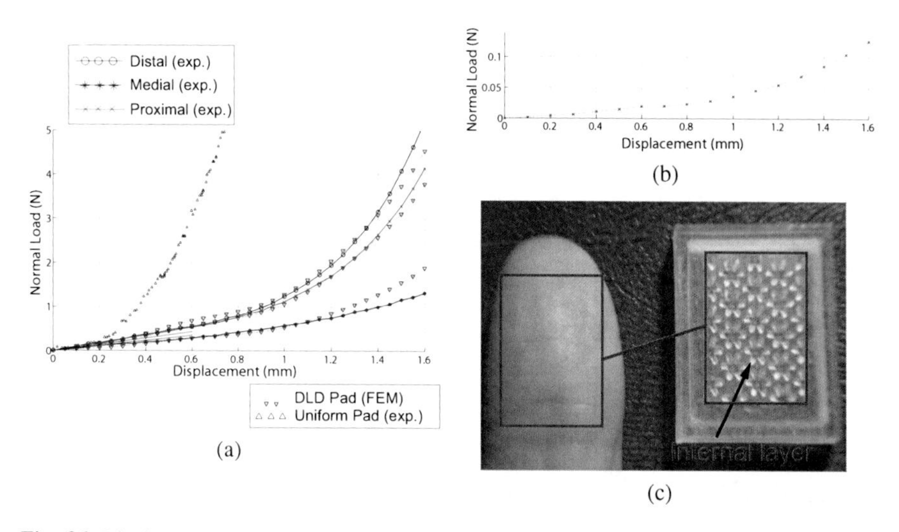

Fig. 36 Displacement (mm) versus normal load (n) for DLD pad and human fingertip (a), TE made of inclined micro-beams for uniform pad (b), Artificial soft pad (c). Experimental (exp.) and FEM results.

to be 3.0 mm (i.e. half the thickness of previously published solutions, see Fig. 32). The TE design is exactly the one depicted in Figs. 34, 36 ($t = 0.5mm, h = 2mm, k = 1mm, \theta = 45°$, l to be designed) and numerical relationship between the applied normal force F and the consequent displacement δ is shown in Fig. 36(b) (FEA results).

Let us consider first the distal phalanx. The contact area to be meshed is a 20mm x 15mm rectangle which is meshed by means of 36 TEs in order to obtain the desired compliance. Such TE presents a surface area of 8.3333 mm^2. Figure 36(a) shows the numerical relationship between the normal load (N) and the resulting displacement (mm) for: 1) the structured pad depicted in Fig. 34(b); 2) a uniform PAD of the same thickness (3mm) made of a softer material (refer to Fig. 32); 3) for the human finger. It can be seen that a 3mm thick structured pad represents a substantial step forward in human finger mimicry in terms of stiffness, when compared to previously published solutions where different materials and higher pad thickness are used. Finally, the first pad prototype is shown in Fig. 36(c). Concerning the medial phalanx the number of TEs is reduced to 30. Concerning the distal phalanges, the overall contact area is split into two 20mm x 15mm rectangles and the number of TEs for each rectangle is 15.

The potentialities of the DLD concept have been experimentally evaluated on hemispherical soft pads shaped over a rigid core [6]. The pad to be mounted on the DEXMART Hand prototype are shaped as in Fig. 36(c) and their physical implementation is shown in Fig. 2(a).

6 Conclusion

What can be summarized from an analysis of the currently available anthropomorphic human-sized robot hands is that a design approach based on conventional mechanics and commercially available actuators and sensors are not suitable for the implementation of these devices, because of the needs in terms of integration among the components, simplification of the overall structure, reduction of the cost and of the power consumption to name a few. The impression is that new approaches should be identified for the design of anthropomorphic dexterous robot hands. In this chapter, possible solutions and innovative technologies and sensors developed within the DEXMART project for improving the design of robot hands have been presented. On the basis of the results of this research activity it is possible to conclude that innovative approaches based on non-conventional structures and on the study of the biological model can significantly simplify the design, enhance the reliability and the performance, reduce the costs of robotic hands. Moreover, innovative actuators and sensors that can be directly integrated and used in anthropomorphic hands are fundamental for achieving the general goal of improving the reliability and the functionality of the device reducing its complexity and the overall costs for the development and the production process. Finally, the results of the experiments performed for the evaluation of the proposed solutions have been presented, aiming at giving practical answers to the outlined problems.

Acknowledgements. The research leading to these results has been supported by the DEXMART Large-scale integrating project, which has received funding from the European Communitys Seventh Framework Programme (FP7/2007-2013) under grant agreement ICT-216239. The authors are solely responsible for its content. It does not represent the opinion of the European Community and the Community is not responsible for any use that might be made of the information contained therein.

References

1. Armstrong-Hèlouvry, B., Dupont, P., Canudas de Wit, C.: A survey of models, analysis tools and compensation methods for the control of machines with friction. Automatica 30, 1083–1138 (1994)
2. Bergström, J., Boyce, M.: Constitutive modeling of the large strain time-dependent behavior of elastomers. Journal of Mechanics and Physics of Solids 46, 931–954 (1998)
3. Berselli, G., Borghesan, G., Brandi, M., Melchiorri, C., Natale, C., Palli, G., Pirozzi, S., Vassura, G.: Integrated mechatronic design for a new generation of robotic hands. In: IFAC Symposium on Robot Control, Gifu (2009)
4. Berselli, G., Piccinini, M., Vassura, G.: On designing structured soft covers for robotic limbs with predetermined compliance. In: ASME International Design Engineering Technical Conference, Montréal (2010)
5. Berselli, G., Piccinini, M., Vassura, G.: Tailoring the viscoelastic properties of soft pads for robotic limbs through purposely designed fluid filled structures. In: IEEE International Conference on Robotics and Automation, Anchorage, AK (2010)

6. Berselli, G., Vassura, G.: Differentiated layer design to modify the compliance of soft pads for robotic limbs. In: IEEE International Conference on Robotics and Automation, Kobe (2009)
7. Berselli, G., Vertechy, R., Vassura, G., Parenti Castelli, V.: Design of a single-acting constant-force actuator based on dielectric elastomers. ASME Journal of Mechanisms and Robotics 1(3) (2009)
8. Biagiotti, L., Lotti, F., Melchiorri, C., Palli, G., Tiezzi, P., Vassura, G.: Development of UB Hand 3: Early results. In: IEEE International Conference on Robotics and Automation, Barcelona (2005)
9. Biagiotti, L., Tiezzi, P., Melchiorri, C., Vassura, G.: Modelling and controlling the compliance of a robotic hand with soft finger-pads. In: IEEE International Conference on Robotics and Automation, New Orleans, LA (2004)
10. Bicchi, A., Tonietti, G.: Fast and soft arm tactics: Dealing with the safety-performance tradeoff in robot arms design and control. IEEE Robotics & Automation Magazine 11(2), 22–33 (2004)
11. Bicchi, A., Tonietti, G., Bavaro, M., Piccigallo, M.: Variable stiffness actuators for fast and safe motion control. In: Dario, P., Chatila, R. (eds.) The Eleventh International Symposium on Robotics Research, pp. 527–536. Springer, Heidelberg (2005)
12. Birglen, L., Laliberté, T., Gosselin, C.: Underactuated Robotic Hands. Springer, Heidelberg (2008)
13. Borghesan, G., Palli, G., Melchiorri, C.: Design of tendon-driven robotic fingers: Modeling and control issues. In: IEEE International Conference on Robotics and Automation, Anchorage, AK (2010)
14. Borghesan, G., Palli, G., Melchiorri, C.: Friction compensation and virtual force sensing for robotic hands. In: IEEE International Conference on Robotics and Automation, Shanghai (2011)
15. Butterfaß, J., Grebenstein, M., Liu, H., Hirzinger, G.: DLR-hand II: Next generation of a dextrous robot hand. In: IEEE International Conference on Robotics and Automation, Seoul (2001)
16. Carrozza, M., Cappiello, G., Stellin, G., Zaccone, F., Vecchi, F., Micera, S., Dario, P.: A cosmetic prosthetic hand with tendon driven under-actuated mechanism and compliant joints: Ongoing research and preliminary results. In: IEEE International Conference on Robotics and Automation, Barcelona (2005)
17. Carrozza, M., Cappiello, G., Stellin, G., Zaccone, F., Vecchi, F., Micera, S., Dario, P.: On the development of a novel adaptive prosthetic hand with compliant joints: Experimental platform and EMG control. In: IEEE/RSJ International Conference on Intelligent Robots and Systems, Edmonton (2005)
18. Cavallo, A., De Maria, G., Natale, C., Pirozzi, S.: Optoelectronic joint angular sensor for robotic fingers. Sensors and Actuators A: Physical 152, 203–210 (2009)
19. Chou, P.C., Hannaford, B.: Measurement and modeling of McKibben pneumatic artificial muscles. IEEE Transactions on Robotics and Automation 12, 90–102 (1996)
20. Cutkosky, M.R., Jourdain, J.M., Wright, P.K.: Skin materials for robotic fingers. In: IEEE International Conference on Robotics and Automation, Raleigh, NC (1987)
21. Cutkosky, M.R., Wright, P.K.: Friction, stability and the design of robotic fingers. International Journal of Robotics Research 5(4), 20–37 (1986)
22. Dalley, S., Wiste, T., Withrow, T., Goldfarb, M.: Design of a multifunctional anthropomorphic prosthetic hand with extrinsic actuation. IEEE/ASME Transactions on Mechatronics 14, 699–706 (2009)
23. D'Amore, A., De Maria, G., Grassia, L., Natale, C., Pirozzi, S.: Silicone-rubber-based tactile sensors for the measurement of normal and tangential components of the contact force. Journal of Applied Polymer Science 122, 3758–3770 (2011)

24. De Maria, G., Natale, C., Pirozzi, S.: Force/tactile sensor for robotic applications. Sensors and Actuators A: Physical 175, 60–72 (2012)
25. Du, H., Lau, G.K., Lim, M.K., Qui, J.: Topological optimization of mechanical amplifiers for piezoelectric actuators under dynamic motion. Smart Materials and Structures 9, 788–800 (2000)
26. Gao, X., Jin, M., Jiang, L., Xie, Z., He, P., Yang, L., Liu, Y., Wei, R., Cai, H., Liu, H., Butterfaß, J., Grebenstein, M., Seitz, N., Hirzinger, G.: The HIT/DLR dexterous hand: Work in progress. In: IEEE International Conference on Robotics and Automation, Taipei (2003)
27. Gardner, M.: Knotted Doughnuts and Other Mathematical Entertainments. Freeman, New York (1986)
28. Gialias, N., Matsuoka, Y.: Muscle actuator design for the act hand. In: IEEE International Conference on Robotics and Automation, New Orleans, LA (2004)
29. Howell, L.L.: Compliant Mechanisms. Wiley, New York (2001)
30. Jacobsen, S., Iversen, E., Knutti, D., Johnson, R., Biggers, K.: Design of the Utah/M.I.T. dexterous hand. In: IEEE International Conference on Robotics and Automation, San Francisco, CA (1986)
31. Jacobsen, S., Wood, J., Knutti, D., Biggers, K.: The UTAH/M.I.T. dextrous hand: Work in progress. International Journal of Robotics Research 3(4), 21–50 (1984)
32. Jacobsen, S.C., Jerrard, R.B., Knutti, D., Carruth, J.: The LADD actuator as a prosthetic muscle. In: International Symposium on External Control of Human Extremities, Dubrovnik (1975)
33. Jung, S., Kang, S., Lee, M., Moon, I.: Design of robotic hand with tendon-driven three fingers. In: International Conference on Control, Automation and Systems, Seoul (2007)
34. Kaneko, M., Wada, M., Maekawa, H., Tanie, K.: A new consideration on tendon-tension control system of robot hands. In: IEEE International Conference on Robotics and Automation, Sacramento, CA (1991)
35. Kaneko, M., Yokoi, K., Tanie, K.: On a new torque sensor for tendon drive fingers. IEEE Transactions on Robotics and Automation 6, 501–507 (1990)
36. Kao, I., Yang, F.: Stiffness and contact mechanics for soft fingers in grasping and manipulation. IEEE Transactions of Robotics and Automation 20, 132–135 (2004)
37. Kasap, S.: Optoelectronics and Photonics: Principles and Practices. Prentice Hall, Englewood Cliffs (2001)
38. Kawasaki, H., Komatsu, T., Uchiyama, K.: Dexterous anthropomorphic robot hand with distributed tactile sensor: Gifu hand II. IEEE/ASME Transactions on Mechatronics 7, 296–303 (2002)
39. Kota, S., Hetrick, J., Li, Z., Saggere, L.: Tailoring unconventional actuators with compliant transmissions: Design methods and applications. IEEE/ASME Transactions on Mechatronics 4, 396–408 (1999)
40. Lee, M.H., Nicholls, H.R.: Tactile sensing for mechatronics–A state of the art survey. Mechatronics 9, 1–31 (1999)
41. Lee, Y., Choi, H., Chung, W., Youm, Y.: Stiffness control of a coupled tendon-driven robot hand. IEEE Control Systems Magazine 14(5), 10–19 (1994)
42. Li, Y., Saitou, K., Kikuchi, N.: Topology optimization of thermally actuated compliant mechanisms considering time-transient effect. Finite Elements in Analysis and Design 40, 1317–1331 (2004)
43. Liu, H., Butterfaß, J., Knoch, S., Meusel, P., Hirzinger, G.: A new control strategy for DLR's multisensory articulated hand. IEEE Control Systems Magazine 19(2), 47–54 (1999)

44. Liu, H., Meusel, P., Hirzinger, G., Jin, M., Liu, Y., Xie, Z.: The modular multisensory DLR-HIT-Hand: Hardware and software architecture. IEEE/ASME Transactions on Mechatronics 13, 461–469 (2008)
45. Lobontiu, N., Paine, J.S.N., Garcia, E., Goldfarb, M.: Corner-filleted flexure hinges. ASME Journal of Mechanical Design 123, 346–352 (2001)
46. Melchiorri, C., Kaneko, M.: Robot hands. In: Siciliano, B., Khatib, O. (eds.) Springer Handbook of Robotics, pp. 345–360. Springer, Heidelberg (2008)
47. Mennitto, G., Buehler, M.: Ladd transmissions: Design, manufacture, and new compliance models. ASME Journal of Mechanical Design 119, 197–203 (1997)
48. Murray, R., Li, Z., Sastry, S.: A Mathematical Introduction to Robotic Manipulation. CRC Press, Boca Raton (1994)
49. Namiki, A., Imai, Y., Ishikawa, M., Kaneko, M.: Development of a high-speed multifingered hand system and its application to catching. In: IEEE/RSJ International Conference on Intelligent Robots and Systems, Las Vegas, NV (2003)
50. Palli, G., Berselli, G., Melchiorri, C., Vassura, G.: Design and modeling of variable stiffness joints based on compliant flexures. In: ASME International Design Engineering Technical Conferences, Montréal (2010)
51. Palli, G., Borghesan, G., Melchiorri, C.: Modeling, identification and control of tendon-based actuation systems. IEEE Transactions on Robotics (2012) (in Press)
52. Palli, G., Melchiorri, C.: Velocity and disturbance observer for non-model based load and friction compensation. In: International Workshop on Advanced Motion Control, Trento (2008)
53. Palli, G., Natale, C., May, C., Melchiorri, C., Würtz, T.: Modeling and control of the twisted string actuation system. IEEE/ASME Transactions on Mechatronics (2012) (in Press)
54. Palli, G., Pirozzi, S.: Force sensor based on discrete optoelectronic components and compliant frames. Sensors and Actuators A: Physical 165, 239–249 (2011)
55. Palli, G., Pirozzi, S.: Miniaturized optical-based force sensors for tendon-driven robots. In: IEEE International Conference on Robotics and Automation, Shanghai (2011)
56. Pawluk, R., Howe, R.D.: Dynamic lumped element response of the human fingerpad. Journal of Biomechanical Engineering 121, 178–183 (1999)
57. Piccinini, M., Berselli, G., Zucchelli, A., Vassura, G.: Predicting the compliance of soft fingertips with differentiated layer design: A numerical and experimental investigation. In: 17th International Conference on Advanced Robotics, München (2009)
58. Pirozzi, S., Grassia, L.: Tactile sensor based on led-phototransistor couples. In: 9th IEEE-RAS International Conference on Humanoid Robots, Paris (2009)
59. Pottebaum, K.L., Beaman, J.J.: A dynamic model of a concentric ladd actuator. ASME Journal of Dynamic Systems, Measurement, and Control 105, 157–164 (1983)
60. Rothling, F., Haschke, R., Steil, J.J., Ritter, H.: Platform portable anthropomorphic grasping with the bielefeld 20-DOF shadow and 9-DOF TUM hand. In: IEEE/RSJ International Conference on Intelligent Robots and Systems, San Diego, CA (2007)
61. Salisbury, J., Craig, J.: Articulated hands: Force control and kinematic issues. International Journal of Robotic Research 1(1), 4–17 (1982)
62. Serina, E., Mockensturm, F., Mote Jr., C., Rempel, D.: A structural model of the forced compression of the fingertip pulp. Journal of Biomechanics 31, 639–646 (1998)
63. Shadow Robot Company: Design of a dextrous hand for advanced CLAWAR applications. In: Climbing and Walking Robots and the Supporting Technologies for Mobile Machines, Catania (2003)
64. Shao, F., Childs, T.H., Henson, B.: Developing an artificial fingertip with human friction properties. Tribology International 42, 1575–1581 (2009)

65. Shimoga, K., Goldenberg, A.: Soft robotic fingertips Part I: A comparison of construction materials. International Journal of Robotics Research 15, 320–334 (1996)
66. Shimoga, K., Goldenberg, A.: Soft robotic fingertips Part II: Modelling and impedance regulation. International Journal of Robotics Research 15, 335–350 (1996)
67. Siciliano, B., Khatib, O. (eds.): Springer Handbook of Robotics. Springer, Heidelberg (2008)
68. Sonoda, T., Godler, I.: A five fingered robotic hand prototype by using twist drive. In: International Symposium on Robotics, München (2010)
69. Sonoda, T., Godler, I.: Multi-fingered robotic hand employing strings transmission named Twist Drive. In: IEEE/RSJ International Conference on Intelligent Robots and Systems, Taipei (2010)
70. Tiezzi, P.: Experimental analysis and modeling of soft pads for application on robotic fingers, Ph.D. thesis, Department of Mechanical Engineering. University of Bologna (2006)
71. Tiezzi, P., Kao, I.: Modeling of viscoelastic contacts and evolution of limit surface for robotic contact interface. IEEE Transastions on Robotics 23, 206–217 (2007)
72. Tiezzi, P., Lotti, F., Vassura, G.: Polyurethane gel pulps for robotic fingers. In: 11th International Conference on Advanced Robotics, Coimbra (2003)
73. Tsai, C.D., Kao, I.: The latency model for viscoelastic contact interface in robotics: Theory and experiments. In: IEEE International Conference on Robotics and Automation, Kobe (2009)
74. Wang, L.X.: Fuzzy systems are universal approximators. IEEE Transactions on Systems, Man, and Cybernetics 7, 1163–1170 (1992)
75. Wu, J., Dong, R., Smutz, W., Rakheja, S.: Dynamic interaction between a fingerpad and a flat surface: Experiments and analysis. Medical Engineering and Physics 25, 397–406 (2003)
76. Xydas, N., Kao, I.: Modeling of contact mechanics and friction limit surface for soft fingers in robotics, with experimental results. International Journal of Robotics Research 18, 941–950 (1999)

Grasping and Control of Multi-Fingered Hands

Luigi Villani, Fanny Ficuciello, Vincenzo Lippiello, Gianluca Palli,
Fabio Ruggiero, and Bruno Siciliano

Abstract. An important issue in controlling a multi-fingered robotic hand grasping an object is the evaluation of the minimal contact forces able to guarantee the stability of the grasp and its feasibility. This problem can be solved online if suitable sensing information is available. In detail, using finger tactile information and contact force measurements, an efficient algorithm is developed to compute the optimal contact forces, assuming that, during the execution of a manipulation task, both the position of the contact points on the object and the wrench to be balanced by the contact forces may change with time. Since manipulation systems can be redundant also if the single fingers are not –due to the presence of the additional degrees of freedom (DOFs) provided by the contact variables– suitable control strategies taking advantage of such redundancy are adopted, both for single and dual-hand manipulation tasks. Another goal pursued in DEXMART is the development of a human-like grasping approach inspired to neuroscience studies. In order to simplify the synthesis of a grasp, a configuration subspace based on few predominant postural synergies of the robotic hand is computed. This approach is evaluated at kinematic level, showing that power and precise grasps can be performed using up to the third predominant synergy.

1 Grasping Force Optimization

The control of a robotic system equipped with multi-fingered hands involves several aspects which range from the synthesis of the optimal grasping contact points and

Luigi Villani · Fanny Ficuciello · Vincenzo Lippiello · Fabio Ruggiero · Bruno Siciliano
PRISMA Lab, Dipartimento di Informatica e Sistemistica,
Università degli Studi di Napoli Federico II, via Claudio 21, 80125 Napoli, Italy
e-mail: {luigi.villani,fanny.ficuciello,vincenzo.lippiello,
fabio.ruggiero,bruno.siciliano}@unina.it

Gianluca Palli
Dipartimento di Elettronica Informatica e Sistemistica,
Alma Mater Studiorum Università di Bologna, viale Risorgimento 2, 40136 Bologna, Italy
e-mail: gianluca.palli@unibo.it

B. Siciliano (Ed.): Advanced Bimanual Manipulation, STAR 80, pp. 219–266.
springerlink.com

grasp planning, to the load sharing and grasp control. With respect to this last, the evaluation of the grasping forces able to guarantee the stability of the grasp and its feasibility, in the face of the external disturbances, represents a crucial problem. The complexity of the problem relies on the necessity of resolving online an optimization problem where both constraints and objective functions are nonlinear, the number of variable and constraints are relatively large, and the grasp configuration and the load wrench may change with time.

The force closure [41] and the optimal grasp configuration selection problems are not considered here, as they would be in charge of the grasp planner. On the other hand, the grasping force optimization problem has been intensively investigated only for relative simple robotic systems and not yet explicitly in the case of bimanual human-like robotic systems. For this last, the computational complexity becomes a major issue to be considered for an efficient online solution.

The nonlinearity of the contact friction models (point contact with friction or soft-finger contact) complicates the solution of the optimal contact force distribution problem. In [26] the friction cone constraints have been formulated in terms of linear matrix inequalities (LMIs), and the grasping optimization problem is addressed as a convex optimization problem involving LMIs with the max–det function as objective function. This problem can be efficiently solved with the interior point algorithm for a small number of fingers.

Starting from the observation that verifying the friction cone constraints is equivalent to testing the positive definiteness of certain symmetric matrices, in [11] the grasp force optimization has been formulated as a convex optimization problem on a Riemannian manifold with linear constraints. Several gradient flow type algorithms have been proposed to provide solutions suitable for real-time applications [12]; to reduce complexity of matrix inversion, the computation of the solution can be split into an on-line phase and an off-line phase, and sparse matrix techniques can be adopted [30]. This technique has been employed and experientially tested with an impedance control approach addressing the regrasping problem for dextrous manipulation tasks [54].

A further improvement has been presented in [27], consisting in a new compact semidefinite representation of the friction cone constraints which allows a significant reduction of the dimension of the optimization problem. Moreover, an estimation technique and a recursion method for selecting the step size in the gradient algorithm are proposed, together with the proof of the quadratic convergence of the algorithm.

In [50] and [53] a method based on the minimization of a cost function, which gives an analytical solution but does not ensure by itself the satisfaction of the friction constraints is presented. An iterative correction algorithm allows modifying this function until the internal forces enter the friction cone, resulting in a fast suboptimal solution suitable for real-time applications. The grasping force optimization problem in the case of power grasp is addressed in [60]. In this case, the optimization problem is formulated as a convex optimization problem involving LMIs similarly to [26], but considering a decomposition of the contact force space into four orthogonal subspaces of active and passive forces.

The method proposed in [11] requires the on-line pseudo-inversion of a constrained matrix whose dimension linearly increases with the number of fingers with a factor that depends on the contact type. By adopting the frictional cone constraint matrix representation proposed in [27], the dimension of the problem decreases and the solution can be computed in real time. However, if torque limits constraints are considered, the complexity of the problem increases more than quadratically with the number of joints, which is higher in a dual-hand system, making it unsuitable for real-time applications. Moreover, all the proposed solutions require, at each iteration, the evaluation of an initial point that satisfies the frictional cone constraints and the joint torque limits. The initial point can be computed with the method proposed in [34], but at the expense of a significant computational effort.

The proposed algorithm is based on the compact formulation of [27] and on the solution of a convex optimization case as in [12], and it extends to bimanual manipulation systems our previous works on single-hand manipulation [32], [33]. The method allows considering also joint torque constraints, with a minimum increase of computation complexity, compatible with real-time constraints. Moreover, the iterative formulation does not require the evaluation at each step of a new initial point. Finally, a sub-optimal single-hand optimization algorithm is proposed to cope with very limited computational hardware availability, and compared with the optimal solution. In particular, a new criterion for load sharing [61], [58], [56] between the hands is here introduced to improve the solution. The feasibility and the effectiveness of this approach have been tested in a simulation scenario where a robotic torso equipped with two dextrous hands is used to empty a half-filled bottle.

1.1 Problem Formulation

Consider a bimanual robotic system equipped with two multi-fingered hands grasping an object with n contacts between the object and the fingertips, the links of the fingers and the palm. Denote the contact wrench of the grasp with $\mathbf{c} = \left[\mathbf{c}_r^{\mathrm{T}}\ \mathbf{c}_l^{\mathrm{T}}\right]^{\mathrm{T}} = \left[\mathbf{c}_1^{\mathrm{T}} \ldots \mathbf{c}_n^{\mathrm{T}}\right]^{\mathrm{T}} \in \mathbb{R}^{nm}$, where $\mathbf{c}_i \in \mathbb{R}^m$ is the wrench vector of the i-th contact with dimension m depending from the adopted contact model, and $\mathbf{c}_r$ and $\mathbf{c}_l$ are the corresponding wrench vectors of all the contact points of the right and left hand, respectively.

The grasping force optimization problem (GFO) consists in finding the set of contact wrenches balancing the generalized external force $\mathbf{h}_e \in \mathbb{R}^6$ acting on the object (including object inertia and weight), which are feasible with respect to the kinematic structure of the hand and to the corresponding joint torque limits, and minimize the overall stress applied the object, i.e, the internal forces. Moreover, to avoid the slippage of the fingers on the object surface, each contact wrench has to be confined within the friction cone.

The balance equation for the generalized forces applied to the object can be written in the form

$$\mathbf{h}_e = \mathbf{G}\mathbf{c}, \tag{1}$$

where $\mathbf{G} = \begin{bmatrix} \mathbf{G}_r \ \mathbf{G}_l \end{bmatrix} \in \mathbb{R}^{6\times nm}$ is the grasp map composed of the the grasp matices of the right and left hand, which is full rank for force-closure grasps [41]. It is assumed that the contact point configurations ensuring the force-closure constraint are assigned at each time by the planning system.

Although several contact models can be used, the two usually adopted are the *point contact with friction* (PCWF) model and the *soft finger contact* (SFC) model.

In the PCWF case, the contact wrench has three DOFs ($m = 3$): the normal component $c_{i,z}$ to the object surface and the two components $c_{i,x}$, $c_{i,y}$ on the tangent plane. The friction constraint is represented by the law

$$\frac{1}{\mu_i^2}\left(c_{i,x}^2 + c_{i,y}^2\right) \leq c_{i,z}^2 \text{ and } c_{i,z} > 0, \tag{2}$$

where μ_i is the friction coefficient at the i-th contact point.

In the SFC case, the contact wrench has an additional DOF $c_{i,t}$ ($m = 4$), corresponding to the torsional component of the moment about the contact normal. In this case, the friction constraint in an elliptic approximation can be expressed in the form

$$\frac{1}{\mu_i}\left(c_{i,x}^2 + c_{i,y}^2\right) + \frac{1}{\mu_{t,i}}c_{i,t}^2 \leq c_{i,z}^2 \text{ and } c_{i,z} > 0, \tag{3}$$

where μ_i and $\mu_{i,t}$ denote the tangential and torsion friction coefficients at the i-th contact point, respectively.

The balance equation for the torques applied to fingers joints of the hand can be written in the form

$$\mathbf{J}^{\mathrm{T}}(\mathbf{q})\mathbf{c} + \boldsymbol{\tau}_e = \boldsymbol{\tau}, \tag{4}$$

where and $\mathbf{J}(\mathbf{q}) = \begin{bmatrix} \mathbf{J}_r^{\mathrm{T}} \ \mathbf{J}_l^{\mathrm{T}} \end{bmatrix}^{\mathrm{T}}$ is the $(nm \times l)$ hands Jacobian matrix, depending on the (l-dimensional) vector $\mathbf{q}$ of the joint variables, being l the total number of the joints, $\boldsymbol{\tau}_e$ is the external torque, including gravity, Coriolis, centripetal and inertia effects at the fingers joints, and $\boldsymbol{\tau}$ is the torque provided by the actuators. For simplicity, it is assumed that $\mathscr{N}(\mathbf{J}^{\mathrm{T}}) = \emptyset$, meaning the absence of structurally dependent forces, namely, contact forces not caused by joint torques but depending on hand mechanics (see, e.g., [41]).

To ensure that the joint actuators are able to provide the required torques, a joint torque constraint must also be considered

$$\boldsymbol{\tau}_L \leq \boldsymbol{\tau} \leq \boldsymbol{\tau}_U, \tag{5}$$

where $\boldsymbol{\tau}_L$ ($\boldsymbol{\tau}_U$) is the lower (upper) joint torque limit.

The simultaneous satisfaction of the force balance equation (1), with the friction constraints (2) and (3), and of the joint torque balance equation (4) with constraint (5), implies that the grasp is stable and feasible. The GFO problem considered here consists in finding the optimal grasp wrench that minimizes the internal forces acting on the object, under the above constraints. The internal forces are contact wrenches that satisfy the friction cone constraints and belong to the null space of the grasp matrix $\mathbf{G}$. These wrenches $\mathbf{c}_{int}$ do not contribute to the balance equation (1), being $\mathbf{G}\mathbf{c}_{int} = 0$, but are used to satisfy the friction cone constraints at the contact points.

1.2 Grasping Constraints and Cost Function

The frictional inequalities (2) and (3) are equivalent to the positive definiteness of the block-diagonal matrix [27]

$$\mathbf{F}(\mathbf{c}) = \mathrm{diag}\left(\mathbf{F}_1(\mathbf{c}_1),\ldots,\mathbf{F}_n(\mathbf{c}_n)\right) > 0, \tag{6}$$

where $\mathbf{F}_i(\mathbf{c}_i)$ is the symmetric (2×2) matrix

$$\mathbf{F}_i(\mathbf{c}_i) = \begin{bmatrix} c_{i,z} + \frac{c_{i,x}}{\mu_i} & \frac{c_{i,y}}{\mu_i} \\ \frac{c_{i,y}}{\mu_i} & c_{i,z} - \frac{c_{i,x}}{\mu_i} \end{bmatrix} \tag{7}$$

in the PCWF case, while it is the Hermitian (2×2) matrix

$$\mathbf{F}_i(\mathbf{c}_i) = \begin{bmatrix} c_{i,z} + \frac{c_{i,x}}{\sqrt{\mu_i}} & \frac{c_{i,y}}{\sqrt{\mu_i}} - j\frac{c_{i,t}}{\sqrt{\mu_{i,t}}} \\ \frac{c_{i,y}}{\sqrt{\mu_i}} + j\frac{c_{i,t}}{\sqrt{\mu_{i,t}}} & c_{i,z} - \frac{c_{i,x}}{\sqrt{\mu_i}} \end{bmatrix}, \tag{8}$$

in the SFC case.

Similarly, the torque limit constraint (5), in view of the torque balance equation (4), is equivalent to the positive definiteness of the diagonal matrix

$$\mathbf{T}(\mathbf{c},\mathbf{q},\tau_e) = \mathrm{diag}\left(\tau_B\right) > 0, \tag{9}$$

where

$$\tau_B = \begin{bmatrix} \tau_{B,L} \\ \tau_{B,H} \end{bmatrix} = \begin{bmatrix} \mathbf{J}^{\mathrm{T}}(\mathbf{q})\mathbf{c} - \tau_L + \tau_e \\ -\mathbf{J}^{\mathrm{T}}(\mathbf{q})\mathbf{c} + \tau_H - \tau_e \end{bmatrix} \tag{10}$$

contains the distances of actuator torques from the lower ($\tau_{B,L}$) and upper ($\tau_{B,H}$) limits, respectively.

Hence, the simultaneous satisfaction of both frictional and joint torque constraints is equivalent to the positive definiteness of the linearly constrained block-diagonal matrix

$$\mathbf{P} = \mathrm{diag}\left(\mathbf{F},\mathbf{T}\right) > 0. \tag{11}$$

Notice that the elements of the matrices $\mathbf{F}$ and $\mathbf{T}$ are linearly dependent, because both depend on $\mathbf{c}$. Moreover, the force balance equation (1) and the torque balance equation (4) corresponds to linear constraints imposed on matrix $\mathbf{P}$.

By denoting with $\mathbf{c}(\mathbf{F})$ the contact wrench vector extracted from the frictional constraint matrix, with $\tau_B(\mathbf{T})$ the vector composed by the diagonal elements of $\mathbf{T}$, and defining vector $\xi(\mathbf{P}) = \left[\mathbf{c}(\mathbf{F})^{\mathrm{T}}\ \tau_B(\mathbf{T})^{\mathrm{T}}\right]^{\mathrm{T}}$, the linear constraints on matrix $\mathbf{P}$ imposed by (1) and (4) can be represented in the following affine general form

$$\mathbf{A}\xi(\mathbf{P}) = \mathbf{b} \tag{12}$$

with

$$\mathbf{A} = \begin{bmatrix} \mathbf{G}\ \mathbf{0}_{6\times 2l} \\ \mathbf{A}_\tau \end{bmatrix} \quad \mathbf{b} = \begin{bmatrix} \mathbf{h}_e \\ \tau_L - \tau_e \\ \tau_H - \tau_e \end{bmatrix}, \tag{13}$$

where $\mathbf{A}_\tau$ is a $(2l \times nm + 2l)$ matrix defined as follows

$$\mathbf{A}_\tau = \begin{bmatrix} \mathbf{J}(\mathbf{q})^{\mathrm{T}} & -\mathbf{I}_l & \mathbf{0}_l \\ \mathbf{J}(\mathbf{q})^{\mathrm{T}} & \mathbf{0}_l & \mathbf{I}_l \end{bmatrix}, \tag{14}$$

being $\mathbf{0}_\times$ the null matrix and $\mathbf{I}_\times$ the identity matrix of the indicated dimensions.

The optimization procedure is based on the minimization of the cost function $\Phi(\mathbf{P}) : \mathscr{P}(r) \to \mathbb{R}$, being $\mathscr{P}(r)$ the set of positive definite symmetric $(r \times r)$ matrices $\mathbf{P} = \mathbf{P}^{\mathrm{T}} > 0$, defined as

$$\Phi(\mathbf{P}) = \mathrm{tr}\left(\mathbf{W}_p\mathbf{P} + \mathbf{W}_b\mathbf{P}^{-1}\right), \tag{15}$$

where $\mathrm{tr}(\cdot)$ denotes the trace operator, $\mathbf{W}_p$ and $\mathbf{W}_b$ are positive definite symmetric matrices. Notice that Φ is a strictly convex twice continuously differentiable function on $\mathscr{P}(r)$ and $\Phi(\mathbf{P}) \to +\infty$ for $\mathbf{P} \to \partial\mathscr{P}(r)$, being $\partial\mathscr{P}(r)$ the boundary of $\mathscr{P}(r)$.

By noting that the sum of the elements of $\mathbf{T}$ (i.e. of τ_B) is constant for each $\mathbf{c}$, because the sum of the two joint torque constraints for the i-th joint is constant and equal to $\tau_{H,i} - \tau_{L,i}$, the diagonal weighting matrix $\mathbf{W}_p = \mathrm{diag}(w_p\mathbf{I}_6, \mathbf{0}_{2l})$, with $w_p > 0$, is considered. In this way, the term $\mathbf{W}_p\mathbf{P}$ weights only the normal forces $\mathbf{c}_{i,z}$ at each contact point, i.e. the pressure forces on the object. If required, different weights can be used allowing higher contact forces for strongest fingers.

The second term $\mathbf{W}_b\mathbf{P}^{-1}$ represents a barrier function, which goes to infinity when $\mathbf{P}$ tends to a singularity, i.e. when friction or torque limits are approached. The barrier weight matrix is also chosen diagonal $\mathbf{W}_b = \mathrm{diag}(\mathbf{W}_{b,F}, \mathbf{W}_{b,T})$, with

$$\begin{aligned} \mathbf{W}_{b,F} &= w_{b,F}\mathrm{diag}\left(\mu_1, \ldots, \mu_n\right) \\ \mathbf{W}_{b,T} &= w_{b,T}\mathrm{diag}\big(\tau_{H,1} - \tau_{L,1}, \ldots, \tau_{H,l} - \tau_{L,l}, \\ &\quad \tau_{H,1} - \tau_{L,1}, \ldots, \tau_{H,l} - \tau_{L,l}\big), \end{aligned} \tag{16}$$

being $w_{b,F} > 0$ and $w_{b,T} > 0$.

Hence, the minimization of the cost function (15) with the linear constraint (12) corresponds to the minimization of the normal contact wrench components applied to the object while satisfying the friction and torque constraints.

1.3 Semidefinite Programming

The minimization problem can be solved using the linearly constrained gradient flow approach on the smooth manifold of positive definite matrices presented in [28], [11]. In particular, it is possible to prove that $\Phi(\mathbf{P})$ presents a unique minimum that can be reached through the linear constrained exponentially convergent gradient flow

$$\xi(\dot{\mathbf{P}}) = \mathbf{Q}\xi(\mathbf{P}^{-1}\mathbf{W}_b\mathbf{P}^{-1} - \mathbf{W}_p), \tag{17}$$

where $\mathbf{Q} = (\mathbf{I} - \mathbf{A}^{\dagger}\mathbf{A})$ is the linear projection operator onto the tangent space of $\mathbf{A}$, and $\mathbf{A}^{\dagger} = \mathbf{A}^{\mathrm{T}}(\mathbf{A}\mathbf{A}^{\mathrm{T}})$ is the pseudo-inverse of $\mathbf{A}$. Consequently, $\mathbf{AQ} = \mathbf{0}$ and $\mathbf{A}\xi(\dot{\mathbf{P}}) = \mathbf{0}$; hence, if the solution satisfies the constraint (12) at $t = 0$, it will satisfy the constraint for all $t > 0$.

A discrete-time version of (17) based on the Euler numerical integration algorithm is

$$\xi(\mathbf{P}_{k+1}) = \xi(\mathbf{P}_k) + \alpha_k\mathbf{Q}\xi(\mathbf{P}_k^{-1}\mathbf{W}_b\mathbf{P}_k^{-1} - \mathbf{W}_p), \tag{18}$$

where the step size α_k is chosen to ensure down hill steps. Notice that the choice of α_k strongly affects the performance of the optimization algorithm. A wrong choice could determine a very slow convergence or the break of the barrier. Several strategies have been proposed for the self-tuning of α_k at each iteration (see [34] for details). The sensitivity to the step size choice can be reduced by adopting a Dikin-type recursive algorithm [12], [19], that leads to the discrete flow

$$\xi(\mathbf{P}_{k+1}) = \xi(\mathbf{P}_k) - \alpha_k\mathbf{Q}\frac{\xi(\mathbf{P}_k^{-1}\mathbf{W}_b\mathbf{P}_k^{-1} - \mathbf{W}_p)}{\|\mathbf{P}_k^{-1}\mathbf{W}_b\mathbf{P}_k^{-1} - \mathbf{W}_p\|_{P_k}}, \tag{19}$$

where $\|\mathbf{X}\|_Y = \mathrm{tr}(\mathbf{Y}^{-1}\mathbf{X}\mathbf{Y}^{-1}\mathbf{X})$, and $0 \le \alpha_k \le 1$ can be evaluated with a bounded line search minimizing $\Phi(\mathbf{P}_{k+1})$.

The online implementation of the proposed algorithm requires the inversion of a $(6+2l)$ square matrix $\mathbf{A}\mathbf{A}^{\mathrm{T}}$ needed for the evaluation of $\mathbf{A}^{\dagger}$ at each iteration, also when the grasping configuration is unchanged, i.e. when $\mathbf{G}$ is constant, due to the variation of $\mathbf{J}(\mathbf{q})$.

Starting from the discrete version of the gradient flow (18), the following new formulation can be derived

$$\mathbf{c}_{k+1} = \mathbf{c}_k + \alpha_k\bar{\mathbf{Q}}\xi(\mathbf{P}^{-1}(\mathbf{c}_k)\mathbf{W}_b\mathbf{P}^{-1}(\mathbf{c}_k) - \mathbf{W}_p), \tag{20}$$

where $\bar{\mathbf{Q}} = (\mathbf{I} - \mathbf{G}^{\dagger}\mathbf{G})[\mathbf{I}_{nm}\mathbf{0}_{2l}](\mathbf{I} - \mathbf{A}_{\tau}^{\dagger}\mathbf{A}_{\tau})$ is the result of the projection onto the null space of matrix $\mathbf{A}_{\tau}$ in (14), which guarantees the coherence of the elements of matrix $\mathbf{P}$, and of the subsequent projection onto the null space of the grasp matrix, ensuring the force balance constraint (1). Therefore, the evaluation of the inverse of a $6+2l$ square matrix is decomposed into the evaluation of the inverse of two matrices of lower dimensions (6 and $2l$, respectively). Moreover, if the grasp configuration remains unchanged, the projector depending on $\mathbf{G}$ can be evaluated off-line. A similar decomposition can be easily achieved for the gradient flow (19).

1.4 Improvements for Real-Time Applications

An iterative technique for the on-line evaluation, at each sampling time, of the initial point —the initial solution $\mathbf{P}_0$ for the optimization gradient flow algorithm— is proposed here, based on the optimal solution at the previous sampling time. The quantities that can vary between successive sampling times are the hand configuration $\mathbf{q}$, the external torque τ_e, and the grasp map $\mathbf{G}$, while they are taken constant during the iterations of the optimization algorithm between two consecutive sampling times (optimization cycle).

To avoid the evaluation of an initial point at each sampling time, the following approach is proposed. Initially, at time t_0, the method proposed in [34] (or an equivalent one) is used to evaluate off-line a first valid initial solution, which is employed for the first optimization cycle. For the next sampling times t_k, the initial point is computed from the optimal solution $\mathbf{c}_{k-1}$ computed at the end of the previous optimization cycle, through the iterative algorithm

$$\begin{aligned} \bar{\mathbf{c}}_j &= (\mathbf{I}-\mathbf{G}_k^{\dagger}\mathbf{G}_k)\bar{\mathbf{c}}_{j-1}+\gamma_j\mathbf{G}_k^{\dagger}\mathbf{h}_{e,k}+(1-\gamma_j)\mathbf{G}_{k-1}^{\dagger}\mathbf{h}_{e,k-1} \\ \bar{\tau}_j &= \mathbf{J}^{\mathrm{T}}(\gamma_j\mathbf{q}_k+(1-\gamma_j)\mathbf{q}_{k-1})\bar{\mathbf{c}}_j+\gamma_j\tau_{e,k}+(1-\gamma_j)\tau_{e,k-1}, \end{aligned} \tag{21}$$

with initial condition $\bar{\mathbf{c}}_0 = \mathbf{c}_{k-1}$, where the subscript k is referred to the current optimization cycle, while the subscript j and the variables with the bar are referred to the iterations within the cycle. The coefficient $\gamma_j \in (0,1]$ is chosen at each iteration according to a monotone sequence, using a simple linear search algorithm, as the maximum value that does not produce invalid solutions ($\mathbf{P}_0 \leq 0$). In the worst case, γ_0 must be set to a value close to zero.

In detail, at each step of the optimization cycle, the first equation of (21) gradually modifies the external wrench component of the current solution until the full external wrench $\mathbf{h}_{e,k}$ is balanced (i.e., $\gamma_j = 1$). Obviously, the optimization cycle cannot be terminated until γ_j does not reach 1. If the solution evaluated at the previous sampling time ($\mathbf{c}_{k-1}$) is sufficiently far from the boundaries (the distance depends also from the weights assigned to $\mathbf{W}_b$), γ_0 can be set to 1 at the first iteration, and thus the initial point has the same internal wrench component of the previous optimal solution. On the other hand, when $\gamma_0 < 1$, the effect of the barrier function produces a new solution that, at each iteration of the optimization cycle, goes away from the boundaries; this guarantees that γ_j increases at each step, until $\gamma_j = 1$. The second equation is required to modify the joint torque with the same rationale of the first equation. In sum, the sequence γ_j produces an effect similar to a low-pass filter on the variation of the solutions between subsequent optimization steps that are recovered directly within the recursive optimization algorithm.

From the practical experience, if the weight $\mathbf{W}_b$ of the barrier function in the cost function (15) is chosen high enough and the sampling period is small, in most cases the last optimal solution is a valid initial solution, i.e. $\gamma_0 = 1$.

Under the reasonable assumption that the solutions of the optimization algorithm evaluated at successive sampling times are quite close, the joint torque constraints can be simplified observing that not all the joint torque constraints can be effective simultaneously. For example, if, for the current optimal solution, the actuator of joint i provides a torque close to the upper bound $\tau_{H,i}$, the constraint on the lower bound $\tau_{L,i}$ can be deactivated at the next sampling time, being negligible the corresponding barrier term in the cost function. More in general, if for a grasp configuration a given contact force is required along a certain direction, it is reasonable to assume that the corresponding joint torques will not change significantly at the next sampling time. Starting from this observation, the number of joint torque constraints can be dynamically reduced at each sampling time, by using the distance of the torque evaluated at the previous sampling time from the lower and upper bounds as the criterion for selecting the constraint (the lower or the upper one) that needs to be activated. Only those constraints with a distance higher than a torque limit threshold, that can be chosen as a fraction $\sigma_\tau > 0$ of the corresponding torque limit, will be activated.

Wherever required, to reduce chattering phenomena during the activation and deactivation of a constraint, that can introduce noise in the solution, a simple double threshold ($\sigma_{\tau,L} > 0$ and $\sigma_{\tau,H} > 0$) with a hysteretic threshold can be employed.

For applications with limited computational resource, a further simplification in the algorithm can be introduced by splitting the bimanual optimization problem into two simpler single-hand problems. In this case, the initial point iterative self-evaluation algorithm presented above can be employed to find the initial common solution. Then two independent optimization procedures can be started separately for each hand, and the corresponding solutions are composed only at the end to achieve a unique wrench vector solution. The price to pay with the simplified algorithm is that the solution is not optimal in a global sense.

A significant improvement in the solution can be reached by considering a suitable weighted pseudo-inverse of the grasp matrix in (21), with the goal of achieving a load sharing between the hands in reason of the actual load of the hand actuators. In detail, at each sampling time, the minimum distance of the joint torques with respect to the corresponding limits is evaluated for each hand, namely $\delta_{\tau,r}$ and $\delta_{\tau,l}$ for the right and for the left hand, respectively. Then a weighting matrix

$$\mathbf{W}_G = \mathrm{diag}\left(\frac{\delta_{\tau,r}+\delta_{\tau,l}}{\delta_{\tau,r}}\mathbf{I}_{n_r m}, \frac{\delta_{\tau,r}+\delta_{\tau,l}}{\delta_{\tau,l}}\mathbf{I}_{n_l m}\right), \tag{22}$$

with n_r and n_l the number of contact points for the right and for the left hand, is adopted for the evaluation of the weighed pseudo-inverse of the grasp matrix.

Fig. 1 The DEXMART Hand prototype.

$$\mathbf{G}^{\#} = \mathbf{W}_G^{-1}\mathbf{G}(\mathbf{G}\mathbf{W}_G^{-1}\mathbf{G}^{\mathrm{T}})^{-1}. \tag{23}$$

With this choice, the quadratic form $\mathbf{c}^{\mathrm{T}}\mathbf{W}_G\mathbf{c}$ is minimized, reducing the load requirement on the hand closest to its torque limits. This approach, as demonstrated in the following case study, can produce a reduction up to 50% of the computational time when a large number of joint torque constraints are active.

1.5 Case Study

The proposed GFO algorithm has been tested in simulation using two models of the DEXMART Hand (see Fig.1), mounted on an anthropomorphic torso, as shown in Fig. 2. It is assumed that the hands grasp a cylinder representing a bottle half filled with water and the task consists in pouring water by reorienting the bottle. The bottle is initially grasped with the main axis aligned to the vertical direction; then the task can be decomposed into three steps:

- a rotation of 135 deg about the horizontal axis through the geometric center of the cylinder is commanded;
- the hand is stopped while some water is poured from the bottle (the mass and inertia of the bottle change accordingly);
- the opposite rotation is commanded to set the bottle back to the initial pose.

A dynamic simulation has been performed using Matlab/Simulink, where the variation of the position of the center of mass of the water and that of its weight have been considered. Figure 2 shows on the right a section of the bottle half filled with water. In the figure, the intensity of the gravity force is proportional to the black vertical arrow applied to the instantaneous center of mass (of length proportional to

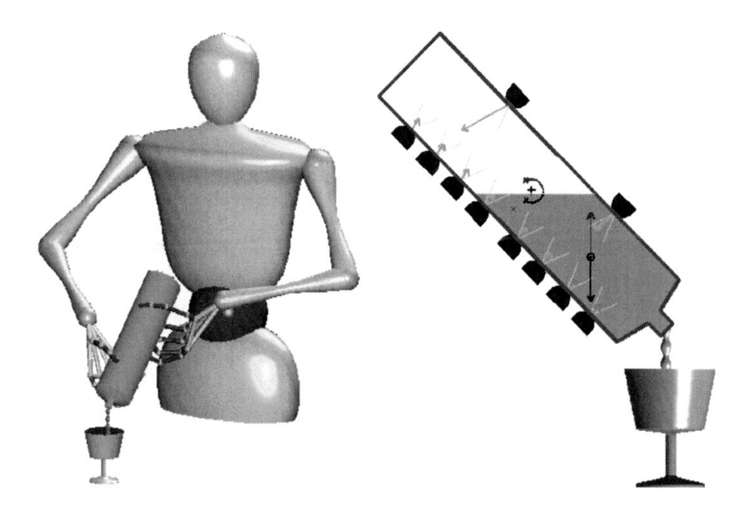

Fig. 2 *Left*: anthropomorphic torso with two DEXMART Hands grasping a bottle. *Right*: section of the grasped bottle with graphical representation of the gravity force and torque (black arrows), of the resultant force and torque applied by the fingers (red arrows), of the optimal contact forces (green arrows if not interested by joint torque constraints, orange arrow otherwise), and of the friction cones (yellow triangles).

the intensity of the force), while the intensity of the gravity torque with respect to the center of the bottle is proportional to the black circular arrow. The red arrows represent the external force and torque balancing the gravity effects and resulting from the contact forces applied by the fingers, represented by green arrows if not interested by joint torque constraints, orange arrow otherwise. The sections of the friction cones in the contact points are colored in yellow. A sequence of significant configurations of the bottle during task execution is shown in Fig. 3.

The effectiveness of the friction and of joint torque limits constraints is shown by considering two different simulations: in the first one only the friction constraint is considered, without any constraint on the joint torque limits, while in the second one different torque limits are set for the fingers. In particular, the thumb actuators are considered stronger than the corresponding actuators of the other fingers of the hand (± 0.5 vs. ± 0.075 Nm), like for the human hand.

In Fig. 4 the trajectory and the areas covered by the contact force vector of each finger in the corresponding contact point during the bottle motion are shown, in blue (red) color for the case without (with) torque constraints. As expected, the frictional constraints are always respected in both simulation cases accordingly to the barrier function considered into the cost function (15).

The time history of the minimum distance of the joint torques for all the actuators from the corresponding limits is shown in Fig. 5, with the red (blue) line refers to the case with (without) torque constraints. The effect of the barrier function settled up also on torques insures the full respect of the adopted limits, without affecting significantly the contact wrenches as shown in Fig. 6.

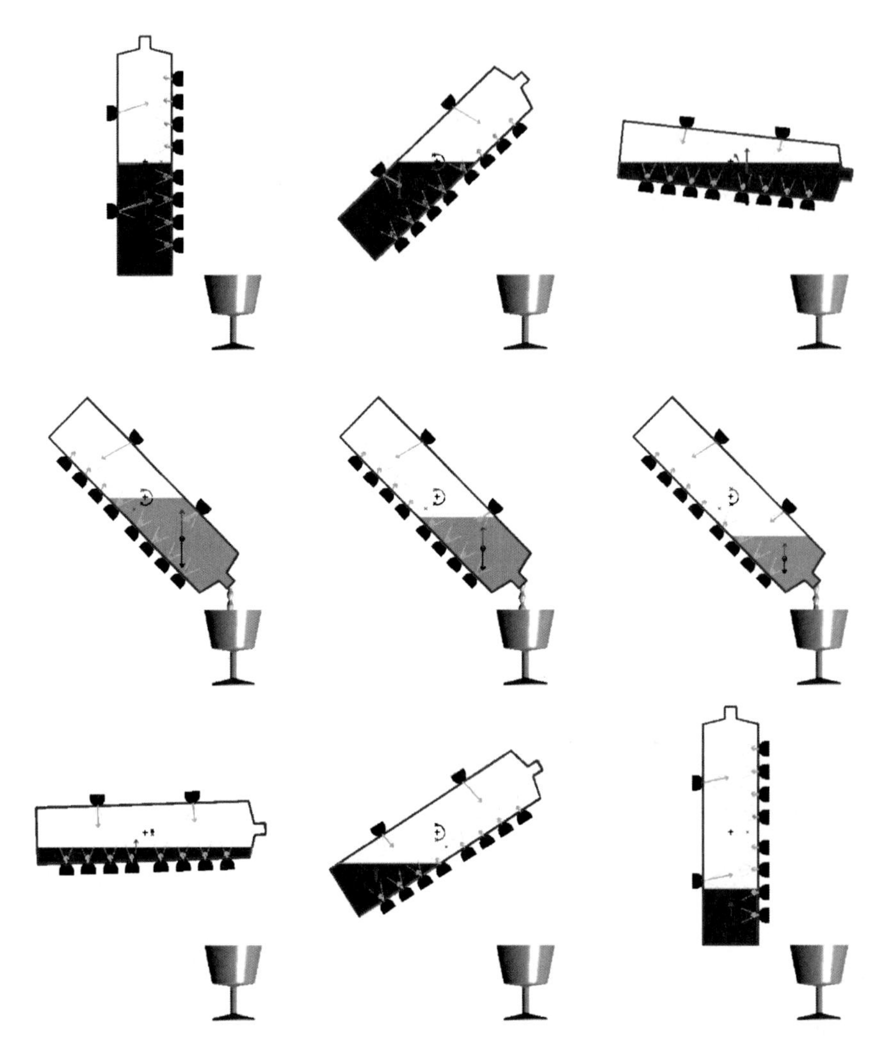

Fig. 3 Sequence of significant configurations of the bottle and of the forces during the task execution.

In Fig. 6 a comparison of the norm of the contact wrenches (on the left) and of the joint torques (on the right) is shown for both simulation cases. The differences for the norm of the torque between the two cases is very limited, while the contact wrenches are improved (smaller in norm), due to a better balancing of the load between the fingers.

The benefits resulting from the adoption of the online joint-torque constraints selection are shown in Fig. 7, where the time history of the computational time effort are represented on the left and the number of employed constraints are represented on the right. To remove the dependence on the employed hardware, all the

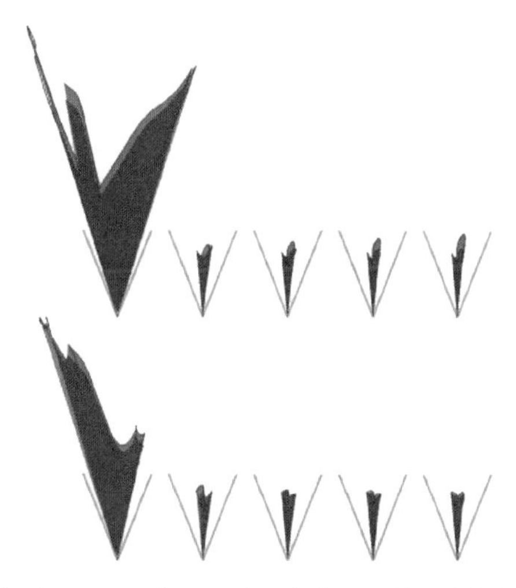

Fig. 4 Areas covered by the contact forces of each finger (*Top*: right hand, *Bottom*: left hand; from the left side: from the thumb to the little finger) without (blue color) and with (red color) torque constraints with respect to the friction cones (represented with green lines).

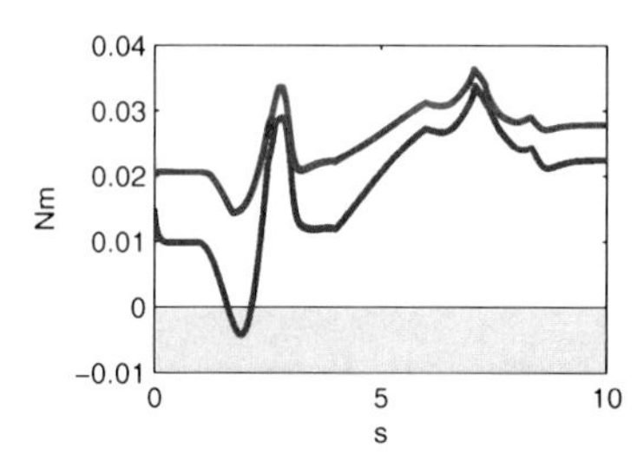

Fig. 5 Time history of the minimum distance of the joint torques from the corresponding limits for all the actuators, in red (blue) color for the case with (without) torque constraints. Negative values (the yellow area) correspond to the violation of one or more joint torque limits.

considered cases are normalized with respect to the maximum value of the fully constrained case (black line) to the value 100 (corresponding about to 23 ms on an Intel Pentium IV at 2.8 Ghz). In particular, four different cases are compared: all constraints (black lines), $\sigma_\tau = 0$ (red lines), $\sigma_\tau = 0.5$ (green lines), $\sigma_\tau = 0.8$ (blue lines), and unconstrained (gray lines), where σ_τ is the threshold for the activation of the joint torque constraints. The achieved reduction of the mean of the computational time varies from a minimum of about 30% for $\sigma_\tau = 0$ to a maximum of about 90% for $\sigma_\tau = 0.8$.

The adoption of the sub-optimal single-hand GFO algorithm can provide a significant reduction on the computational time with respect to the optimal one, as shown in Fig. 7 for the case of all joint torque constraint simultaneously active. However, the drawback is that the sub-optimal solution has reduced performance in terms of both the norm of contact wrenches and of joint torque (see Fig. 8). Consequently,

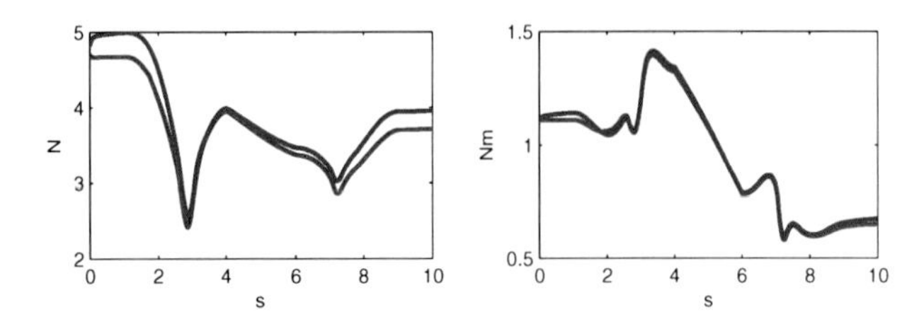

Fig. 6 Time history of the norm of the contact wrenches (on the left) and of the joint torques (on the right), in red (blue) color for the case with (without) torque constraints.

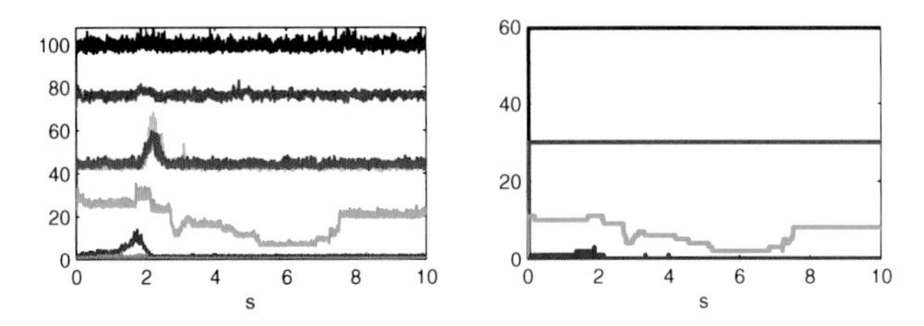

Fig. 7 Time history of the normalized computational-time effort (left) and of the number of employed joint torque constraints (right) for the cases with all constraints (black), $\sigma_\tau = 0$ (red), $\sigma_\tau = 0.5$ (green), $\sigma_\tau = 0.8$ (blue), unconstrained (gray), and single-hand local optimization with (violet) and without (cyan) weighted pseudo-inverse of the grasp matrix.

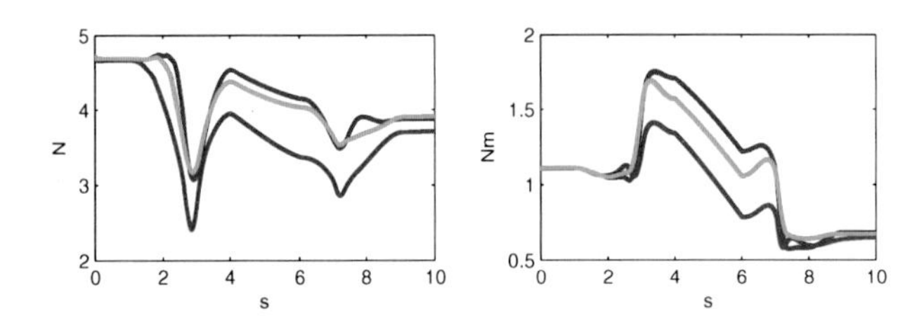

Fig. 8 Time history of the norm of the contact wrenches (on the left) and of the joint torques (on the right) for the cases of local single-hand optimization without (blue) and with (green) weighted pseudo-inverse of the grasp matrix, and global optimization (red).

also the distance with respect to the joint torque limits result reduced significantly, as shown in Fig. 9), but without violating the imposed constrained. As shown in these figures, the adoption of the weighted pseudo-inverse of the grasp matrix in (23) can improve the achieved solution resulting in a well-shared load between the two hands. This behavior is mainly due to the reduction of the DOFs available to the optimization algorithm considering separately the the two hands instead of both together.

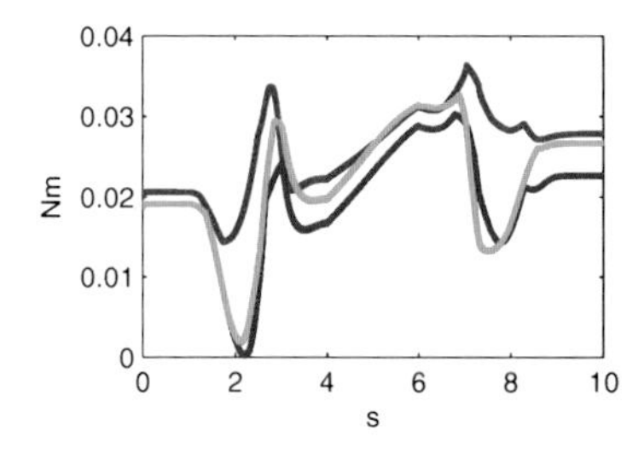

Fig. 9 Time history of the minimum distance of the joint torques from the corresponding limits for all the actuators for the cases of local single-hand optimization without (blue) and with (green) weighted pseudo-inverse of the grasp matrix, and global optimization (red).

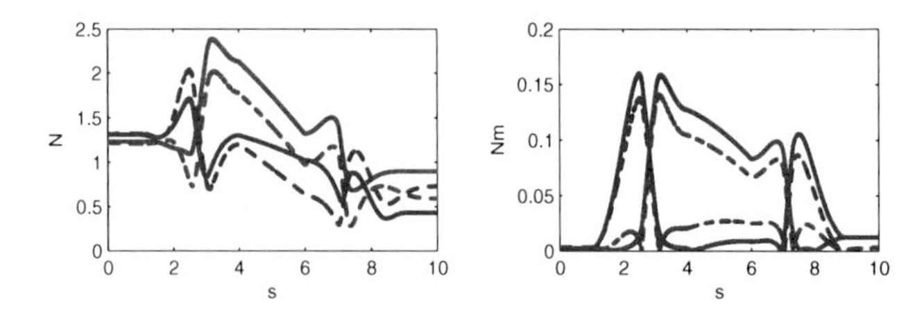

Fig. 10 Time history of the norm of the load force (on the *left*) and moments (on the *right*) for the right (red) and left (blue) hand in the cases of local (continuous lines) and global (dashed lines) method.

On top of Fig. 10 the time history of the normalized load-sharing coefficients $\delta_{\tau,r}/(\delta_{\tau,r}+\delta_{\tau,l})$ (red) and $\delta_{\tau,l}/(\delta_{\tau,r}+\delta_{\tau,l})$ (blue) employed in (23) is shown, while the time history of the norm of the load force and moments for the right (red) and left (blue) hand in the cases of sub-optimal (continuous lines) and optimal (dashed lines) method are shown in Fig. 10. As expected, the whole balancing of effort between the hands is degraded with respect to the optimal solution (dashed lines), but the adopted load sharing method allows an online load repartition according to the current load capability of each hand.

2 Kinematic Control with Force Feedback

Dual-arm/hand object manipulation with multi-fingered hands is a challenging task, especially in service robotics applications, but it has not investigated as extensively as it should deserve. In order to achieve the desired motion of the manipulated object, arms and fingers should operate in a coordinated fashion. In the absence of physical interaction between the fingers and the object, simple motion synchronization shall be ensured. Further, the execution of object grasping or manipulation requires controlling also the interaction forces to ensure grasp stability [43], [49].

From a kinematics point of view, an object manipulation task can be assigned in terms of the motion of the fingertips and/or in terms of the desired object motion. The planner (or the controller) has to map the desired task into the corresponding joint trajectories of the fingers and the arms, thus requiring the solution of an inverse kinematics problem.

In this work, starting from the framework presented in [31], a kinematic model for object manipulation using a dual-arm/hand robotic system is derived, which allows computing the object pose from the joint variables of each arm and each finger (active joints), as well as from a set of contact variables, modelled as passive joints [40]. Suitable conditions are derived ensuring that a given motion can be imposed to the object using only the active joints. Exploiting also the information provided by force sensors mounted inside the fingertips, a two-stage control scheme is proposed so as to achieve the desired object motion and to maintain the desired contact normal forces.

The kinematic redundancy of the system, deriving also from the presence of the passive joints, is suitably exploited to satisfy a certain number of secondary tasks with lower priority, aimed at ensuring grasp stability and manipulation dexterity —without violating system constraints— besides the main task corresponding to the desired object motion. To this aim, a prioritized task sequencing with smooth transitions between the different tasks [36] is employed.

At the best of authors knowledge, the focus of previous works on kinematics of multi-fingered manipulation was on constrained kinematic control [25], [40], or manipulability analysis [8], without considering redundancy resolution and the benefits of integrating a force feedback in a kinematic control loop. The effectiveness of the proposed approach is demonstrated in simulation by considering an object exchange task for a planar bimanual system.

2.1 Kinematic Model

Consider a bimanual manipulation system, e.g., the humanoid manipulator of Fig. 11 composed by a three DOFs torso and two DLR manipulators (each with seven DOFs). The direct kinematics can be computed as reported in [55], by introducing a frame Σ_b fixed with the base of the torso, two frames, Σ_r and Σ_l, attached at the base of the right and left arm, respectively, and two frames, Σ_{rh} and Σ_{lh}, attached to the palms of the right and left hand, respectively. Moreover, assuming that each arm ends with a robotic hand composed by N fingers, it is useful to introduce a frame Σ_{rf_i} (Σ_{lf_i}), attached to the distal phalanx of finger i ($i = 1 \ldots N$) of the right (left) hand.

The pose of Σ_{rf_i} with respect to the base frame Σ_b can be represented by the well known (4×4) homogeneous transformation matrix $\mathbf{T}^b_{rf_i}(\mathbf{R}^b_{rf_i}, \mathbf{o}^b_{rf_i})$, where $\mathbf{R}^b_{rf_i}$ is the (3×3) rotation matrix expressing the orientation of Σ_{rf_i} with respect to the base frame and $\mathbf{o}^b_{rf_i}$ is the (3×1) position vector of the origin of Σ_{rf_i} with respect to the base frame. Hence, the direct kinematics can be expressed as

$$\mathbf{T}^b_{rf_i} = \mathbf{T}^b_r(\mathbf{q}_t)\mathbf{T}^r_{rh}(\mathbf{q}_{rh})\mathbf{T}^{rh}_{rf_i}(\mathbf{q}_{rf_i}) \tag{24}$$

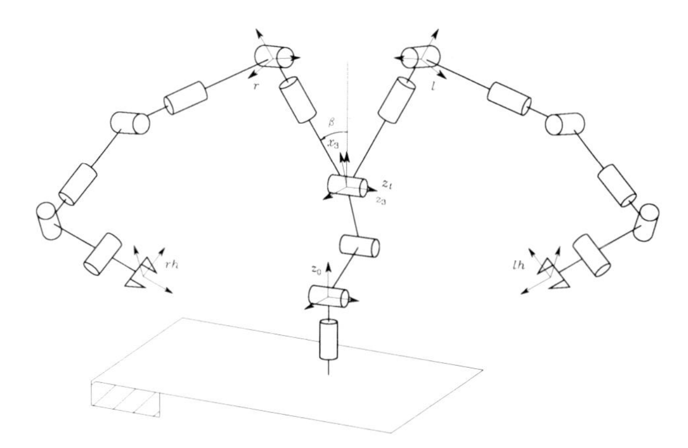

Fig. 11 Kinematic structure of a humanoid manipulator with torso and arms inspired to the DLR Justin.

where $\mathbf{T}_r^b$ is the matrix relating the frame at the basis of the right arm to the base frame (which depends, in turn, on the torso joint vector, $\mathbf{q}_t$), $\mathbf{T}_{rh}^r(\mathbf{q}_{rh})$ is the matrix relating the right palm frame to the base frame of the right arm (which depends, in turn, on the joint vector of the right arm, $\mathbf{q}_{rh}$), and $\mathbf{T}_{rf_i}^{rh}$ is the matrix relating the frame attached to finger i to the palm frame of the right hand (which depends, in turn, on the joint vector $\mathbf{q}_{rf_i}$, where the fingers are assumed to be identical). An equation similar to (24) holds for the left hand fingers, with subscript l in place of subscript r.

Due to the branched structure of the manipulator, the kinematic equations of both the right and the left arm depend on the joint vector $\mathbf{q}_t$ of the torso and, thus, they are not independent. Without loss of generality, hereafter it is assumed that the torso is motionless, i.e., $\mathbf{q}_t$ is constant; therefore, the kinematics of the right and of the left hand can be considered separately. Hence, in the sequel, the superscripts r and l will be omitted and will be used explicitly only when it is required to distinguish between the right and the left arm.

The velocity of frame Σ_{f_i} with respect to the base frame can be represented by the (6×1) twist vector $\mathbf{v}_{f_i} = [\dot{\mathbf{o}}_{f_i}^T \ \omega_{f_i}^T]^T$, where ω_{f_i} is the angular velocity, such that $\dot{\mathbf{R}}_{f_i} = \mathbf{S}(\omega_{f_i})\mathbf{R}_{f_i}$, with $\mathbf{S}(\cdot)$ the skew-symmetric operator representing the vector product. The superscript b, denoting the base frame, has been omitted to simplify notation.

The differential kinematics equation relating the joint velocities to the velocity of frame Σ_{f_i} can be written as

$$\upsilon_{f_i} = \begin{bmatrix} \mathbf{J}_{P_i}(\mathbf{q}_i) \\ \mathbf{J}_{O_i}(\mathbf{q}_i) \end{bmatrix} \dot{\mathbf{q}}_i = \mathbf{J}_{F_i}(\mathbf{q}_i)\dot{\mathbf{q}}_i, \tag{25}$$

where $\mathbf{q}_i^T = \left[\mathbf{q}_h^T \ \mathbf{q}_{f_i}^T\right]^T$ and $\mathbf{J}_{F_i}$ is the Jacobian of the arm, ending with finger i.

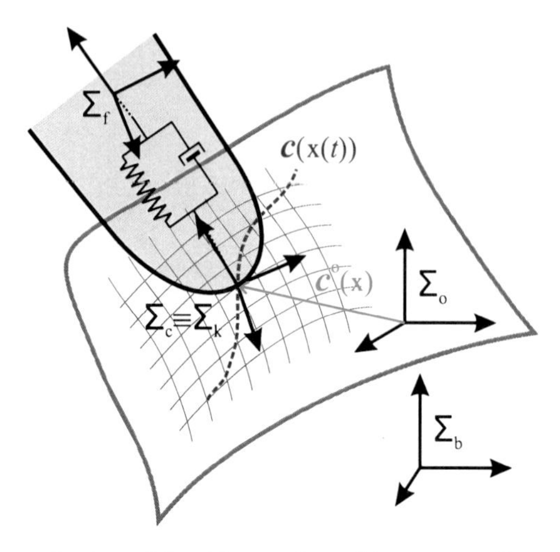

Fig. 12 Local parametrization of the object surface with respect to Σ_O.

Therefore, the differential kinematics equation of the whole arm-hand system, considering the N fingers as end-effectors, can be written in the form

$$\widetilde{\mathbf{v}}_f = \mathbf{J}(\mathbf{q})\dot{\mathbf{q}}, \tag{26}$$

where $\widetilde{\mathbf{v}}_f^{\mathrm{T}} = \left[\mathbf{v}_{f_1}^{\mathrm{T}} \cdots \mathbf{v}_{f_N}^{\mathrm{T}}\right]^{\mathrm{T}}$, $\mathbf{q}^{\mathrm{T}} = \left[\mathbf{q}_h^{\mathrm{T}}\ \mathbf{q}_1^{\mathrm{T}} \cdots \mathbf{q}_N^{\mathrm{T}}\right]^{\mathrm{T}}$, and $\mathbf{J}$ is the Jacobian of the overall arm-hand system.

Assuming that the hand grasps a rigid object, to derive the kinematic mapping between the joint variables of the arm-hand system and the pose (position and orientation) of the object, it is useful introducing an object frame Σ_o attached to the object, usually chosen with the origin in the object center of mass. Let $\mathbf{R}_o$ and $\mathbf{o}_o$ denote respectively the rotation matrix and the position vector of the origin of Σ_o with respect to the base frame, and let $\mathbf{v}_o$ denote the velocity twist vector.

Grasping situations may involve moving rather than fixed contacts: often, both the object and the robotic fingers are smooth surfaces, and manipulation involves rolling and/or sliding of the fingertips on the object surface, depending on the contact type. If the fingers and object shapes are completely known, the contact kinematics can be described introducing contact coordinates defined on the basis of a suitable parametrization of the contact surfaces [39], [41].

In this work, it is assumed that the fingertips are sharp (i.e. they end with a point, denoted as tip point) and covered by an elastic pad. The elastic contact is modeled by introducing a finger contact frame, Σ_{k_i}, attached to the soft pad and with the origin in the tip point $\mathbf{o}_{k_i}$, and a spring-damper system connecting $\mathbf{o}_{k_i}$ with the origin of frame Σ_{f_i}, attached to the rigid part of the finger (see Fig. 13) and with the same orientation of Σ_{k_i}. The displacement between Σ_{f_i} and Σ_{k_i}, due to the elastic contact force, can be computed as

$$\mathbf{o}_{f_i} - \mathbf{o}_{k_i} = (l_i - \Delta l_i)\mathbf{R}_o\hat{\mathbf{n}}^o(\xi), \tag{27}$$

where l_i and $0 \leq \Delta l_i \leq l_i$ are the rest position and the compression of the spring, respectively, and $\hat{\mathbf{n}}^o$ is the vector representing the outward normal to the object's surface at the contact point, referred to Σ_o.

Let Σ_{c_i} be the contact frame attached to the object, with the origin at contact point $\mathbf{o}_{c_i}$. Notice that, instantaneously, the object contact point $\mathbf{o}_{c_i}$ and the finger contact point $\mathbf{o}_{k_i}$ are coincident. One of the axes of Σ_{c_i}, e.g., the Z axis, is assumed to be the outward normal to the tangent plane to the object surface at the contact point.

It is assumed that, at least locally, the position of the contact point with respect to the object frame $\mathbf{o}^o_{o,c_i} = \mathbf{o}^o_{c_i} - \mathbf{o}^o_o$ can be parameterized in terms of a coordinate chart $\mathbf{c}^o_i : U_i \subset \mathbb{R}^2 \mapsto \mathbb{R}^3$ which maps a point $\xi_i = [u_i \; v_i]^{\mathrm{T}} \in U_i$ to the point $\mathbf{o}^o_{o,c_i}(\xi_i)$ of the surface of the object.

Assuming that $\mathbf{c}^o_i$ is a diffeomorphism and that the coordinate chart is orthogonal and right-handed, the contact frame Σ_{c_i} can be chosen as a Gauss frame [39], where the the relative orientation expressed by the rotation matrix $\mathbf{R}^o_{c_i}$ is computed as a function of the orthogonal tangent vectors $\mathbf{c}^o_{u_i} = \partial \mathbf{c}^o_i / \partial u_i$ and $\mathbf{c}^o_{v_i} = \partial \mathbf{c}^o_i / \partial v_i$ [31].

Consider the contact kinematics from the object point of view. Let $\mathbf{c}^o_i(\xi_i(t))$ denote a curve on the surface of the object, with $\xi_i(t) \in U$ (see Fig. 13). The corresponding motion of Σ_{c_i} with respect to the base frame can be determined as a function of: object motion, geometric parameters of the object and the curve geometric features. Namely, the velocity of the contact frame can be expressed as

$$\upsilon_{c_i} = \begin{bmatrix} \dot{\mathbf{o}}_{c_i} \\ \omega_{c_i} \end{bmatrix} = \mathbf{G}^{\mathrm{T}}_{\xi_i}(\xi_i)\upsilon_{o_i} + \mathbf{J}_{\xi_i}(\xi_i)\dot{\xi}_i, \tag{28}$$

where $\mathbf{G}_{\xi_i}(\xi_i)$ and $\mathbf{J}_{\xi_i}(\xi_i)$ are respectively (6×6) and (6×2) full rank matrices, whose expressions can be found in [31].

Consider now the contact kinematics from the fingers point of view. The contact can be modeled with an unactuated 3-DOF ball and socket kinematic pair centered at the origin $\mathbf{o}_{k_i}$ of Σ_{k_i}, fixed to the soft pad of the finger; the origin may also move on the surface, if sliding is allowed. Therefore, the relative orientation $\mathbf{R}^{k_i}_{c_i}$ of Σ_{c_i} with respect to Σ_{k_i} can be computed in terms of a suitable parametrization of the ball and socked joint, e.g., Euler angles.

A vector $\theta_i = \begin{bmatrix} \theta_{i1} \; \theta_{i2} \; \theta_{i3} \end{bmatrix}^{\mathrm{T}}$ of XYZ Euler angles can be considered, and thus $\mathbf{R}^{k_i}_{c_i} = \mathbf{R}^{k_i}_{c_i}(\theta_i)$. Singularities occurs for $\theta_{2i} = \pm\pi/2$, but they do not correspond to physical singularities of the kinematic pair.

Notice that, in the presence of a contact force, because of the tip elasticity, frame Σ_{k_i} translates from the finger frame Σ_{f_i} according to (27), but the orientation does not change. Therefore, $\mathbf{R}^{k_i}_{c_i} = \mathbf{R}^{f_i}_{c_i}$. Moreover, the angular velocity of Σ_{c_i} relative to Σ_{f_i} can be expressed as $\omega^{f_i}_{f_i,c_i} = \mathbf{H}(\theta_i)\dot{\theta}_i$, where $\mathbf{H}$ is a transformation matrix depending on the joint parameterization. In view of the decomposition $\omega_{c_i} = \omega_{f_i} + \mathbf{R}_{f_i}(\mathbf{q}_i)\omega^{f_i}_{f_i,c_i}$, and from (25), the angular velocity of Σ_{c_i} can be computed also as a function of joint and contact variables in the form

$$\omega_{c_i} = \mathbf{J}_{O_i}(\mathbf{q}_i)\dot{\mathbf{q}}_i + \mathbf{R}_{f_i}(\mathbf{q}_i)\mathbf{H}(\theta_i)\dot{\theta}_i, \tag{29}$$

with $\mathbf{J}_{O_i}$ defined in (25). Moreover, since the origins of Σ_{c_i} and Σ_{k_i} coincide, the following equalities hold:

$$\begin{aligned}\mathbf{o}_{c_i} &= \mathbf{o}_{k_i} = \mathbf{o}_{f_i} - (l_i - \Delta l_i)\mathbf{R}_o\hat{\mathbf{n}}_i^o(\xi_i)\\ \dot{\mathbf{o}}_{c_i} &= \mathbf{J}_{P_i}(\mathbf{q}_i)\dot{\mathbf{q}}_i + \dot{\Delta l}_i\mathbf{R}_o\hat{\mathbf{n}}_i^o(\xi_i)\\ &\quad + (l_i - \Delta l_i)\mathbf{S}(\mathbf{R}_o\hat{\mathbf{n}}_i^o(\xi_i))\omega_o - (l_i - \Delta l_i)\mathbf{R}_o\frac{\partial\hat{\mathbf{n}}_i^o(\xi_i)}{\partial\xi_i}\dot{\xi}_i,\end{aligned} \tag{30}$$

with $\mathbf{J}_{P_i}$ defined in (25). Using (29) and (30), the velocity of the contact frame can be expressed as

$$\begin{aligned}\upsilon_{c_i} &= \mathbf{J}_{F_i}(\mathbf{q})\dot{\mathbf{q}} + \mathbf{J}_{\theta_i}(\theta_i,\mathbf{q}_i)\dot{\theta}_i + \mathbf{J}_{\Delta l_i}(\xi_i)\dot{\Delta l}_i\\ &\quad - \mathbf{J}'_{\xi_i}(\xi_i,\Delta l_i)\dot{\xi}_i - \mathbf{G}^T_{\Delta l_i}(\xi_i,\Delta l_i)\upsilon_o,\end{aligned} \tag{31}$$

where $\mathbf{J}_{\theta_i}$ is a (6×3) full rank matrix, whose detailed expression can be found in [31], $\mathbf{J}_{\Delta l_i}$ is a (6×1) vector

$$\mathbf{J}_{\Delta l_i} = \begin{bmatrix}\mathbf{R}_o\hat{\mathbf{n}}_i^o(\xi_i)\\ \mathbf{0}\end{bmatrix},$$

$\mathbf{J}'_{\xi_i}$ is a (6×2) full rank matrix

$$\mathbf{J}'_{\xi_i} = \begin{bmatrix}(l - \Delta l_i)\mathbf{R}_o\dfrac{\partial\hat{\mathbf{n}}_i^o(\xi_i)}{\partial\xi_i}\\ \mathbf{0}\end{bmatrix},$$

and $\mathbf{G}_{\Delta l_i}$ is the (6×6) matrix

$$\mathbf{G}_{\Delta l_i} = \begin{bmatrix}\mathbf{0} & \mathbf{0}\\ (\Delta l_i - l_i)\mathbf{S}(\mathbf{R}_o\hat{\mathbf{n}}_i^o(\xi_i)) & \mathbf{0}\end{bmatrix}.$$

Hence, from (28) and (31), the contact kinematics of finger i has the form

$$\mathbf{J}_{F_i}(\mathbf{q}_i)\dot{\mathbf{q}}_i + \mathbf{J}_{\eta_i}(\eta_i,\mathbf{q}_i,\Delta l_i)\dot{\eta}_i + \mathbf{J}_{\Delta l_i}(\xi)\dot{\Delta l}_i = \mathbf{G}_i^{\mathrm{T}}(\eta_i,\Delta l_i)\upsilon_o, \tag{32}$$

where $\eta_i = \begin{bmatrix}\xi_i^{\mathrm{T}} & \theta_i^{\mathrm{T}}\end{bmatrix}^{\mathrm{T}}$ is the vector of contact variables, $\mathbf{J}_{\eta_i} = \begin{bmatrix}-(\mathbf{J}_{\xi_i} + \mathbf{J}'_{\xi_i}) & \mathbf{J}_{\theta_i}\end{bmatrix}$ is a (6×5) full rank matrix, and $\mathbf{G}_i = \mathbf{G}_{\xi_i} + \mathbf{G}_{\Delta l_i}$ is a (6×6) full rank grasp matrix. This equation can be interpreted as the differential kinematics equation of an "extended" finger corresponding to the kinematic chain including the arm and finger joint variables (active joints) and the contact variables (passive joints), from the base frame to the contact frame [40].

It is worth noticing that equation (32) involves all the 6 components of the velocity, differently from the grasping constraint equation usually considered (see, e.g., [41]), which contains only the components of the velocities that are transmitted by the contact. The reason is that the above formulation takes into account also the velocity components not transmitted by contact i, parameterized by the contact variables and lying in the range space of $\left[\mathbf{J}_{\eta_i}\ \mathbf{J}_{\Delta l_i}\right]$. As a consequence, $\mathbf{G}_i$ is always a full-rank matrix.

Depending on the considered contact type [49], some of the parameters of ξ_i and θ_i are constant. Hence, assuming that the contact type remains unchanged during the task, the variable parameters at each contact point are grouped in an $(n_{c_i} \times 1)$ vector η_i of contact variables, with $n_{c_i} \leq 5$.

Differently form the classical grasp analysis, in this work the elasticity of the soft pad has been explicitly modeled (although using a simplified model). This means that the force along the normal to the contact surface is always of elastic type. The quantity Δl_i, at steady state, is related to the normal contact force f_{ni} by the equation $\Delta l_i = f_{ni}/k_i$, being k_i the elastic constant of the soft pad of finger i.

Object dynamic manipulation is, in general, a difficult task, since the number of the control variables (the active joints) is lower than the number of configuration variables (active and passive joints). However, in some particular situations, it is possible to simplify the analysis, considering only the kinematics of the system.

To this purpose, assume that force sensors are available on the fingertips and a force control strategy is employed to ensure a desired constant contact forces f_{di} along the direction normal to the contact point. Therefore, $\Delta l_i = \Delta l_{di} = f_{di}/k_i$ can be assumed to be fixed ($\dot{\Delta l}_i = 0$) and equation (32) can be rewritten as

$$\mathbf{J}_{F_i}(\mathbf{q}_i)\dot{\mathbf{q}}_i + \mathbf{J}_{\eta_i}(\eta_i, \mathbf{q}_i, \Delta l_i)\dot{\eta}_i = \mathbf{G}_i^{\mathrm{T}}(\eta_i, \Delta l_i)\upsilon_o. \tag{33}$$

On the basis of (33), it is possible to make a kinematic classification of the grasp [49].

A grasp is *redundant* if the null space of the matrix $\left[\mathbf{J}_{F_i}\ \mathbf{J}_{\eta_i}\right]$ is non-null, for at least one finger i. In this case, the mapping between the joint variables of "extended" finger i and the object velocity is many to one: motions of active and passive joints of the extended finger are possible when the object is locked.

A grasp is *indeterminate* if the intersection of the null spaces of $[-\mathbf{J}_{\eta_i}\ \mathbf{G}_i^{\mathrm{T}}]$, for all $i = 1, \ldots, N$, is non-null. In this case, motions of the object and of the passive joints are possible when the active joints of all the fingers are locked.

It is worth noticing that, also in the case of redundant and indeterminate grasps, for a given object pose and fingers configuration, the value of the contact variables is uniquely determined. More details can be found in [31].

2.2 Control Scheme with Redundancy Resolution

In the case of a kinematically determinate and, possibly, redundant grasp, a two-stage control scheme is proposed for the dual arm-hand manipulation system. The first stage is an inverse kinematics scheme with redundancy resolution, which computes the joint references for the active joints corresponding to a desired object's motion —assigned in terms of the homogeneous transformation matrix $\mathbf{T}_d$ and the corresponding twist velocity vector $\mathbf{v}_{o_d}$— and to the desired normal contact force $\mathbf{f}_d^T = [f_{d1} \cdots f_{dN}]$. The second stage is a parallel control composed of a PD position controller and a PI tip force controller, ensuring the desired object motion and desired contact forces on the basis of the previously computed joint references.

Namely, in ideal conditions, the joint references computed by the kinematic stage ensure tracking of the desired object motion, with the desired contact forces. In the presence of modeling errors and parameters uncertainty, the contact forces may differ from those planned. Using the force sensors at the fingertips, a force control strategy is adopted to ensure the desired contact force by modifying the joint references computed by the inverse kinematics stage. In principle, the joint references of the overall manipulation system could be involved; however, it is reasonable to design a force controller acting only on the joints of the fingers.

In order to derive the equations of the first stage, starting from (32), it is useful to write the differential kinematic equations of the whole (right or left) arm-hand system as

$$\mathbf{J}(\mathbf{q})\dot{\mathbf{q}} + \mathbf{J}_\eta(\eta, \mathbf{q}, \Delta\mathbf{l})\dot{\eta} = \mathbf{G}^{\mathrm{T}}(\eta, \Delta\mathbf{l})\widetilde{\mathbf{v}}_o, \tag{34}$$

where $\mathbf{J}$ is the Jacobian of the arm-hand system defined in (26), $\mathbf{J}_\eta$ is a block diagonal matrix $\mathbf{J}_\eta = \mathrm{diag}\{\mathbf{J}_{\eta_1}, \cdots, \mathbf{J}_{\eta_N}\}$ corresponding to the vector of passive joints $\eta^{\mathrm{T}} = [\eta_1^{\mathrm{T}} \cdots \eta_N^{\mathrm{T}}]^{\mathrm{T}}$, $\mathbf{G}$ is the block diagonal grasp matrix $\mathbf{G} = \mathrm{diag}\{\mathbf{G}_1, \cdots, \mathbf{G}_N\}$, $\Delta\mathbf{l}^{\mathrm{T}} = [\Delta l_1 \cdots \Delta l_N]^{\mathrm{T}}$ and $\widetilde{\mathbf{v}}_o^{\mathrm{T}} = [\mathbf{v}_o^{\mathrm{T}} \cdots \mathbf{v}_o^{\mathrm{T}}]^{\mathrm{T}}$.

From (34), the following closed-loop inverse kinematics algorithm can be derived:

$$\begin{bmatrix} \dot{\mathbf{q}}_d \\ \dot{\eta}_d \end{bmatrix} = \widetilde{\mathbf{J}}^{\dagger}(\mathbf{q}_d, \eta_d, \Delta\mathbf{l}_d)\mathbf{G}^{\mathrm{T}}(\widetilde{\mathbf{v}}_{o_d} + \mathbf{K}_o\widetilde{\mathbf{e}}_o) + \mathbf{N}_o\sigma, \tag{35}$$

where $\widetilde{\mathbf{J}} = [\mathbf{J}\ \mathbf{J}_\eta]$, the symbol $\dagger$ denotes a right (weighted) pseudo-inverse, $\widetilde{\mathbf{v}}_{o_d}^{\mathrm{T}} = [\mathbf{v}_{o_d}^{\mathrm{T}} \cdots \mathbf{v}_{o_d}^{\mathrm{T}}]^{\mathrm{T}}$, $\mathbf{K}_o$ is a diagonal and positive definite matrix gain, $\widetilde{\mathbf{e}}_o^{\mathrm{T}} = [\mathbf{e}_{o_1}^{\mathrm{T}} \cdots \mathbf{e}_{o_N}^{\mathrm{T}}]^{\mathrm{T}}$, being $\mathbf{e}_{o_i}$ the pose error between the desired and the current object pose computed on the basis of the direct kinematics of the extended finger i, and $\mathbf{N}_o = \mathbf{I} - \widetilde{\mathbf{J}}^{\dagger}\widetilde{\mathbf{J}}$ is a projector in the null space of the Jacobian matrix $\widetilde{\mathbf{J}}$. The quantity $\Delta\mathbf{l}_d$ in (35) is the vector collecting the finger soft pad deformations $\Delta l_{di} = f_{di}/k_i$ corresponding to the desired contact force f_{di}.

Equation (35) is used to compute the joint reference vector $\mathbf{q}_d$ for the controller of the second stage.

In view of the above considerations, any kind of joint motion control can be adopted for the arms of the bimanual manipulation system, receiving as input the joint references computed by the inverse kinematics scheme. In this chapter, the joint torques for finger i are set according to the parallel force/position control law

$$\tau_i = \mathbf{J}_i^T(\mathbf{q}_i)\left(k_P\Delta\mathbf{p}_i + \mathbf{f}_{di} + k_F\Delta\mathbf{f}_{ni} + k_I\int_0^t \Delta\mathbf{f}_{ni}\mathrm{d}\tau\right) - k_d\dot{\mathbf{q}}_i + \mathbf{g}_i(\mathbf{q}_i) \tag{36}$$

where $\mathbf{g}_i(\mathbf{q}_i)$ is the vector of the gravity torque of finger i, $\Delta\mathbf{p}_i$ denotes the position error of finger i between the desired value computed through direct kinematics starting from $\mathbf{q}_{d_i}$ and the current one, and $\Delta\mathbf{f}_{ni}$ is the projection of the force error along the normal to the object surface at the contact point. The above control law regulates the contact force to the desired value at the expense of a position error (i.e., a displacement of the positions of the fingers with respect to the palm), in the presence of uncertainties.

Since the system may be highly redundant, multiple tasks could be fulfilled, provided that they are suitably arranged in a priority order, according to the *augmented projection method* [1]. Consider m secondary tasks, each expressed by a task function $\sigma_{t_h}(\widetilde{\mathbf{q}})$ $(h = 1,\ldots,m)$, where $\widetilde{\mathbf{q}} = \begin{bmatrix}\mathbf{q}_d^T & \eta_d^T\end{bmatrix}^T$. According to the *augmented projection method* [1], the control law (35) can be replaced by

$$\dot{\widetilde{\mathbf{q}}}=\widetilde{\mathbf{J}}^{\dagger}(\widetilde{\mathbf{q}},\Delta\mathbf{l}_d)\mathbf{G}^{\mathrm{T}}(\widetilde{\mathbf{v}}_{o_d}+\mathbf{K}_o\widetilde{\mathbf{e}}_o)+\sum_{h=1}^{m}\mathbf{N}(\mathbf{J}_{t_h}^A)\mathbf{J}_{t_h}^{\dagger}\mathbf{K}_{t_h}\mathbf{e}_{t_h}, \tag{37}$$

where $\mathbf{J}_{t_h}$ is the Jacobian of the hth task, $\mathbf{J}_{t_h}^A$ is the augmented Jacobian, given by

$$\mathbf{J}_{t_h}^A(\widetilde{\mathbf{q}},\Delta\mathbf{l}_d)=\left[\widetilde{\mathbf{J}}^T(\widetilde{\mathbf{q}},\Delta\mathbf{l}_d)\;\mathbf{J}_{t_1}^T(\widetilde{\mathbf{q}})\ldots\mathbf{J}_{t_{h-1}}^T(\widetilde{\mathbf{q}})\right]^T.$$

$\mathbf{N}(\mathbf{J}_{t_h}^A)$ is a null projector of the matrix $\mathbf{J}_{t_h}^A$, $\mathbf{K}_{t_h}$ is a positive definite gain matrix and $\mathbf{e}_{t_h} = \sigma_{t_{hd}} - \sigma_{t_h}$ is the task error, being $\sigma_{t_{hd}}$ the desired value of the h-th task variable.

The augmented projection method can be also adopted to fulfill mechanical or environmental constraints, such as joint limits and obstacle avoidance (other fingers or the grasped object). To this aim, each constraint can be described by means of a cost function, $\mathscr{C}(\widetilde{\mathbf{q}})$, increasing when the manipulator comes close to violate the constraint. In order to minimize the cost function, the manipulator could be moved according to the opposite of the gradient $-\nabla_{\widetilde{q}}^T\mathscr{C}(\widetilde{\mathbf{q}})$, that could be considered as a fictitious force moving the manipulator away from configurations violating the constraints. In order to include the constraints in (37), an overall cost function $\mathscr{C}_\Sigma$ given by

$$\mathscr{C}_\Sigma(\widetilde{\mathbf{q}})=\sum_{c_s}\gamma_{c_s}\mathscr{C}_{c_s}(\widetilde{\mathbf{q}}), \tag{38}$$

is introduced, where γ_{c_s} and $\mathscr{C}_{c_s}$ are a positive weight and a cost function, respectively, referred to the c_sth constraint.

2.3 Task Sequencing

If the system comes close to violate a constraint, a high level supervisor has to remove some secondary tasks and relax enough DOFs to fulfill the constraint [36]. To manage in a correct way removal/insertion of tasks from/into the stack (task sequencing), a suitable task supervisor can be designed, based on a three layers architecture: the lower layer computes the motion variables on the basis of a stack of active tasks; the intermediate layer determines which tasks must be removed from the stack in order to respect the constraints; the upper layer verifies if the previously removed tasks can be pushed back in the stack.

A task must be removed from the stack when the predicted value of the overall cost function at the next time step is above a suitable defined threshold, $\overline{\mathscr{C}}$. Let T be the sampling time adopted to implement the control law and κT the actual time, the configuration at time instant $(\kappa+1)T$ can be estimated as follows

$$\widehat{\widetilde{\mathbf{q}}}(\kappa+1) = \widetilde{\mathbf{q}}(\kappa) + T\dot{\widetilde{\mathbf{q}}}(\kappa). \tag{39}$$

Hence, a task must be removed from the stack if

$$\mathscr{C}_\Sigma\left(\widehat{\widetilde{\mathbf{q}}}(\kappa+1)\right) \geq \overline{\mathscr{C}}. \tag{40}$$

Once it has been ascertained that a task must be removed from the stack, the problem is to detect which task has to be removed. To the purpose, several criteria have been proposed in [36], with the aim of verifying the conflict between the constraints and each task. In this chapter, the overall cost function gradient is projected in the null space of the task Jacobian, i.e.,

$$\mathscr{P}_{t_h} = \left\|\mathbf{N}\left(\mathbf{J}_{t_h}\right)\left(-\nabla_{\widetilde{q}}^T\mathscr{C}_\Sigma\right)\right\| \quad h = 1,\ldots,m; \tag{41}$$

the task corresponding to the minimum of $\mathscr{P}_{t_h}$ is then removed, since its projection into the null-space of $\mathbf{J}_{t_h}$ should be, ideally, zero to ensure constraint fulfillment.

The tasks removed by the second layer must be reinserted in the stack as soon as possible, provided that the constraints will not be violated. A prediction of the $\mathscr{C}_\Sigma$ evolution at the next time step has to be evaluated by considering the effect of each task currently out of the stack, i.e.

$$\widehat{\widetilde{\mathbf{q}}}_{t_h}(\kappa+1) = \widetilde{\mathbf{q}}(\kappa) + \mathbf{J}_{t_h}^{\dagger}\mathbf{e}_{t_h}(\kappa). \tag{42}$$

Therefore, let $\underline{\mathscr{C}} < \overline{\mathscr{C}}$ be a suitably chosen threshold; a task is pushed back in the stack if

$$\mathscr{C}_\Sigma\left(\widehat{\widetilde{\mathbf{q}}}_{t_h}(\kappa+1)\right) \leq \underline{\mathscr{C}}. \tag{43}$$

Task sequencing might cause discontinuities in the commanded joint velocities due to the change of active tasks in the stack. For each task a variable gain ρ_{t_h} is introduced to achieve a smooth behavior of the controller output

$$\rho_{t_h}(t) = \begin{cases} 1-\mathrm{e}^{-\mu(t-\tau)} & \text{if the } h\text{-th task is in the stack} \\ \mathrm{e}^{-\mu(t-\tau')} & \text{if the } h\text{-th task is out of the stack,} \end{cases}$$

where τ and τ' are the time instant in which the task is inserted in the stack and the time instant in which it is removed, respectively, and $1/\mu$ is a time constant.

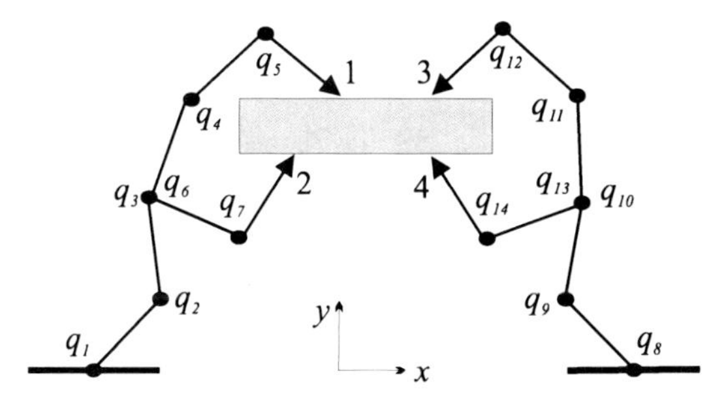

Fig. 13 Manipulation system.

Hence, the first stage control law can be written in its complete form

$$\dot{\tilde{\mathbf{q}}} = \tilde{\mathbf{J}}^{\dagger}(\tilde{\mathbf{q}}, \Delta \mathbf{l}_d)\mathbf{G}^{\mathrm{T}}(\tilde{\mathbf{v}}_{O_d} + \mathbf{K}_o \tilde{\mathbf{e}}_o) + \sum_{h=1}^{m} \rho_{t_h} \mathbf{N}(\mathbf{J}_{t_h}^{A})\mathbf{J}_{t_h}^{\dagger}\mathbf{K}_{t_h}\mathbf{e}_{t_h} - k_{\nabla}\mathbf{N}(\mathbf{J}_{t_{m+1}}^{A})\nabla_{\tilde{q}}^{T}\mathscr{C}_{\Sigma},$$

where k_{∇} is a positive gain.

2.4 *Case Study*

The presented control scheme has been tested on a manipulation system grasping a certain object, represented in Fig. 14, composed by two identical planar grippers, each with two branches and 7 DOFs, resulting in a total of $N = 4$ fingers and 14 active joints. The idea is that of performing an object exchange.

In its initial configuration, it is assumed that the system grasps the object with only tips 1 and 2, which are in a force closure condition, since the contact normal forces are acting on the same straight line [41]. Tips 3 and 4 approach the object until they reach a condition in which all the tips are in contact with the object. The main task consists in keeping the object still, while Tips 3 and 4 move in order to achieve a force closure condition upon the object in a dexterous configuration, without violating a certain number of limits and constraints. Then, Fingers 1 and 2 can leave the object, simulating in this way an hand-to-hand object passing.

The force control loop ensures that the planned forces are applied on the object. In this case study, the desired forces for Tips 3 and 4 are negligible, since they have to slide on the object's surface so as to reconfigure themselves to reach force closure condition. The desired forces for Tips 1 and 2 are dynamically planned, on the basis of the current value of the forces exerted by the fingers, in order to produce zero net force and moment on the object and to balance disturbances caused by movements of the other two fingertips.

A sequence of snapshots representing the described task are shown in Fig. 14. It can be noticed that, in the final configuration (fifth snapshot), Fingers 3 and 4 are in a force closure condition, since the normals at the contact points act on the same straight line.

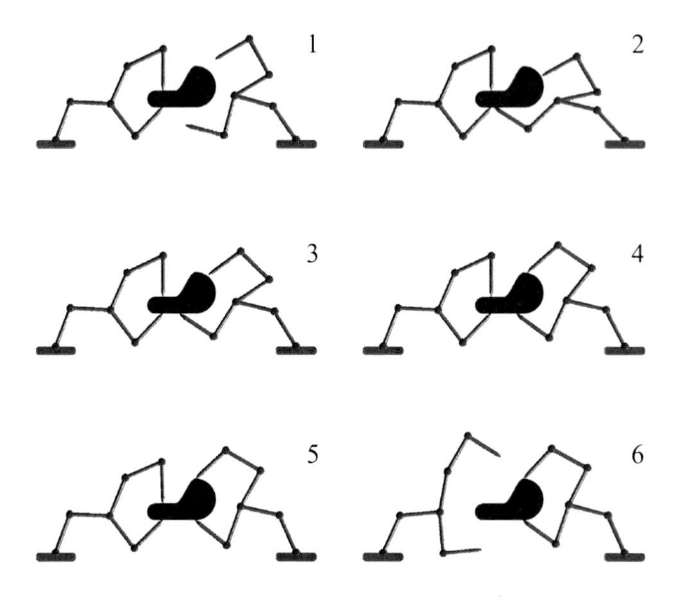

Fig. 14 Snapshots describing the case study.

Four different subtasks have been considered: the first two, aimed at choosing the optimal contact points, are related to the grasp quality; the others regard the manipulability and the distance between the palm and the grasped object.

Unit frictionless equilibrium. The grasp quality can be guaranteed by moving the contact points on the object surface until the unit frictionless equilibrium is reached. This condition is a special case of a force-closure grasp; it is satisfied when two positive indices, called frictionless force (ε_f) and moment (ε_m) residuals, are zero: [16], [45]

$$\varepsilon_f = \frac{1}{2}\mathbf{f}^T\mathbf{f} \qquad \mathbf{f} = \sum_{i=1}^{4} \hat{\mathbf{n}}_i^o \tag{44}$$

$$\varepsilon_m = \frac{1}{2}\mathbf{m}^T\mathbf{m} \qquad \mathbf{m} = \sum_{i=1}^{4} \mathbf{c}_i^o \times \hat{\mathbf{n}}_i^o\,, \tag{45}$$

where $i = 1,\ldots,4$, and where $\hat{\mathbf{n}}_i^o(\xi_i)$ and $\mathbf{c}_i^o(\xi_i)$ are the surface normal and the position of the ith contact point, respectively, both referred to the object frame. It has been shown that, for two or more contact points, unit frictionless equilibrium is a force closure condition for any nonzero friction coefficient [45], [46].

Manipulability. In order to keep the manipulator far from singularities, a manipulability index of each finger can be considered. In detail, the following manipulability measure, which vanishes at a singular configuration, is adopted for the i-th finger [55].

$$w_i(q_i) = \sqrt{det\left(J_i(q_i)J_i^T(q_i)\right)} \quad i = 1,\ldots,4. \tag{46}$$

The considered task function is then

$$\sigma_{w_i} = \begin{cases} \frac{1}{2}(\overline{w}_i - w_i(q_i))^2 & \text{if} \, w_i(q_i) < \overline{w}_i \\ 0 & \text{otherwise,} \end{cases} \tag{47}$$

where $\overline{w}_i$ is a threshold for the task activation. The desired value, $\sigma_{w_i d}$, is zero.

Distance between palm and object. Consider the position $\mathbf{p}_c^o$ of the palm centroid in the object frame and a suitably chosen surface $\mathscr{S}$ surrounding the object characterized by the equation $\mathscr{F}(\mathbf{p}^o) = 0$. When the centroid is inside the surface $\mathscr{S}$, a collision can occur; therefore, the centroid must be moved on the boundary, i.e, in a position such that $\mathscr{F}(\mathbf{p}_c^o) = 0$. Hence, the task function is the following

$$\sigma_P(p_c^o) = \begin{cases} \mathscr{F}(p_c^o) & \text{if the centroid is inside } \mathscr{S} \\ 0 & \text{otherwise.} \end{cases} \tag{48}$$

In the following the two considered constraints are described.

Joint-limit avoidance. A physical constraint to the motion of the system is imposed by the mechanical joint limits. The system configuration is considered safe if $q_j \in [\underline{q}_j, \overline{q}_j]$, for $j = 1, \ldots, 14$, with $\underline{q}_j$ and $\overline{q}_j$ suitable chosen values far enough from the limits. The cost function, directly defined in the joint space, is the following:

$$\mathscr{C}_{JL}(\mathbf{q}) = \sum_{j=1}^{14} c_j(q_j), \tag{49}$$

$$c_j(q_j) = \begin{cases} k_j e^{\delta(q_j - \underline{q}_j)^2} - 1 & \text{if } q_j \le \underline{q}_j \\ 0 & \text{if } \underline{q}_j < q_j \le \overline{q}_j \\ k_j e^{\delta(q_j - \overline{q}_j)^2} - 1 & \text{if } q_j > \overline{q}_j, \end{cases}$$

where k_j and δ are positive constants.

Collision avoidance. In order to avoid collisions between the fingers, it is imposed that the distance between the fingers be larger than a safety value, d_s; hence, if $d_{ii'}$ denotes the distance between the ith and the i'th finger, the following cost function can be formalized

$$\mathscr{C}_{CA}(\widetilde{\mathbf{q}}) = \sum_{i,i'} c_{ii'}(\widetilde{\mathbf{q}}), \tag{50}$$

where the sum is extended to all the couples of fingers,

$$c_{ii'}(d_{ii'}) = \begin{cases} k_{ii'} \dfrac{d_s - d_{ii'}}{d_{ii'}^2} & \text{if } d_{ii'} \le d_s \\ 0 & \text{if } d_{ii'} > d_s, \end{cases} \tag{51}$$

and $k_{ii'}$ is a positive gain.

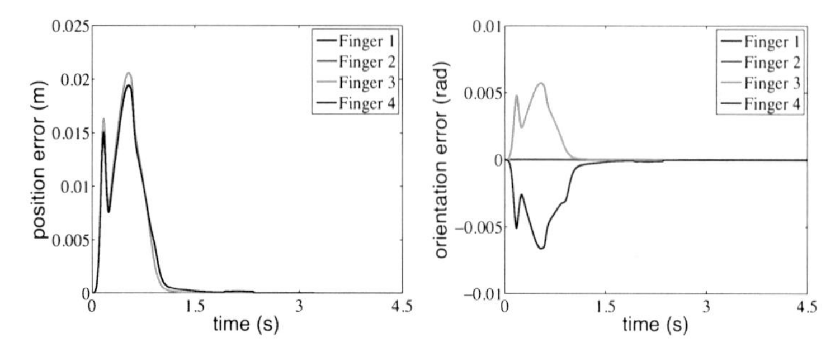

Fig. 15 Object's pose errors for each finger.

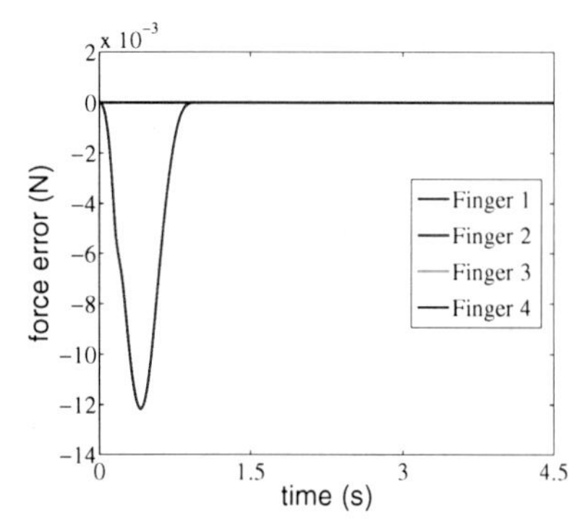

Fig. 16 Finger force errors.

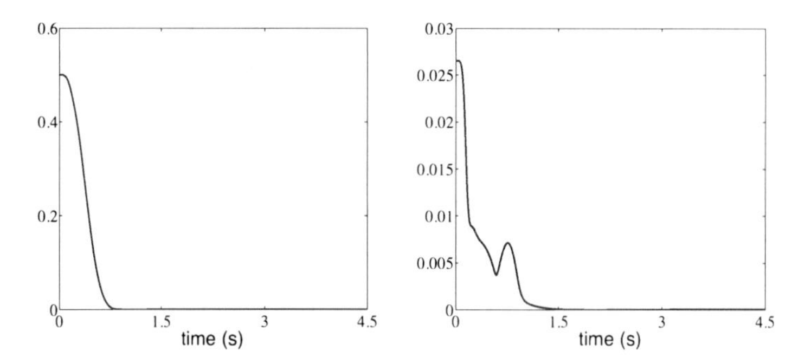

Fig. 17 Force and moment residuals.

The parameters of the elastic contact are: $5 \cdot 10^4$ N/m for the springs elastic coefficients, 5 Ns^2/ m for the springs damper coefficients and $l_i = 5 \cdot 10^{-3}$ m for the springs rest condition. The parameters used to define the subtasks are chosen as follows: $\overline{w}_i = 2.55$ for the manipulability subtask, $\underline{q}_j = -110^o, \overline{q}_j = 110^o, k_j = 5, \delta = 2$ for the joint-limit avoidance and $k_{ii'} = 1, d_s = 5$ cm for the collision avoidance. In the system of Fig. 13 the palm is represented by the ramification point of the right manipulator. The task has a duration of 4.5 s; a Runge-Kutta integration method, with a step size of 2 ms, has been used to simulate the system.

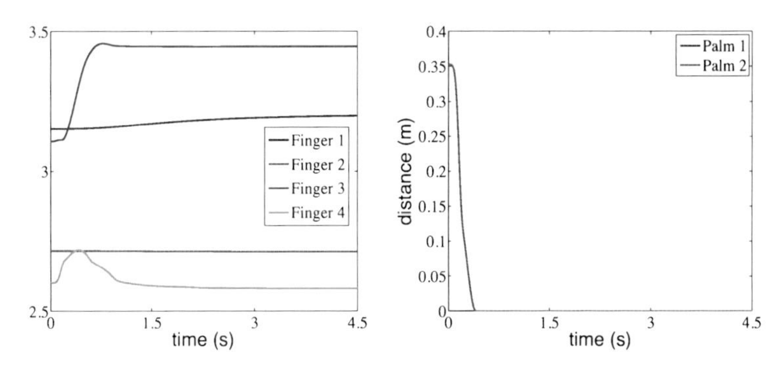

Fig. 18 Subtask functions.

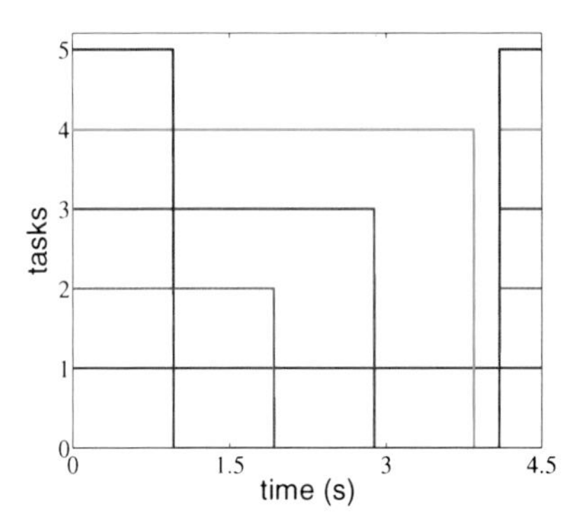

Fig. 19 Time history of the stack status: 1 is the main task, corresponding to keep the object fixed; 2 and 3 are the force and moment residual tasks, respectively; 4 is the manipulability task; 5 is the task about the distance between palms and object.

Figure 15 shows the time history of the norm of the object's pose error for each finger (position on the left, orientation on the right). Figure 16 shows the evolution of the force error for each finger: in detail, Finger 1 is much more affected by the motion of Fingers 3 and 4 than that of Finger 2; the desired value for the normal force at Tips 3 and 4 is very small and it is impossible to see remarkable variations in the time history.

Figure 17 shows the force and moment residuals, ε_f and ε_m, respectively. Since both residuals asymptotically converge to zero, it is clear that Fingers 3 and 4 reach a force closure condition.

Figure 18 shows the time history of the manipulability measure (left) and the distance from the palm function σ_P in (48) (right). The manipulability measure of each finger is above the limit value $\overline{w}_i$, while σ_P is zero when the task is not activated, since the palm is sufficiently far from the object.

Finally, Figure 19 depicts the time history of the stack status during the simulation. It can be noticed that the main task is never removed from the stack, while the other tasks are removed when some constraints are near to be violated. When the system is in a safe condition with respect to the constraints, the tasks are re-inserted in the stack maintaining their previous priorities. Notice that the label assigned at each task denotes its priority in the stack.

3 Control Using Postural Synergies

Recent works on the application of human hand postural synergies to a robotic hand demonstrate that understanding human prehension is a promising way to simplify and optimize the control of multiple-DOF limbs. In order to interact with humans directly, the robots of the future will require enhanced manipulation capabilities similar to those of human beings. For this purpose, complex dexterous hands with advanced sensorimotor skills and human-like kinematics are needed. The human hand is an excellent example of dexterous bio-mechanical architecture with versatile capabilities to perform different kinds of tasks. The undergoing research in the field aims at the reproduction of human's abilities not only by means of anthropomorphic design but also by the adoption of human-inspired control strategies. To this purpose postural synergies have been identified to be the key strategies in planning and control of the robotic and prosthetic hands of the future.

The studies on grasp taxonomy carried out by scientists such as Napier [42], Cutkosky [17] and Iberall [29] aim at defining which fingers (and which parts of the fingers) are used by humans to generate forces on the grasped object. According to this classification, the hand configuration during grasping operation can be decomposed in a limited set of basic postures. On the other hand, recent advances in neuroscience have shown that control of the human hand during grasping is dominated by movements in a continuous configuration space of highly reduced dimensionality with respect to the number of DOFs [52], [37].

In this work, the eigengrasps of the DEXMART Hand (Fig. 1) have been derived using the principal component analysis (PCA) by considering a set of 36 hand configurations among precision, intermediate and power grasps of common (for humans) objects contained in a comprehensive human grasp taxonomy [51]. A method is proposed to derive the postural synergies of the DEXMART Hand from experiments and the kinematic patterns connected to the three predominant eigenpostures. Moreover, the temporal variation of the three synergies weights is exploited for real-time execution of the grasps.

From the data analysis it can be argued that the introduction of the third predominant synergy significantly improves the grasp synthesis and performance, especially with regard to the improvement of the adduction/abduction motion of the thumb. Experimental results show that grasp planning and control of the DEXMART Hand performed by using the three predominant postural synergies allows not only reproducing the set of postures adopted to derive the eigengrasps with a high level of fidelity, but also synthesizing and performing a wide set of grasps, namely precision, intermediate and power grasps of objects with different shapes and dimensions. Indeed, in order to prove the efficiency of the method, the synthesis of new grasp/object pairs not contained in the reference set of postures used for PCA has been realized. The intent here is to imitate the typical attitude of humans in manipulation as well as to cover the entire grasp variety in recently proposed taxonomies, e.g. [20]. All the data from experiments are obtained using a prototype of the DEXMART Hand without sensors. Thus, the motivation of this study is to test the method for deriving synergies and the potential of the anthropomorphic design of the DEXMART Hand to work efficiently in a synergies based framework for grasp planning and prehension control during reach to grasp.

3.1 Related Work

Recently, the studies on kinematic synergies have collected the interest of many researchers not only belonging to the field of neuroscience but also working on control theory and mechanical design of artificial hands. Preliminary studies on the human hand had pointed out that the combination of tendon coupling and muscle activation patterns exhibited by humans lead to significant joint coupling and inter-finger coordination, or, in other words, to postural synergies, that are evidence of simplified control schemes occurring at neurological level for the organization of the hand movements.

In [52] the PCA has been used to calculate the postural synergies from real-world data collected on a variety of human hand postures by means of a data glove. Moreover, the authors show that a wide set of hand postures during grasping operation evolves continuously within a linear space spanned by few postural synergies that account for most of the hand configurations variance, without distinguishing between power and precision grasps. In [37] it is shown that even if higher principal components account for a small percentage of the variance, they give critical details not only for the static grasp when the hand adapts to the object shape, but also for the act of preshaping during the grasp. In [15], [13], [14] the authors extend the concept of postural synergies to robotic hands showing how a similar dimensionality reduction can be used to derive comprehensive planning and control algorithms that produce stable grasps for a number of different robot hand models. Synergies have been used to solve the dimensionality reduction problem in control and coordination of a 16-DOF underactuated prosthetic hand prototype (CyberHand), in order to perform the three prehensile forms mostly used in activities of daily living [38].

Other applications have been made in order to simplify the design and analysis of robotic hand structures [10]. In [47] the authors investigate how the number and types of synergies are related to the possibility of controlling the contact forces and the object motion in grasping and manipulation tasks. In [22], using the definition of force-closure for underactuated hands and the definition of grasping force optimization, the authors investigate the role of different postural synergies in the ability of obtaining force closure grasps and the quality of the grasps in two case studies addressing a precision and a power grasp. The manipulability analysis has been extended to synergy-actuated hands, where compliance is utilized in order to solve the force distribution problem [48]. The authors introduce new manipulability indices which take into account underactuation and compliance. A modified model of synergies including the mechanical compliance of the hand's musculotendinous system has been proposed in [7] in order to account in the synergy model for the force distribution in the actual grasp. In [59] a synergy impedance controller has been derived and implemented on the DLR hand.

Recent work on mapping synergies from the human hand to the robot hand has been addressed in [24]. The proposed mapping strategy between the synergies of a paradigmatic human hand and a robotic hand is carried out in the task space and it is based on the use of a virtual sphere. This approach has the advantages to be independent of the robotic hand and depends only on the specific operation, and thus it can be used for robotic hands with very dissimilar kinematics. In [23] three synergies from data on human grasping experiments have been extracted and mapped to a robotic hand. Then a neural network with the features of the objects and the coefficients of the synergies has been trained and employed to control robot grasping. Neural networks have been utilized also in other papers in order to simulate temporal coordination of human reaching and grasping. In [57] the neural network model includes a synergistic control of the whole fingers during prehension and the design of a library of hand gestures.

3.2 *Postural Synergies of the DEXMART Hand*

Postural synergies describe patterns occurring at the joint displacement level. In [52], the authors measure a set of static human hand postures by recording 15 joint angles and, by means of the PCA, they show that the first two principal components account for >80% of the hand postures. Thus the use of the principal components, also called postural synergies, holds great potential for robot hands control, implying a substantial reduction of the grasp synthesis problem dimension with respect to the case of considering the entire number of DOFs of the robotic hand.

Drawing inspiration from the studies on the human hand motion, and since the DEXMART Hand presents human-like kinematics, we have found a set of principal components of the DEXMART Hand configuration space. The study of the two predominant synergies of the DEXMART Hand was carried out in [21], where experimental results showed that it is possible to obtain grasp synthesis for a large set of objects in the case of both precision and power grasps.

More recently, the third synergy has been experimentally obtained and evaluated. It has been shown that, by exploiting the third synergy, the movement of adduction/abduction of the thumb covers the whole range of joint limits without violating the limits of the other joints. This improves the correct opposition of the thumb and allows synthesizing and executing more precisely complex grasps and reproducing the set of postures adopted to derive the eigengrasps with higher accuracy with respect to the case in which only two synergies are used.

The first two synergies found for the DEXMART Hand account for >77% of the hand postures, thus matching quite well the results reported in [52]. Therefore, since the three predominant postural synergies account for >85% of the hand postures, a robot hand control strategy that uses also the third synergy will significantly improve the grasping performance, as experiments reported in this work show.

The DEXMART Hand kinematics is rather close to that of the human hand. Hence, with the aim of deriving the PCA, a set of grasps similar to those illustrated in [51] has been considered. The choice of the reference set of postures has been made by taking into account all the most common human grasps considered in the grasp taxonomy literature. This set is composed of grasps of objects such as spheres of different dimensions involving a different number of fingers in both power and precise grasp configuration. Cylindrical grasps have been considered as well, distinguishing also between different positions of the thumb. Moreover, several configurations for precise grasps with index and thumb opposition as well as intermediate side grasps have been included. Following the taxonomy adopted in [20], a comprehensive hierarchical human grasp classification used for the PCA is reported in Fig. 20. Furthermore, a suitable number of open-hand configurations with different positions of the thumb and of the adduction/abduction fingers joint has been added in order to find synergies that allow the hand to moves continuously also toward open-hand configurations which are equally important to reach and grasp the objects. A total amount of $n = 36$ hand configurations have been evaluated to derive the fundamental eigenpostures. Each grasp configuration of the reference set of postures has been experimentally reproduced with the DEXMART Hand as close as possible to a natural human-like grasp, and the vector $\mathbf{c}_i \in \mathbb{R}^{15}$ of the joint angle values corresponding to each reproduced grasp has been measured. Once the set of the DEXMART Hand configurations matrix $\mathbf{C} = \{\mathbf{c}_i \mid i = 1 \ldots n\}$ has been built, the vector $\bar{\mathbf{c}}$ representing the mean hand position in the grasp configurations space (zero-offset position) and the matrix $\mathbf{F} = \{\mathbf{c}_i - \bar{\mathbf{c}} \mid i = 1 \ldots n\}$ of the grasp offsets with respect to the mean configuration have been computed. The PCA has then been performed on $\mathbf{F}$ and a base matrix $\mathbf{E}$ of the postural synergies subspace has been found. The PCA can be performed by diagonalizing the covariance matrix of $\mathbf{F}$ as

$$\mathbf{F}\mathbf{F}^T = \mathbf{E}\mathbf{S}^2\mathbf{E}^T. \tag{52}$$

The $(h \times h)$ orthogonal matrix $\mathbf{E}$ gives the directions of variance of the data, and the diagonal matrix $\mathbf{S}^2$ is the variance in each direction sorted in decreasing magnitude, i.e. the element on the diagonal represents the eigenvalue of the covariance matrix.

To verify the effectiveness of the synergy-based modeling approach, the percentage σ of the total variance of the data described by the first j-th principal components can be obtained by means of the following equation

$$\sigma_j = \sum_{k=0}^{j} \mathbf{s}_k \Big/ \sum_{k=0}^{15} \mathbf{s}_k \tag{53}$$

where $\mathbf{s}_k$ is the k-th element of the diagonal of the matrix $\mathbf{S}^2$. Since the three principal components account for >85% of the postures ($\sigma_3 = 0.8503$), the posture matrix $\mathbf{C}$ can be reconstructed with good accuracy by adopting the matrix

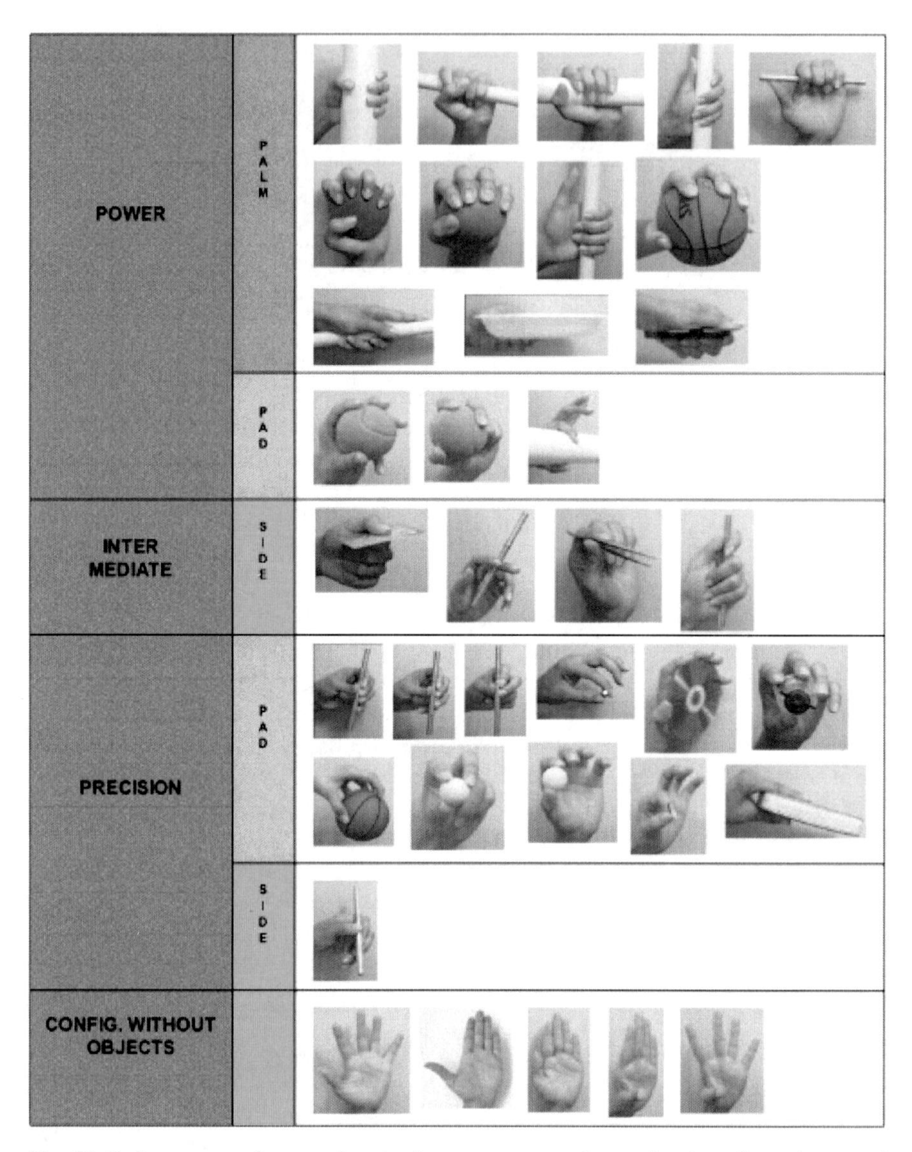

Fig. 20 Reference set of comprehensive human grasps and open-hand configurations used for PCA.

$$\hat{\mathbf{E}} = [\mathbf{e}_1\ \mathbf{e}_2\ \mathbf{e}_3] \tag{54}$$

composed of the three principal components of $\mathbf{E}$ as a base of the robotic hand configuration space, thus allowing the control of the robotic hand motion in a configuration space of highly reduced dimensions with respect to the DOFs of the hand itself. Each hand grasp posture $\mathbf{c}_i$ can be obtained by a suitable selection of the weights

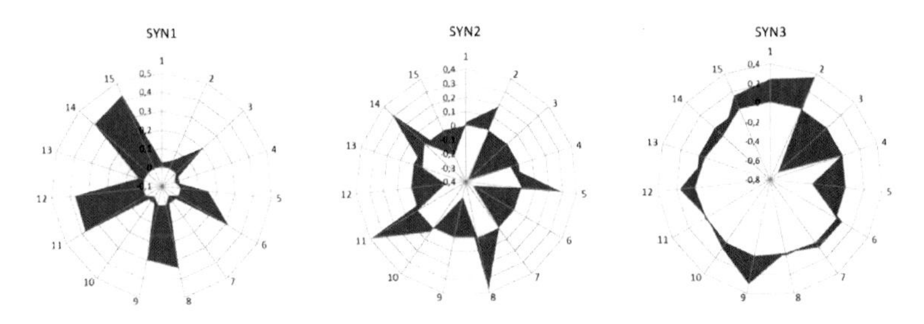

Fig. 21 In this figure, the angular change in degrees for each joint due to a positive unitary variation in α_1, α_2, and α_3 for the first three synergies is represented. The adduction/abduction, proximal and medial flexion joints are indicated from 1 to 3 for the thumb, from 4 to 6 for the index finger, from 7 to 9 for the middle finger, from 10 to 12 for the ring finger and finally from 13 to 15 for the little finger.

$[\alpha_1\ \alpha_2\ \alpha_3]^T \in \mathbb{R}^3$ of the postural synergies. Therefore, the projection $\hat{\mathbf{c}}_i$ of each robotic hand configuration $\mathbf{c}_i$ on the postural synergies subspace can be evaluated as

$$\hat{\mathbf{c}}_i = \bar{\mathbf{c}} + \hat{\mathbf{E}} \begin{bmatrix} \alpha_{1,i} \\ \alpha_{2,i} \\ \alpha_{3,i} \end{bmatrix}. \tag{55}$$

In the following, the three fundamental synergies derived for the DEXMART Hand, i.e. the robotic hand motions spanned by $\mathbf{e}_1$, $\mathbf{e}_2$ and $\mathbf{e}_3$ respectively, are briefly described, referring to the minimum and maximum configuration of each synergy as the hand configurations obtained by means of, respectively, the minimum and maximum value of the corresponding synergy weights without violating the joint limits [21]. When the weights of the synergies are zero, the hand posture corresponds to the zero-offset position $\bar{\mathbf{c}}$. The vectors of the three synergies and the zero-offset vector of the DEXMART Hand are reported in Tab. 1.

The circular graphs represented in Fig. 21 are a useful tool for identifying the joints whose rotations are more involved in each synergy. From left to right, the angular variations in degrees for each joint due to a unitary variation of the corresponding synergy weight is represented for the first, the second and the third synergy. It is easy to observe how the adduction/abduction thumb joint motion (Joint 1) is more involved in the third synergy rather than in the first two. Moreover, in the third synergy the movement of the index and of the thumb are more engaged than for the other fingers. This justifies the use of the third synergy in order to grasp objects more precisely, especially for precision grasps and intermediate side grasps, where the position of the thumb and of the index is crucial, as the experiments reported in Sect. 3.4 demonstrate.

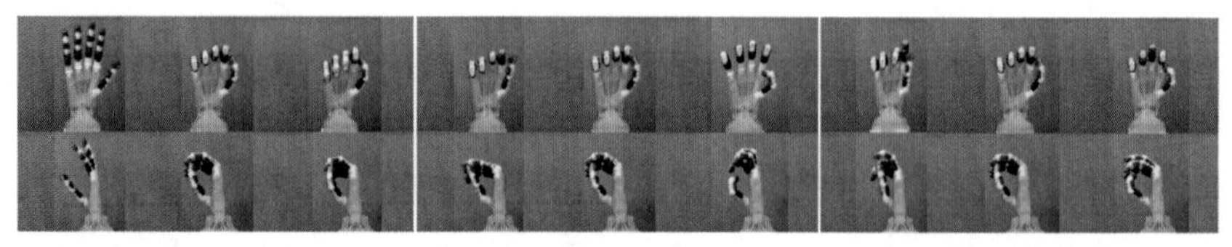

(a) First postural synergy. (b) Second postural synergy. (c) Third postural synergy.

Fig. 22 Representation of the DEXMART Hand postural synergies. On the top of each figure, from left to right, a sequence of hand postures going form the minimum to the maximum configuration are represented. On the bottom the lateral views are reported.

3.3 Control with Postural Synergies

In order to perform the desired grasp, the value of the three eigenpostures weights $[\alpha_1\ \alpha_2\ \alpha_3]^T$ are computed by projection of the desired grasp posture in the synergies subspace:

$$\begin{bmatrix} \alpha_{1,i} \\ \alpha_{2,i} \\ \alpha_{3,i} \end{bmatrix} = \hat{\mathbf{E}}^{\dagger}\left(\mathbf{c}_i - \bar{\mathbf{c}}\right) \tag{56}$$

where $\hat{\mathbf{E}}^{\dagger}$ is the Moore-Penrose pseudo-inverse of the base matrix $\hat{\mathbf{E}}$. It is straightforward to note that the motions shown in Fig. 22(a), 22(b) and 22(c), derived by considering separately the three synergies, are obtained from (55) by assuming $\alpha_2 = 0$ and $\alpha_3 = 0$ for the first synergy, $\alpha_1 = 0$ and $\alpha_3 = 0$ for the second synergy and finally $\alpha_1 = 0$ and $\alpha_2 = 0$ for the third synergy.

The temporal value of the weights α_1, α_2, α_3 during grasp operations has to be chosen in such a way that, starting from the zero-offset position $\bar{\mathbf{c}}$ (i.e. $\alpha_1 = \alpha_2 = \alpha_3 = 0$), the hand opens during the reach in preparation for object grasp, and then closes reaching a suitable shape determined from (56) and depending on the original grasp configuration $\mathbf{c}_i$ for the considered object. In the open-hand configuration, namely $\mathbf{c}_0$, all the flexion joint angles are close to zero, and the corresponding values of α_1, α_2 and α_3 can be determined from (56) by posing $\mathbf{c}_i = \mathbf{c}_0$.

The intermediate values of the synergy weights have been determined by assuming a suitable time interval for the grasp operation (six seconds for the whole reach to grasp phase, three seconds for both the opening and closing phases) and by linear interpolation of the α_1, α_2 and α_3 values in the three reference configurations $\{\bar{\mathbf{c}},\ \hat{\mathbf{c}}_0,\ \hat{\mathbf{c}}_i\}$.

Three synergies, shown in Figs. 22(a), 22(b) and 22(c) respectively, are now analyzed in detail. With reference to the first postural synergy (column $\mathbf{e}_1$ in Tab. 1), in the minimum configuration the proximal and medial flexion joint angles of all the fingers are all almost zero and increase their value during the motion toward the maximum configuration. The adduction/abduction movements are not very involved in this synergy. In Fig. 22(a) the minimum, zero-offset and maximum configuration in frontal and lateral view of the first postural synergy are represented. The second postural synergy (column $\mathbf{e}_2$ in Tab. 1) is characterized by a movement in opposite

Table 1 First three eigenpostures and zero offset vectors of the DEXMART Hand postural synergies subspace (data in degrees).

		$\mathbf{e}_1$	$\mathbf{e}_2$	$\mathbf{e}_3$	$\bar{\mathbf{c}}[deg]$
Thumb	adduction/abduction	0.0282	0.0235	-0.2454	-0.833
	proximal	0.0674	0.1874	-0.3639	20.5
	medial	0.2004	-0.3853	0.6991	34.7
Index	adduction/abduction	-0.0266	-0.0647	0.0228	2.92
	proximal	0.1575	0.2893	0.3648	34.9
	medial	0.3220	-0.3494	-0.0735	50.5
Middle	adduction/abduction	-0.0404	0.0069	-0.0675	-0.694
	proximal	0.3405	0.3794	0.0304	41.4
	medial	0.2999	-0.2948	-0.3034	42.2
Ring	adduction/abduction	-0.0374	0.0343	-0.0778	-1.11
	proximal	0.3775	0.3977	0.0200	45.5
	medial	0.3766	-0.2568	-0.1675	49.2
Little	adduction/abduction	0.0364	-0.0738	0.0720	0.694
	proximal	0.3892	0.3213	0.1273	48.7
	medial	0.4235	-0.2026	-0.1491	51.7

directions of the proximal and medial flexion joints. In this synergy, the adduction/abduction movements of all the fingers are more involved with respect to the first synergy for the index and the little finger. In Fig. 22(b) the minimum, zero-offset and maximum configurations of the second postural synergy are depicted in frontal and lateral views. In the third postural synergy (column $\mathbf{e}_3$ in Tab. 1) the movement involves especially the index and the thumb. Thanks to this synergy, the movement of adduction/abduction of the thumb covers the whole joint range without violating other joint limits. This characteristic is crucial because the correct index/thumb opposition allows increasing the grasp accuracy, and thus achieving more stable grasps. This justifies the use of three predominant synergies for the hand control in order to improve the grasp performance. Finally, the excursion of the angles of adduction/abduction of the middle and ring fingers are quite involved in this synergy, more than in the first two. In Fig. 22(c) the minimum, zero-offset and maximum configuration in frontal and lateral views of the third postural synergy are represented.

3.4 Experimental Evaluation

The hand controller developed in the Matlab/Simulink environment is based on the RTAI-Linux realtime operating system. The Matlab Realtime Workshop toolbox has been used for the automatic generation of the real-time application of the DEXMART Hand controller. The user interface to the real-time application has been implemented by means of the Simulink External Mode capabilities, for which the RTAI-Linux support has been purposely developed.

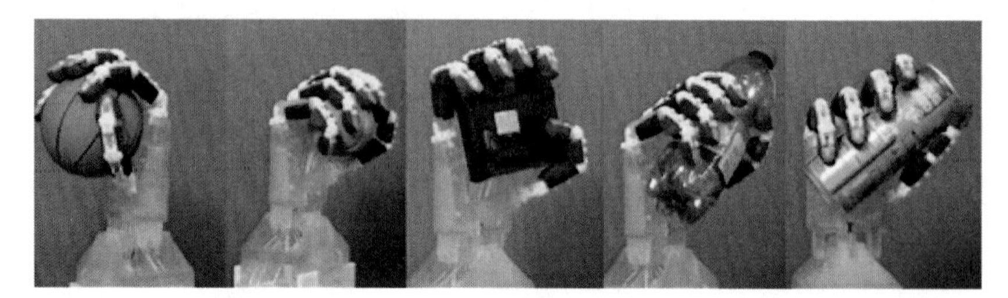

Fig. 23 Reproduced power grasps from the reference set of postures using the first three synergies.

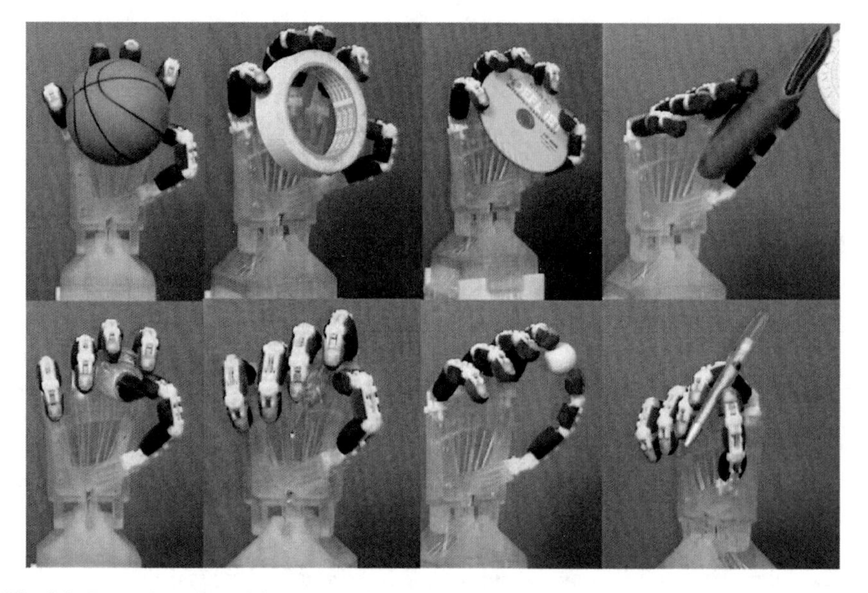

Fig. 24 Reproduced precision grasps from the reference set of postures using the first three synergies.

In the experiments, starting from the zero-offset position, the hand moves continuously in the synergies configuration subspace and goes in an open-hand configuration. Then, it closes reaching a configuration that depends on the particular grasp to be performed. During the closing phase, the weights of the three postural synergies are obtained by linear interpolation from those corresponding to the open-hand configuration to those suitable values unique for each object and computed using (56).

Experimental results reveal that, by using the three predominant eigengrasps, it is possible to reproduce several grasp configurations more precisely than in the case of using two synergies only [21].

The linear combination of the three synergies allows a power grasp of both cylinders and spheres of different dimensions by means of suitable opposition of the thumb, see Fig. 23.

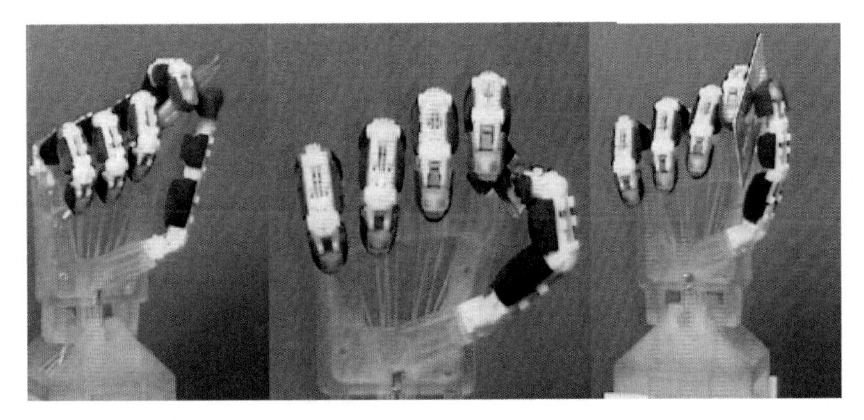

Fig. 25 Reproduced intermediate side grasps from the reference set of postures using the first three synergies.

In Fig. 24, the performance of precise grasp operations is reported considering different objects, achieving opposition of the thumb and index as well as precise grasps using from two to five fingers for prismatic and circular objects.

The reproduced intermediate side grasps from the reference set of postures is depicted in Fig. 25. All the evaluations of the experiments data have been carried out with the aid of the following tables. Table 2 reports the first three synergy weights computed by projection of the reference set of postures in the synergies subspace. In Tab. 3 the absolute value of the angle error of the adduction/abduction joint of the thumb obtained comparing the reference set of configurations and the reproduced configurations using two and three synergies are reported. Finally, the average of the joint errors in the case of using two and three synergies are reported in Tab. 4. The average of the joint errors is computed using the Euclidean norm

$$e = \frac{||\mathbf{c_i} - \widehat{\mathbf{c_i}}||}{15}.$$

By observing the first image from the left (pen, Configuration C27 in Tab. 2) of Fig. 25, it is interesting to note that this posture is very close to the minimum configuration of the third synergy; indeed the weight of the third synergy is high with respect to the other grasps, and thus the use of the third synergy is essential for this performance.

By looking at Tab. 3, it is possible to note that the use of the three predominant synergies reduces the error on the angular position of the adduction/abduction thumb joint for almost all the 36 configurations with respect to the case in which only two synergies are used.

In Tab. 3, the grasp configurations executed using only two synergies are marked with a star, while the new grasp configurations that have been performed successfully adding the third synergy are marked with a diamond. This table shows that, by introducing the third predominant synergy, the joint angle error of the thumb is reduced for almost all the grasps configurations marked with a star, except for

C10 (box), C16 (credit card) and C19 (pen). Nevertheless, a global improvement obtained using also the third synergy is evident observing the average joint angle errors reported in Tab. 4. Only for the configuration C14 no improvement has been obtained and this is confirmed by the very small value of the third synergy weight; this means that the third synergy gives almost no contribution to the variance of this posture. The improvement on the adduction/abduction thumb joint angle using the

Table 2 Synergy weights of the grasps from the reference set of postures.

Conf.	C1	C2⋆	C3	C4⋆	C5	C6	C7	C8
α_1	0.90	63.7	64.7	38.0	21.9	42.7	53.1	68.2
α_2	-3.47	28.1	-24.5	36.5	-33.6	-5.76	-8.21	-8.66
α_3	7.06	3.61	7.95	-21.2	-65.9	-2.33	-4.68	-8.56
Conf.	C9⋄	C10⋆	C11⋆	C12⋆	C13⋄	C14⋆	C15	C16⋆
α_1	-30.8	-4.00	58.8	12.3	-7.89	-18.9	8.86	63.7
α_2	-12.0	-117	9.39	-63.5	-59.0	8.47	1.36	39.3
α_3	21.4	-2.30	44.3	-14.0	27.6	0.15	-23.7	-8.38
Conf.	C17	C18	C19⋆	C20⋆	C21	C22	C23⋄	C24
α_1	13.7	42.7	67.4	-29.8	-18.4	-30.4	82.8	39.8
α_2	44.2	-43.2	28.4	122	-0.103	-30.2	2.58	23.3
α_3	-37.6	-13.2	16.0	25.3	17.7	11.2	31.7	18.3
Conf.	C25⋄	C26	C27⋄	C28	C29	C30	C31⋆	C32
α_1	35.4	65.3	75.9	81.8	-103	-120	26.3	-134
α_2	-17.8	1.82	19.8	26.4	-28.0	29.5	-0.933	4.10
α_3	27.6	18.8	-68.0	-20.3	27.8	18.7	31.0	-9.15
Conf.	C33	C34	C35	C36				
α_1	-134	-127	-134	-135				
α_2	4.20	18.3	4.96	3.73				
α_3	-10.2	-36.5	-5.04	-5.26				

third synergy is very clear at least for the configurations marked with a diamond. For what concerns Configuration C27 (pen, intermediate side grasp), the improvement can be seen mainly in the error average and it is spread on the thumb and index joints. These results show that the use of the third synergy allows grasping objects more precisely, especially when the position of the thumb and index is crucial, as in the case of precision grasps. The confirmation of this is given by the observation that the configurations marked with a diamond correspond to precision grasps, except for C27.

In Fig. 26, the distribution of the synergy weights adopted for executing the grasping experiments is shown in the space of the three predominant eigengrasps, and an example of a complete hand trajectory computed by linear interpolation of the synergies weights in the three reference configurations for the grasp of a generic object (a CD) is reported by the red dashed line. For the sake of clarity, only the weights of the grasps obtained during the experiments are reported, and only some of them are named. In this figure the full bullets represent the final configuration

Table 3 Adduction/abduction thumb joint angle error (in degrees) in the reproduced grasps from the reference set of postures obtained using two and three synergies.

Conf.	C1	C2⋆	C3	C4⋆	C5	C6	C7	C8
two syn	9.11	11.6	9.59	8.89	11.0	9.76	9.53	9.11
three syn	7.38	10.7	11.5	3.69	5.16	9.19	8.38	7.01

Conf.	C9⋄	C10⋆	C11⋆	C12⋆	C13⋄	C14⋆	C15	C16⋆
two syn	8.02	6.30	11.0	12.0	7.56	8.83	10.5	1.88
three syn	2.77	6.87	0.17	8.54	0.77	8.80	4.73	3.94

Conf.	C17	C18	C19⋆	C20⋆	C21	C22	C23⋄	C24
two syn	9.41	9.35	8.26	11.2	8.65	7.60	11.6	10.8
three syn	0.186	12.6	12.2	4.97	4.30	4.84	3.79	6.33

Conf.	C25⋄	C26	C27⋄	C28	C29	C30	C31⋆	C32
two syn	9.74	8.95	8.23	7.91	5.59	6.48	9.89	14.5
three syn	2.98	13.6	8.46	2.93	1.22	1.88	2.28	12.3

Conf.	C33	C34	C35	C36				
two syn	14.5	14.0	5.50	5.45				
three syn	12.0	5.02	6.74	6.74				

Table 4 Average joint angle errors (in degrees) in the reproduced grasps from the reference set of postures obtained using two and three synergies.

Conf.	C1	C2⋆	C3	C4⋆	C5	C6	C7	C8
two syn	1.81	1.84	1.83	2.70	4.89	2.25	2.24	2.12
three syn	1.75	1.82	1.75	2.30	2.16	2.24	2.22	2.05

Conf.	C9⋄	C10⋆	C11⋆	C12⋆	C13⋄	C14⋆	C15	C16⋆
two syn	2.21	1.38	3.48	1.90	2.64	1.63	3.62	1.56
three syn	1.69	1.37	1.84	1.65	1.89	1.63	3.26	1.46

Conf.	C17	C18	C19⋆	C20⋆	C21	C22	C23⋄	C24
two syn	4.11	2.01	3.20	2.19	2.06	3.78	2.77	2.34
three syn	3.26	1.81	3.02	1.40	1.68	3.70	1.79	1.99

Conf.	C25⋄	C26	C27⋄	C28	C29	C30	C31⋆	C32
two syn	2.95	2.31	4.83	4.27	3.29	4.05	2.60	2.00
three syn	2.31	1.94	1.66	4.05	2.72	3.86	1.57	1.90

Conf.	C33	C34	C35	C36				
two syn	1.92	4.99	1.99	1.70				
three syn	1.80	4.35	1.97	1.66				

weights corresponding to the grasps performed also in the previous work [21]. The triangles represent the final configuration weights corresponding to the objects that have been grasped thanks to the use of the third synergy also. Finally, the circles represent the final synergy weights corresponding to the synthesized grasps of different grasp/object pairs (Fig. 28(b)) not included in the table of Fig. 20 and obtained by projection in the synergies subspace of the desired configuration of the hand. This

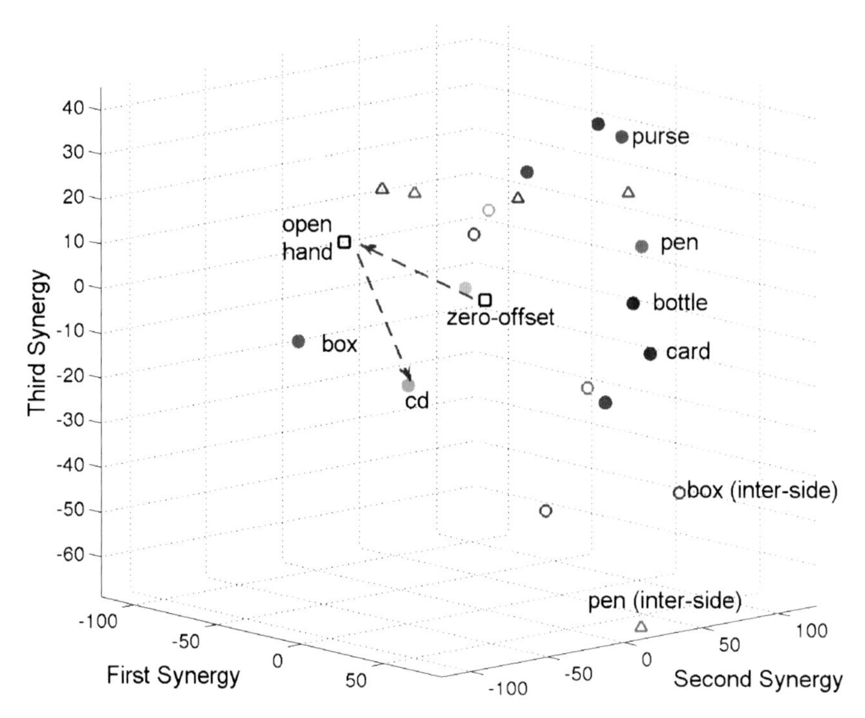

Fig. 26 The distribution of hand postures in the space of the first three postural synergies is represented, distinguishing between the object graspable also using the first two synergies only (full bullets), the objects graspable thanks to the use of the third synergy (triangles), and the synthesized grasps obtained by projection (circles).

desired configuration is obtained experimentally, moving singularly the joints in order to realize the desired grasp.

In Fig. 27 two grasp configurations executed using both two and three synergies are represented. From left to right the first pictures of the ball and of the CD are the ones executed using three synergies. From these pictures, the improvement on the position of the thumb can be noticed thanks to the introduction of the third synergy in the hand control.

The grasps realized above show that through the use of three synergies we can reproduce the matrix of the reference set of postures (Fig. 20) with accuracy greater than 85%.

The idea now is to extend the method in order to grasp any object not necessarily contained in the reference table. The advantage of the synergies subspace is that of simplifying grasp synthesis of common objects using also complex hand shapes typical of human manipulation.

To accomplish this goal, we have selected grasps of five common objects in Fig. 28(a). The object/grasp pairs have been chosen so as to cover the entire variety in recently proposed taxonomy [20], namely a power palm grasp, a power pad grasp, an intermediate side grasp, a precision pad grasp, and a precision side

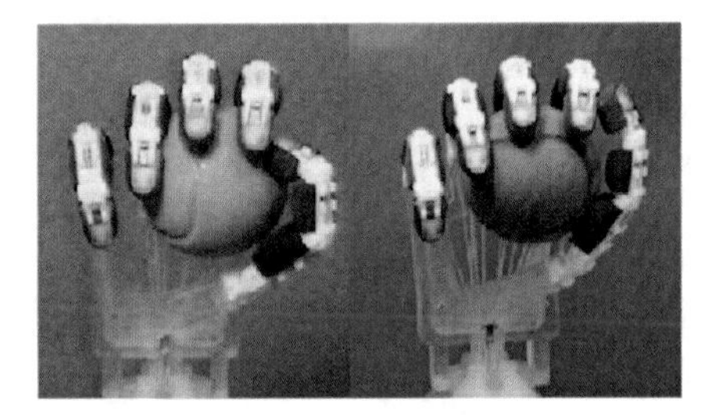 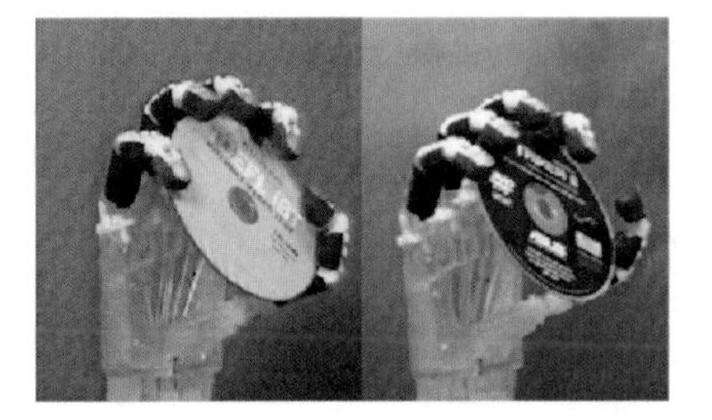

Fig. 27 Comparison between two grasps configuration executed using both two and three synergies. From left to right the first pictures of the ball and of the CD are the ones executed using three synergies.

POWER		INTERMEDIATE	PRECISION	
PALM	PAD	SIDE	PAD	SIDE

(a) Reference set of human grasps.

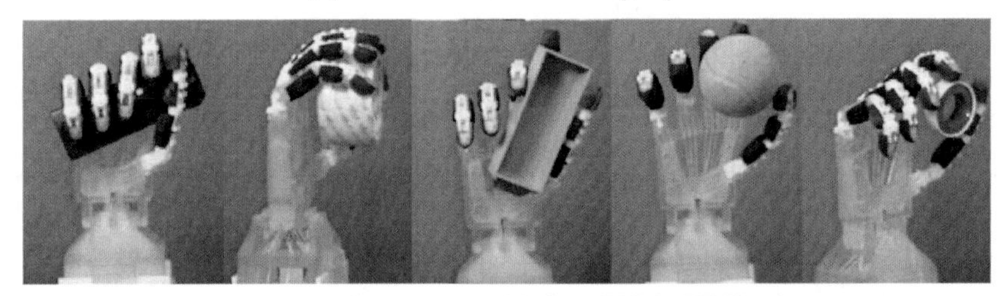

(b) Synthesized grasps with the DEXMART Hand.

Fig. 28 On the top the reference set of human grasps, not included in the PCA analysis, covering the entire variety of grasps in recently proposed taxonomies, is depicted. On the bottom the synthesized grasps realized with the DEXMART Hand using synergies subspace projection is shown. A power palm grasp, a power pad grasp, an intermediate side grasp, a precision pad grasp and a precision side grasp are reported from left to right.

grasp. Moreover, the choices have been made with the intent to imitate the typical modality of human manipulation.The selected grasp configurations are unusual for a typical application of robotic manipulation as they are carried out by a robotic hand with high dexterity and anthropomorphism. In fact, they constitute a high standard for robotic hands currently on the market and designed for applications of service robotics and prosthetics. On the other hand, the synthesis of this kind of grasps involves complex problems of planning and control. The high number of DOFs and complex kinematics similar to that of the human hand is essential for this specific type of tasks. Despite this, the DOFs have to be managed.

Through a teaching-by-showing technique, the use of synergies allows reducing significantly synthesis complexity. By singularly guiding the finger joints, the hand has been carried to the desired grasp configuration and the joint displacement vector has been recorded. Through the projection of the desired configuration in the synergies subspace we have derived their weights in the final configuration $\hat{\mathbf{c}}_i$. The weights are subsequently used in the control algorithm that performs motion during reach to grasp and realize the desired final grasp configuration, see Fig. 28(b).

Table 5 Synergy weights of the grasps in Fig. 28(b) (from left to right).

Conf.	G1	G2	G3	G4	G5
α_1	84.3	-6.61	16.7	-14.4	42.8
α_2	34.8	8.96	20.6	7.93	18.5
α_3	-37.3	18.6	-47.1	12.9	-16.2

The weights shown in Tab. 5 indicate the contribution rate of each synergy to achieve the final configuration. Referring to what we have previously argued about the use of the third synergy, for the grasp represented in the third image of Fig. 28(b) (G3), the weight of the third synergy greatly influences the success of the grasp, see Tab. 5.

The experimental results demonstrate a good choice of the reference set of postures to retrieve the synergies subspace, and confirm that the planning/control method based on synergies can work efficiently for every object and grasp choice throughout a complete taxonomy. During reach to grasp, the hand behaves like a human hand reaching impressive human like shape using a small effort in planning.

4 Conclusion

In this chapter, some issues related to the control of anthropomorphic sensorised hands have been addressed. First, the problem of computing online the optimal contact forces to grasp an object has been considered, assuming that these forces may vary during task execution, thanks to the availability of force/torque sensors at the fingertips of the DEXMART Hand. The proposed algorithm takes into account the maximum joint torques that can be provided by the fingers and has been extended to bimanual manipulation tasks with a limited increase of the computational complexity, thanks to a novel load sharing technique. The other interesting feature of anthropomorphic manipulation, especially in the bimanual case, is the availability of redundant degrees of freedom. These have been suitably exploited to design a multi-priority control approach that allows satisfying a certain number of secondary tasks, aimed at ensuring grasp stability and manipulation dexterity, besides the main task corresponding to the desired object motion. Finally, anthropomorphism has been exploited for the development of a human-like grasping approach based on the synergic motions that can be observed in the human hand. In detail, the synthesis and control of grasping for the DEXMART Hand have been simplified by computing a reduced configuration subspace based on few predominant postural synergies. This approach has been evaluated at kinematic level, showing that both power and precise grasps can be performed using up to the third predominant synergy.

Acknowledgements. The research leading to these results has been supported by the DEXMART Large-scale integrating project, which has received funding from the European Communitys Seventh Framework Programme (FP7/2007-2013) under grant agreement ICT-216239. The authors are solely responsible for its content. It does not represent the opinion of the European Community and the Community is not responsible for any use that might be made of the information contained therein.

The authors would like to thank Fabrizio Caccavale, Giuseppe Muscio and Francesco Pierri for their contribution in the design and implementation of the control approach with redundancy resolution.

References

1. Antonelli, G.: Stability analysis for prioritized closed-loop inverse kinematic algorithms for redundant robotic systems. IEEE Transactions on Robotics 25, 985–994 (2009)
2. Berselli, G., Borghesan, G., Brandi, M., Melchiorri, C., Natale, C., Palli, G., Pirozzi, S., Vassura, G.: Integrated mechatronic design for a new generation of robotic hands. In: IFAC Symposium on Robot Control, Gifu (2009)
3. Berselli, G., Piccinini, M., Palli, G., Vassura, G.: Engineering design of fluid-filled soft covers for robotic contact interfaces: Guidelines, nonlinear modelling, and experimental validation. IEEE Transactions on Robotics 27, 436–449 (2011)
4. Berselli, G., Piccinini, M., Vassura, G.: On designing structured soft covers for robotic limbs with predetermined compliance. In: ASME International Design Engineering Technical Conferences, Montréal (2010)
5. Berselli, G., Vassura, G.: Differentiated layer design to modify the compliance of soft pads for robotic limbs. In: IEEE International Conference on Robotics and Automation, Kobe (2009)
6. Biagiotti, L., Lotti, F., Melchiorri, C., Palli, G., Tiezzi, P., Vassura, G.: Development of UB Hand 3: Early results. In: IEEE International Conference on Robotics and Automation, Barcelona (2005)
7. Bicchi, A., Gabiccini, M., Santello, M.: Modelling natural and artificial hands with synergies. Philosophical Transactions of the Royal Society B: Biological Sciences 366, 3153–3161 (2011)
8. Bicchi, A., Prattichizzo, D.: Manipulability of cooperative robots with unactuated joints and closed-chain mechanisms. IEEE Transactions on Robotics and Automation 16, 336–345 (2000)
9. Borghesan, G., Palli, G., Melchiorri, C.: Design of tendon-driven robotic fingers: Modelling and control issues. In: IEEE International Conference on Robotics and Automation, Anchorage, AK (2010)
10. Brown, C., Asada, H.: Inter-finger coordination and postural synergies in robot hands via mechanical implementation of principal components analysis. In: IEEE/RSJ Interational Conference on Intelligent Robots and Systems, San Diego, CA (2007)

11. Buss, M., Hashimoto, H., Moore, J.B.: Dextrous hand frasping force optimization. IEEE Transactions on Robotics and Automation 12, 406–418 (1996)
12. Buss, M., Faybusovich, L., Moore, J.B.: Dikin-type algorithms for dextrous grasping force optimization. International Journal of Robotics Research 17, 831–839 (1998)
13. Ciocarlie, M., Allen, P.: Hand posture subspaces for dexterous robotic grasping. International Journal of Robotics Research 28, 851–867 (2009)
14. Ciocarlie, M., Allen, P.: On-line interactive dexterous grasping. In: 6th International Conference on Haptics: Perception, Devices and Scenarios, Madrid (2008)
15. Ciocarlie, M., Goldfeder, C., Allen, P.: Dimensionality reduction for hand-independent dexterous robotic grasping. In: IEEE/RSJ International Conference on Intelligent Robots and Systems, San Diego (2007)
16. Coelho, J., Grupen, R.: A control basis for learning multifingered grasps. Journal of Robotic Systems 14, 545–557 (1997)
17. Cutkosky, M.: On grasp choice, grasp models, and the design of hands for manufacturing tasks. IEEE Transactions on Robotics and Automation 5, 269–279 (1989)
18. DEXMART Project website, http://www.dexmart.eu/
19. Faybusovich, L.: Dikin's algorithm for matrix linear programming problems. In: Henry, J., Yvon, J.-P. (eds.) System Modelling and Optimization, pp. 237–247. Springer, New York (1994)
20. Feix, T., Pawlik, R., Schmiedmayer, H., Romero, J., Kragic, D.: The generation of a comprehensive grasp taxonomy. In: Robotics: Science and Systems, Workshop on Understanding the Human Hand for Advancing Robotic Manipulation, Seattle, WA (2009)
21. Ficuciello, F., Palli, G., Melchiorri, C., Siciliano, B.: Experimental evaluation of postural synergies during reach to grasp with the UB Hand IV. In: IEEE/RSJ International Conference on Intelligent Robots and Systems, San Francisco, CA (2011)
22. Gabiccini, M., Bicchi, A.: On the role of hand synergies in the optimal choice of grasping forces. In: Robotics: Science and Systems, Zaragoza (2010)
23. Geng, T., Lee, M., Hulse, M.: Transferring human grasping synergies to a robot. Mechatronics 284, 272–284 (2011)
24. Gioioso, G., Salvietti, G., Malvezzi, M., Prattichizzo, D.: Mapping synergies from human to robotic hands with dissimilar kinematics: An object based approach. In: IEEE International Conference on Robotics and Automation, Workshop on Manipulation Under Uncertainty, Shanghai (2011)
25. Han, L., Trinkle, J.C.: The instantaneous kinematics of manipulation. In: IEEE International Conference on Robotics and Automation, Leuven (1998)
26. Han, L., Trinkle, J.C., Li, Z.X.: Grasp analysis as linear matrix inequality problems. IEEE Transactions on Robotics and Automation 16, 663–674 (2000)
27. Helmke, U., Hüper, K., Moore, J.B.: Quadratically convergent algorithms for optimal dexterous hand grasping. IEEE Transactions on Robotics and Automation 18, 138–146 (2002)
28. Helmke, U., Moore, J.B.: Optimization and Dynamic Systems. Springer, New York (1993)
29. Iberall, T.: Human prehension and dexterous robot hands. International Journal of Robotics Research 16, 285–299 (1997)
30. Li, Z.X., Quin, Z., Jiang, S., Han, L.: Coordinated motion generation and real-time grasping force control for multifingered manipulation. In: IEEE International Conference on Robotics and Automation, Leuven (1998)
31. Lippiello, V., Ruggiero, F., Villani, L.: Exploiting redundancy in closed-loop inverse kinematics for dexterous object manipulation. In: International Conference on Advanced Robotics, Munich (2009)

32. Lippiello, V., Siciliano, B., Villani, L.: Online dextrous-hand grasping force optimization with dynamic torque constraints selection. In: IEEE International Conference on Robotics and Automation, Shanghai (2011)
33. Lippiello, V., Siciliano, B., Villani, L.: A grasping force optimization algorithm with dynamic torque constraints selection for multi-fingered robotic hands. In: American Control Conference, San Francisco, CA (2011)
34. Liu, G., Xu, J., Li, Z.: On geometric algorithms for real-time grasping force optimization. IEEE Transactions on Control System Technology 12, 843–859 (2004)
35. Lotti, F., Vassura, G.: A novel approach to mechanical design of articulated fingers for robotic hands. In: IEEE/RSJ International Conference on Intelligent Robots and Systems, Lausanne (2002)
36. Mansard, N., Chaumette, F.: Task sequencing for high-level sensor-based control. IEEE Transactions on Robotics and Automation 23, 60–72 (2007)
37. Mason, C., Gomez, J., Ebner, T.: Hand synergies during reach-to-grasp. Journal of Neurophysiology 86, 2896–2910 (2001)
38. Matrone, G., Cipriani, C., Secco, E., Magenes, G., Carrozza, M.: Principal components analysis based control of a multi-dof underactuated prosthetic hand. Journal of NeuroEngineering and Rehabilitation 7(16), 1–16 (2010)
39. Montana, D.: The kinematics of contact and grasp. International Journal of Robotics Research 7(3), 17–32 (1988)
40. Montana, D.: The kinematics of multi-fingered manipulation. IEEE Transactions on Robotics and Automation 11, 491–503 (1995)
41. Murray, R., Li, Z.X., Sastry, S.: A Mathematical Introduction to Robotic Manipulation. CRC Press, Boca Raton (1994)
42. Napier, J.: The prehensile movements of the human hand. Journal of Bone and Joint Surgery 38-B, 902–913 (1956)
43. Okamura, A.M., Smaby, N., Cutkosky, M.R.: An overview of dextrous manipulation. In: IEEE International Conference on Robotics and Automation, San Francisco, CA (2000)
44. Palli, G., Borghesan, G., Melchiorri, C.: Modelling, identification and control of tendon-based actuation systems. IEEE Transactions on Robotics 27, 1–14 (2011)
45. Platt, R., Fagg, A.H., Grupen, R.: Null-space grasp control: Theory and experiments. IEEE Transactions on Robotics 26, 282–295 (2010)
46. Ponce, J., Sullivan, S., Sudsang, A., Boissonnat, J., Merlet, J.: On computing four-finger equilibrium and force-closure grasps of polyhedral objects. International Journal of Robotic Research 16, 11–35 (1996)
47. Prattichizzo, D., Malvezzi, M., Bicchi, A.: On motion and force controllability of grasping hands with postural synergies. In: Robotics: Science and Systems, Zaragoza (2010)
48. Prattichizzo, D., Malvezzi, M., Gabiccini, M., Bicchi, A.: On the manipulability ellipsoids of underactuated robotic hands with compliance. Robotics and Autonomous Systems 60, 337–346 (2012)
49. Prattichizzo, D., Trinkle, J.C.: Grasping. In: Siciliano, B., Khatib, O. (eds.) Springer Handbook of Robotics, pp. 671–700. Springer, Heidelberg (2008)
50. Remond, C., Perdereau, V., Drouin, M.: A multi-fingered hand control structure with on-line grasping force optimization. In: IEEE/ASME International Conference on Advanced Intelligent Mechatronics, Como (2001)
51. Romero, J., Feix, T., Kjellstrom, H., Kragic, D.: Spatio-temporal modelling of grasping actions. In: IEEE/RSJ International Conference on Intelligent Robots and Systems, Taipei (2010)
52. Santello, M., Flanders, M., Soechting, J.: Postural hand synergies for tool use. Journal of Neuroscience 18, 10105–10115 (1998)

53. Saut, J.P., Remond, C., Perdereau, V., Drouin, M.: Online computation of grasping force in multi-fingered hands. In: IEEE/RSJ International Conference on Intelligent Robots and Systems, Edmonton (2005)
54. Shlegl, T., Buss, M., Omata, T., Schmidt, G.: Fast dextrous regrasping with optimal contact force and contact sensor-based impedance control. In: IEEE International Conference on Robotics and Automation, Seoul (2001)
55. Siciliano, B., Sciavicco, L., Villani, L., Oriolo, G.: Robotics: Modelling, Planning and Control. Springer, London (2009)
56. Uchiyama, M.: A unified approach to load sharing, motion decomposing, and force sensing of dual arm robots. In: Miura, H., Arimoto, S. (eds.) Robotics Research: The Fifth International Symposium, pp. 225–232. MIT Press, Cambridge (1994)
57. Vilaplana, J., Coronado, J.: A neural network model for coordination of hand gesture during reach to grasp. Neural Networks 19, 12–30 (2006)
58. Walker, I.D., Marcus, S.I., Freeman, R.A.: Distribution of dynamic loads for multiple cooperating robot manipulators. Journal of Robotic Systems 9, 35–47 (1989)
59. Wimboeck, T., Jan, B., Hirzinger, G.: Synergy-level impedance control for a multifingered hand. In: IEEE/RSJ International Conference on Intelligent Robots and Systems, San Francisco, CA (2011)
60. Xu, J., Lou, Y., Li, Z.: Grasping force optimization for whole hand grasping. In: IEEE/RSJ International Conference on Intelligent Robots and Systems, Beijing (2006)
61. Zheng, Y.F., Luh, J.Y.S.: Optimal load distribution for two industrial robots handling a single object. In: IEEE International Conference on Robotics and Automation, Philadelphia, PA (1988)